Jose A. Lozano, Pedro Larrañaga, Iñaki Inza, Endika Bengoetxea (Eds.)

Towards a New Evolutionary Computation

Studies in Fuzziness and Soft Computing, Volume 192

Editor-in-chief
Prof. Janusz Kacprzyk
Systems Research Institute
Polish Academy of Sciences
ul. Newelska 6
01-447 Warsaw
Poland
E-mail: kacprzyk@ibspan.waw.pl

Jose A. Lozano

Pedro Larrañaga

Iñaki Inza

Endika Bengoetxea (Eds.)

Towards a New Evolutionary Computation

Advances in the Estimation of Distribution Algorithms

 Springer

Jose A. Lozano
Pedro Larrañaga
Iñaki Inza
University of the Basque Country
Department of Computer Science and
Artificial Intelligence
Apartado de correos 649
20080 Donostia-San Sebastian
Spain
E-mail: lozano@si.ehu.es
       pedro.larranaga@ehu.es
       inza@si.ehu.es

Endika Bengoetxea
Intelligent Systems Group
Department of Architecture and
Computer Technology
University of the Basque Country
20080 Donostia-San Sebastián
Spain
E-mail: endika@si.ehu.es

ISSN print edition: 1434-9922
ISSN electronic edition: 1860-0808

ISBN 978-3-642-06704-4   e-ISBN 978-3-540-32494-2

Springer is a part of Springer Science+Business Media
springer.com
© Springer-Verlag Berlin Heidelberg 2006
Softcover reprint of the hardcover 1st edition 2006

Printed on acid-free paper

Preface

Estimation of Distribution Algorithms (EDAs) are a set of algorithms in the Evolutionary Computation (EC) field characterized by the use of explicit probability distributions in optimization. Contrarily to other EC techniques such as the broadly known Genetic Algorithms (GAs) in EDAs, the crossover and mutation operators are substituted by the sampling of a distribution previously learnt from the selected individuals.

Since they were first termed by Mühlenbein and Paaß (1996) and the seminal papers written three years later by Etxeberria and Larrañaga (1999), Mühlenbein and Mahnig (1999) and Pelikan et al. (1999), EDAs have experienced a high development that has transformed them into an established discipline within the EC field. Evidence of its establishment is the great number of papers on EDAs published in the main EC conferences and in EC-related journals, as well as the tutorials given in the PPSN, CEC and GECCO conferences.

The work developed in the field since our first edited book (Larrañaga and Lozano (2002)), has motivated us to compile a subset of the great advances on EDAs in this new volume. We hope this will attract the interest of new researchers in the EC field as well as in other optimization disciplines, and that it becomes a reference for all of us working on this topic.

The twelve chapters of this book can be divided into those that endeavor to set a sound theoretical basis for EDAs, those that broaden the methodology of EDAs and finally those that have an applied objective.

In the theoretical field, Ochoa and Soto abound on the relation between the concept of entropy of a distribution and EDAs. Particularly, the authors design benchmark functions for EDAs based on the principle of maximum entropy. The concept of entropy is also applied by Ocenasek to base a stopping criterion for EDAs in discrete domains. The author proposes to end the algorithm at the time point when the generation of new solutions becomes ineffective.

Methodological contributions in the field of continuous optimization are carried out by Ahn et al. The authors define the Real-coded Bayesian Optimization Algorithm, an algorithm that endeavors to convey the good properties of BOA to the continuous domain. Hansen presents a comparison of the CMA (Covariance Matrix Adaption) of evolution strategies with EDAs defined in continuous domains.

The extension of the EDAs framework to broader scopes is performed by Yanai and Iba, Bosman and Thierens and Madera et al. Yanai and Iba introduce EDAs in the context of Genetic Programming. In this context the probability distribution of programs is estimated by using a Bayesian network. Bosman and Thierens extend their IDEA algorithm to the problem of multi-objective optimization. They show how the use of a mixture model of univariate components allows for wide–spread exploration of a multi–objective front. The parallelization of EDAs is deal with by Madera et al. The authors propose several island models for EDAs.

Other two works on the methodological arena are those of Robles et. al. and Miquelez et al. In the view of the great practical success attained by hybrid algorithms, Robles et al. propose several ideas to combine EDAs with GAs in order for the hybrid to share the good points of both GAs and EDAs. Miquelez et al. design a sub-family of EDAs in which Bayesian classifiers are applied in optimization problems. Using the classification labels, a Bayesian classifier is built instead of a common Bayesian network.

Finally, the book contains some concrete examples on using and adapting the EDA framework to the characteristics of complex practical applications. An example of this is presented by Saeys et al. who apply the algorithm in a feature ranking problem in the context of the biological problem of acceptor splice site prediction. They obtain an ordering of the genes from the estimated distribution of an EDA. Flores et al. use EDAs to induce linguistic fuzzy rule systems in prediction problems. The authors integrate EDAs in the recently proposed COR methodology which tries to take advantage of the cooperation among rules. Finally the quadratic assignment problem is tackled by Zhang et al. The authors use an EDA couple with a 2-opt local algorithm. A new operator "guided mutation" is used to generate the individuals.

We would finally like to thank all the contributors of this book for their effort in making it a good and solid piece of work. We are also indebted to the Basque country government for supporting by means of the SAIOTEK S-PE04UN25 and ETORTEK-BIOLAN grants.

Spain
August 2005

Jose A. Lozano
Pedro Larrañaga
Iñaki Inza
Endika Bengoetxea

Contents

Real-coded Bayesian Optimization Algorithm

The CMA Evolution Strategy: A Comparing Review

**Estimation of Distribution Algorithm with 2-opt Local Search
for the Quadratic Assignment Problem**
Qingfu Zhang, Jianyong Sun, Edward Tsang and John Ford 281

List of Contributors

Chang Wook Ahn
Department of Information and
Communications
Gwangju Institute of Science and
Technology (GIST). Korea
cwan@evolution.re.kr

Enrique Alba
Department of Languages and Computer
Science
Málaga University. Spain
eat@lcc.uma.es

Endika Bengoetxea
Intelligent Systems Group
Department of Computer Architecture
and Technology
The University of the Basque Country
Spain
endika@si.ehu.es

Peter A.N. Bosman
National Research Institute for
Mathematics and Computer Science
The Netherlands
Peter.Bosman@cwi.nl

Sven Degroeve
Department of Plant Systems Biology
Ghent University.
Belgium
sven.degroeve@psb.ugent.be

M. Julia Flores
Departamento de Informática
Universidad de Castilla-La Mancha
Spain
Julia.Flores@uclm.es

John Ford
Department of Computer Science
University of Essex
U.K.
fordj@essex.ac.uk

José A. Gámez
Departamento de Informática
Universidad de Castilla-La Mancha
Spain
Jose.Gamez@uclm.es

David E. Goldberg
Department of General Engineering
University of Illinois at
Urbana-Champaign
USA
deg@uiuc.edu

Nikolaus Hansen
CoLab Computational Laboratory, ETH
Zürich, ICoS Institute of Computational
Science, ETH Zürich
Switzerland
nikolaus.hansen@inf.ethz.ch

Vanessa Herves
Department of Computer Architecture
and Technology
Universidad Politécnica de Madrid
Spain
vherves@fi.upm.es

Hitoshi Iba
Dept. of Frontier Informatics, Graduate
School of Frontier Sciences
The University of Tokyo
Japan
iba@iba.k.u-tokyo.ac.jp

Pedro Larrañaga
Intelligent Systems Group
Department of Computer Science
and Artificial Intelligence
The University of the Basque Country
Spain
ccplamup@si.ehu.es

Julio Madera
Department of Computing
Camagüey University
Cuba
jmadera@inf.reduc.edu.cu

Teresa Miquélez
Intelligent Systems Group
Department of Computer Architecture
and Technology
The University of the Basque Country
Spain
teresa@si.ehu.es

Jiri Ocenasek
Computational Laboratory (CoLab)
Swiss Federal Institute of Technology
Switzerland
jirio@inf.ethz.ch

Alberto Ochoa
Institute of Cybernetics
Mathematics and Physics
Cuba
ochoa@icmf.inf.cu

Jose M. Peña
Department of Computer Architecture
and Technology
Universidad Politécnica de Madrid
Spain
jmpena@fi.upm.es

María S. Pérez
Department of Computer Architecture
and Technology
Universidad Politécnica de Madrid
Spain
mperez@fi.upm.es

José M. Puerta
Departamento de Informática
Universidad de Castilla-La Mancha
Spain
Jose.Puerta@uclm.es

R. S. Ramakrishna
Department of Information and
Communications
Gwangju Institute of Science and
Technology (GIST). Korea
rsr@gist.ac.kr

Victor Robles
Department of Computer Architecture
and Technology
Universidad Politécnica de Madrid
Spain
vrobles@fi.upm.es

Marta Soto
Institute of Cybernetics
Mathematics and Physics
Cuba
mrosa@icmf.inf.cu

Yvan Saeys
Department of Plant Systems Biology
Ghent University
Belgium
yvan.saeys@psb.ugent.be

Jianyong Sun
Department of Computer Science
University of Essex
U.K.
jysun@essex.ac.uk

Dirk Thierens
Institute of Information and Computing
Sciences
Utrecht University
The Netherlands
Dirk.Thierens@cs.uu.nl

Edward Tsang
Department of Computer Science
University of Essex
U.K.
edward@essex.ac.uk

Yves Van de Peer
Department of Plant Systems Biology
Ghent University
Belgium
yves.vandepeer@psb.ugent.be

Kohsuke Yanai
Department of Frontier Informatics
Graduate School of Frontier Sciences
The University of Tokyo
Japan
yanai@iba.k.u-tokyo.ac.jp

Qingfu Zhang
Department of Computer Science
University of Essex
U.K.
qzhang@essex.ac.uk

Linking Entropy to Estimation
of Distribution Algorithms

Alberto Ochoa and Marta Soto

Institute of Cybernetics, Mathematics and Physics, Cuba
{ochoa,mrosa}@icmf.inf.cu

Summary. This chapter presents results on the application of the concept of entropy to estimation of distribution algorithms (EDAs). Firstly, the Boltzmann mutual information curves are introduced. They are shown to contain a lot of information about the difficulty of the functions. Next, a design method of discrete benchmark functions is presented. The newly developed approach allows the construction of both single and random classes of functions that obey a given collection of probabilistic constraints. This application and the next – the construction of low cost search distributions – are based on the principle of maximum entropy. The last proposal is the linear entropic mutation (LEM), an approach that measures the amount of mutation applied to a variable as the increase of its entropy. We argue that LEM is a natural operator for EDAs because it mutates distributions instead of single individuals.

1 Introduction

Entropy is a measure of the uncertainty of a random variable, whereas mutual information measures the reduction of the entropy due to another variable. These are fundamental quantities of information theory [3], the building blocks of a field that overlaps with probability theory, statistical physics, algorithmic complexity theory and communication theory, among others disciplines.

In this chapter, we explore several novel uses of the concept of entropy in evolutionary optimization. In particular, we investigate intersections of information theory and the field of estimation of distribution algorithms (EDAs) [26].

A major challenge of evolutionary optimization is the preservation of the right balance between exploitation and exploration. From an entropic point of view, exploitation can be seen as a low-entropy search, whereas exploration is better understood as a high-entropy search. This occurs both at the system and variable levels. At the system level, we see how the joint entropy is reduced as the run approaches the optimum. At the variable level, the mutual information comes into play, the reduction in uncertainty of a variable due to the remainder variables is an indicator of what kind of entropic balance should be enforced at that point. These are just few evidences about the fact that entropy is at the heart of the dynamics of artificial evolution. This has been a major motivation of our work. We believe that EDAs will profit from greater efforts in this area of research. Keeping in mind these arguments, in this chapter we approach the following issues:

- A method for analysing the difficulty of the functions (Sect. 3).
- A design method of benchmark functions (Sect. 4).
- A method for learning low cost maximum-entropy distributions (Sect. 5).
- An entropic approach to mutation (Sect. 6).

Nowadays, simulation is a fundamental tool of verification, validation and comparison of evolutionary algorithms. For EDAs, the design of benchmark functions should emphasize, in the first place, the complexity of the probabilistic structure of the search distributions. We have developed a method, which gives the designer the possibility of specifying a collection of probabilistic constraints that have to be fulfilled by the search distributions. The method is connected to the concept of entropy because it constructs a maximum entropy distribution that satisfies the given constraints.

A good design method should be accompanied by a good analysis method. We introduce a new approach for function complexity analysis in the context of EDA optimization. Our approach investigates the mutual information of Boltzmann distributions as a function of the temperature parameter.

A critical problem of learning search distributions in an EDA, is the sample complexity. Large sample sizes mean large number of function evaluations. The challenge is to reduce the number of evaluations, without damaging the effectiveness and efficiency of the search. We use the concept of entropy to achieve this goal; the true search distribution is substituted by a maximum entropy approximation, which can be reliably computed with less population size.

EDAs have to approach mutation from a distribution perspective, in contrast with the genotype perspective of GAs. While a GA mutates single individuals, an EDA must mutate distributions. We have developed an approach that uses the concept of entropy to fulfill this requirement. The relation between entropy and mutation is quite intuitive: when a random variable is mutated, a certain degree of randomness is added to it. Therefore, it seems reasonable to measure the amount of mutation applied to a variable as the increase of its entropy.

The outline of this contribution is as follows. Section 2 presents the background material. Then we discuss the above problems in Sects. 3-6. Finally, the conclusions are given.

2 Background

This section introduces the general notation of the chapter. It also gives a short introduction to the theories that underlie our main results.

2.1 General Notation

In this chapter, X_i represents a scalar random variable and $p(x_i) = p(X_i = x_i)$ its probability mass function with $x_i \in \mathfrak{X} = \{0, 1, \ldots, K\}$. Note that $p(x_i)$ and

$p(x_j)$ refer to two different random variables, and have in fact different probability mass functions, $p(X_i = x_i)$ and $p(X_j = x_j)$, respectively. Similarly, $\boldsymbol{X} = (X_1, X_2, \ldots, X_n)$ denotes a n-dimensional random variable, $\boldsymbol{x} = (x_1, x_2, \ldots, x_n)$ is a configuration and $p(x_1, x_2, \ldots, x_n)$ represents a joint probability mass. The notation $\boldsymbol{X}_a$ and $\boldsymbol{x}_a$ is used to denote sub-vectors of $\boldsymbol{X}$ and $\boldsymbol{x}$ with indexes from $a \subset \{1, \ldots, n\}$. $p(\boldsymbol{x}_a) = \sum_{x_i, i \notin a} p(\boldsymbol{x})$ and $p(\boldsymbol{x}_a \mid \boldsymbol{x}_b) = p(\boldsymbol{x}_a, \boldsymbol{x}_b) / p(\boldsymbol{x}_b)$ define marginal and conditional distributions, respectively. $p(a)$ or p_a are used to denote $p(\boldsymbol{x}_a)$.

2.2 Boltzmann Estimation of Distribution Algorithms

At the center of most of the ideas and results of this chapter, lies the Boltzmann distribution. Some authors have considered it as the corner stone of the theory of estimation of distribution algorithms [19, 24, 25]. We believe that this chapter is new evidence that supports this way of thinking.

Definition 1 *For $\beta \geq 0$ define the Boltzmann distribution of a function $f(\boldsymbol{x})$ as*

$$p_{\beta, f}(\boldsymbol{x}) := \frac{e^{\beta f(\boldsymbol{x})}}{\sum_{\boldsymbol{y}} e^{\beta f(\boldsymbol{y})}} = \frac{e^{\beta f(\boldsymbol{x})}}{Z_f(\beta)}$$

where $Z_f(\beta)$ is the partition function.

We also use $Z_{\beta, f}$, but to simplify the notation β and f can be omitted. If we follow the usual definition of the Boltzmann distribution, then $-f(\boldsymbol{x})$ is called the free energy and $1/\beta$ the temperature of the distribution. The parameter β is usually called the inverse temperature.

Closely related to the Boltzmann distribution is Boltzmann selection:

Definition 2 *Given a distribution $p(\boldsymbol{x})$ and a selection parameter γ, Boltzmann selection calculates a new distribution according to*

$$p^s(\boldsymbol{x}) = \frac{p(\boldsymbol{x}) e^{\gamma f(\boldsymbol{x})}}{\sum_{\boldsymbol{y}} p(\boldsymbol{y}) e^{\gamma f(\boldsymbol{y})}}$$

Boltzmann selection is important because the following holds [25]:

Theorem 1 *Let $p_{\beta, f}(\boldsymbol{x})$ be a Boltzmann distribution. If Boltzmann selection is used with parameter γ, then the distribution of the selected points is again a Boltzmann distribution with*

$$p^s(\boldsymbol{x}) = \frac{e^{(\beta + \gamma) f(\boldsymbol{x})}}{\sum_{\boldsymbol{y}} e^{(\beta + \gamma) f(\boldsymbol{y})}}$$

The Boltzmann estimation of distribution algorithm (BEDA) was introduced in [25] on the basis of the above. Here, it is shown as Algorithm 1. BEDA is an algorithm with good theoretical properties, it has even a convergence proof. However, in the form in which it is shown in algorithm 1, it is just a conceptional algorithm.

Algorithm 1 BEDA – Boltzmann Estimation of Distribution Algorithm

---

Step 1 $t \leftarrow 0, \beta(t) \leftarrow 0$ and $p(xx, t) = \frac{1}{Z_{\beta(t),f}}$

Step 2 $t \leftarrow t + 1, \Delta\beta(t) \leftarrow \beta(t) - \beta(t - 1)$ and

$$p(\boldsymbol{x}, t + 1) \leftarrow \frac{p(\boldsymbol{x}, t)e^{\Delta\beta(t)f(\boldsymbol{x})}}{\sum_{\boldsymbol{y}} p(\boldsymbol{y}, t)e^{\Delta\beta(t)f(\boldsymbol{y})}} \tag{1}$$

Step 3   If the stopping criterion is not reached, go to step 2.

The reasons are twofold: the exponential complexity of the denominator of (1) and the lack of a method for updating $\Delta\beta(t)$.

The next lemma solves the second problem. The reader is referred to [19] for details.

Lemma 1 $\Delta\beta(t) = c/\sqrt{Var_f(\beta(t))}$ *leads to an annealing schedule where the average fitness, $W_f(\beta(t))$, increases approximately proportional to the standard deviation:*

$$W_f(\beta(t + 1)) - W_f(\beta(t)) \approx c\sqrt{Var_f(\beta(t))}$$

where c is a constant and $Var_f(\beta(t)) = \sigma_f^2(\beta(t))$ is the variance of the fitness function. This annealing schedule has been called standard deviation schedule (SDS).

The exponential complexity of computing the partition function can be avoided if the Boltzmann distribution is approximated with a tractable distribution. There are several ways of accomplishing this approximation [23]. However, for the purposes of this chapter it is enough to restrict ourselves to the special case covered by the factorization theorem [25].

The factorization theorem defines how and under what conditions the search distributions associated to discrete functions can be factorized. The factorization follows the structure of the function and is only exact if the function obeys certain structural constraints.

Definition 3 *Let $s_i \subseteq \{1, \ldots, n\}$ $(1 \le i \le m)$ be index-sets and let $f^{(i)}$ be functions depending only on the variables X_j $(j \in s_i)$. Then, $f(\boldsymbol{x}) = \sum_{i=1}^{m} f^{(i)}(\boldsymbol{x}_{s_i})$ is an additive decomposition of the fitness function $f(\boldsymbol{x})$.*

Definition 4 *Given $s_1, \ldots, s_m$, the sets d_i, b_i and c_i $(i = 1, \ldots, m)$ are defined as follows: $d_0 := \emptyset$, $d_i := \bigcup_{j=1}^{i} s_j$, $b_i := s_i \setminus d_{i-1}$ and $c_i := s_i \cap d_{i-1}$.*

Theorem 2 *(Factorization theorem) For $\beta \ge 0$, let $p_{\beta,f}(\boldsymbol{x})$ be a Boltzmann distribution of a function $f(\boldsymbol{x})$, and $f(\boldsymbol{x}) = \sum_{i=1}^{m} f^{(i)}(\boldsymbol{x}_{s_i})$ be an additive decomposition. If $d_m = \{1, \ldots, n\}$ and the following holds*

$$\begin{aligned}
&\forall i \in \{1, \ldots, m\}, b_i \neq 0 \\
&\forall i \ge 2, \exists j < i \quad such\,that\ c_i \subseteq s_j
\end{aligned} \tag{2}$$

then

$$p_{\beta,f}(\boldsymbol{x}) = \prod_{i=1}^{m} p\left(\boldsymbol{x}_{b_i} \mid \boldsymbol{x}_{c_i}\right) \tag{3}$$

The proof can be found in [25]. Assumption 2 is called the running intersection property [16].

In the simulations of this chapter we use mainly two algorithms: the factorized distribution algorithm (FDA) and the Boltzmann FDA. Both algorithms use the factorization (3) as a model of the search distributions. However, while the FDA uses truncation selection, the BFDA uses Boltzmann selection with SDS.

The following lemma is relevant to this chapter [19].

Lemma 2 *BFDA is invariant under linear transformation of the fitness function with a positive factor.*

2.3 Factorizations

As was said in the previous section, the factorization of probability distributions is a major concern of EDA researchers. In this chapter, Bayesian factorizations are specially relevant. They are connected with the concept of Bayesian network.

A Bayesian network (BN) [30, 31] is a directed acyclic graph containing nodes, representing the variables, and arcs, representing probabilistic dependencies among nodes. For any node (variable) X_i, and set of parents π_{X_i}, the Bayesian network specifies a conditional probability distribution $p(x_i \mid \pi_{x_i})$.

There are single-connected – no more than one undirected path connects two nodes – and multiple-connected BNs. The single-connected BNs are also called polytrees. In a polytree, a node may have several parents and many roots. Trees are special class of polytrees, which have at most one parent and one root. Polytrees describe higher-order interactions than trees, while retaining many of their computational advantages. In a polytree, structures like $X \rightarrow Z \leftarrow Y$ are often called head-to-head patterns. This type of pattern makes X and Y conditionally dependent given Z, which cannot be represented by a tree.

A junction tree [10, 14, 16] is an undirected tree, where each node contains a set of variables. The junction tree satisfies the *junction property*: for any two nodes a and b and any node h on the unique path between a and b, $a \cap b \subseteq h$. The arcs between the nodes are labelled with the intersection of the adjacent nodes; usually, they are called *separating sets* or *separators*.

Junction trees are important for inference and sampling because they have tractable algorithms for these tasks. Given a BN, it is possible to construct at least one junction tree. The reader is referred to [10, 14, 16] for a complete discussion on the issue.

2.4 Entropy and Mutual Information

The entropy $H(X)$ of a discrete random vector X is defined in [3] by

$$H(X) = -\sum_{x \in X} p(x) \log p(x) \tag{4}$$

Note that entropy is a functional of the distribution of X. It does not depend on the actual values taken by the random variable, but only on the probabilities. This means that $H(X)$ is a shortcut for $H(p(X))$. The logarithm in (4) is to the base two and entropy is expressed in bits. We use the convention that $0 \log 0 = 0$.

For a binary variable X, such that $p(X = 1) = p$, we have

$$H(X) = H(p(X)) = H(p) := -p \log p - (1-p) \log (1-p) \tag{5}$$

The entropy of a binary variable is a nonnegative, symmetric and concave function of the distribution. It has the maximum at the point $(0.5, 1)$ and it is zero for $p \in \{0, 1\}$.

The following theorem will be useful later on.

Theorem 3 *(Independence bound on entropy [3]). Let $p(\boldsymbol{x})$ be any joint probability mass of a set of discrete random variables $\boldsymbol{X} = (X_1, X_2, \ldots, X_n)$, then*

$$H(\boldsymbol{X}) \leq \sum_{i=1}^{n} H(X_i)$$

with equality if and only if the variables are independent.

The concepts of marginal and conditional mutual information will be intensively used in the chapter. The mutual information, $I(X, Y)$, is the reduction in the uncertainty of X due to the knowledge of Y. The conditional mutual information, $I(X, Y | Z)$, represents the reduction in the uncertainty of X due to the knowledge of Y given Z. The following theorem connects entropy and mutual information.

Theorem 4 *Between mutual information and entropy the following holds [3]:*

$$I(X, Y) = H(X) + H(Y) - H(X, Y) \tag{6}$$

$$I(X, Y | Z) = H(X | Z) - H(X | Y, Z) \tag{7}$$

The Maximum-Entropy Principle

The maximum-entropy principle (MEP) plays an important role in this chapter. It is used to build probability mass functions that fulfill a collection of marginal constraints. The ideas behind this concept can be shortly explained as follows.

Frequently, partial prior information is available outside of which it is desired to use a prior that is as non-informative as possible. For example, suppose some prior marginal distributions are specified, and among prior distributions with these marginals the most non-informative distribution is sought [12, 13]. If we have the joint distribution with the maximum-entropy of all the joints that fulfill a given collection of marginals, choosing a joint with less entropy amounts to add some information that is not justified by the constraints.

The iterative proportional fitting (IPF) algorithm can be used to find the maximum-entropy distribution [11, 12, 18, 32]. The proof that IPF converges against the maximum-entropy solution can be found in [4]. Unfortunately, the naive implementation of the IPF takes exponential time and space. Therefore, it is not suitable for computing distributions with many variables.

For large distributions, an efficient implementation of the maximum-entropy algorithm was developed in [15, 21]. The general idea is to improve the performance of IPF by combining it with the junction tree technique. It consists of performing IPF locally on the nodes and passing messages to the neighboring nodes. It has been proved that this converges to the unique maximum-entropy solution, so it is equivalent to IPF. The reader is referred to [29] for details on the implementation of the method for computing maximum-entropy distributions of polytrees.

3 Mutual Information and Functions Difficulty

This section presents preliminary ideas about a novel method for analysing the complexity of functions for evolutionary algorithms. The corner stone of the approach is the concept of mutual information, which is studied through its relation with selection.

3.1 Boltzmann Mutual Information Curves

The Goldberg's Deceptive3 function belongs to the class of the so called deceptive problems [6, 7] that are those having local optima which are easier to find than global optima. Deceptive problems contain deceptive attractors, which mislead the algorithm to search for sub-optima because their basins of attraction are much larger than the ones favoring global optima. Often, deceptiveness is considered a challenge to search algorithms. However, deception is a relative category that emerges solely in the context of the relationship problem-algorithm. In other words, a problem may be deceptive for one algorithm, but not for another.

Deception has been intensively studied in the context of genetic algorithms. In [6, 7, 9], the authors described ways to construct deceptive functions and gave sufficient conditions for deception. Figure 1 (left) shows the usual way of describing deceptive problems as a function of unitation. Note, the deep valley separating the optimum from the sub-optimum and the different sizes of their attractors.

In this section, we introduce a new method for analysing the function complexity in the context of EDA optimization. Our approach investigates the mutual information of Boltzmann distributions as a function of the parameter β. Given a function f, this method computes the Boltzmann distribution $p_{f,\beta}$ for $\beta > 0$. Then, it computes the marginal and the conditional mutual information on any sub-set of variables. We show that the Boltzmann mutual information curves, $I(\beta)$, contain a lot of information about the complexity of the function.

Table 1 shows the function Deceptive3 and its Boltzmann distribution for $\beta = 10.49$. On the other hand, Fig. 1 (right) presents the marginal mutual information

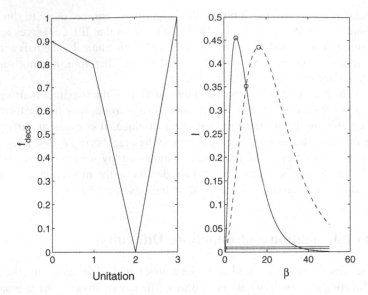

Fig. 1. Explaining the complexity of Goldberg's Deceptive3 function: (*left*) unitation approach – the optimum is isolated and separated from the sub-optima by a deep valley (*right*) mutual information approach – marginal (*dashed line*) and conditional (*solid line*)

Table 1. Goldberg's Deceptive3 function and its Boltzmann distribution for $\beta = 10.49$. At this value, $I(X, Y) = I(X, Y \mid Z)$

$x_3 x_2 x_1$	$f_{dec3}(\mathbf{x})$	$p_{\beta=10.49}(\mathbf{x})$	$x_3 x_2 x_1$	$f_{dec3}(\mathbf{x})$	$p_{\beta=10.49}(\mathbf{x})$
000	0.9	0.2038	100	0.8	0.0714
001	0.8	0.0714	101	0	0
010	0.8	0.0714	110	0	0
011	0	0	111	1	0.5820

and the conditional mutual information. Note that all edges have the same marginal and conditional values of mutual information, i.e. the function is symmetric. This property of the Deceptive3 simplifies its analysis.

To begin with, we recall a result that was presented in [35], which states that the difference between conditional and marginal mutual information is invariant to permuting the variables. Remarkably, the result holds for any three sets of variables X_a, X_b and X_c.

Proposition 1 *(Whittaker [35, Proposition 4.5.1]) Suppose that the partitioned random vector (X_a, X_b, X_c) has a joint density function f_{abc}. The difference between the divergence against the conditional independence of X_a and X_b given X_c and the marginal independence of X_a and X_b is invariant to permuting the symbols X_a, X_b and X_c.*

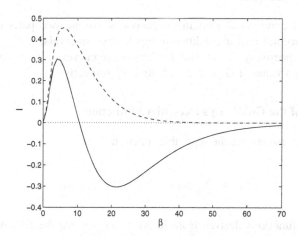

Fig. 2. Conditional information $I(X,Y\,|\,Z)$ *(dashed line)* and $G(X,Y,Z) = I(X,Y\,|\,Z) - I(X,Y)$ *(solid line)*

The above difference is denoted by $G(a,b,c)$. As a consequence of the proposition 1, the curve $G(a,b,c)$ and the three conditional information curves also contain all the marginal mutual information. Therefore, we also use pictures like Fig. 2 as tools for analysing the complexity of functions. In our framework, we refer to these curves as Boltzmann-mutual-information curves or simply Boltzmann-information curves.

From an evolutionary point of view, the Boltzmann-information curves show how selection influences the strength of the dependencies among the variables of the problem. If the algorithm uses Boltzmann selection as is the case of BEDAs, then β directly measures the selection pressure. Although for other selection schemes the connection is not direct, the information gathered from curves is still useful.

The curves are continuous, monotonously increasing up to their maximum values and decreasing to zero as β increases. This simple observation has an important implication for learning: there is a strong correlation between mutual information values at consecutive steps of the evolution.

Note in Fig. 1 (right), the horizontal lines at $I \approx 0.0069$ and $I \approx 0.0107$; they are thresholds for marginal and conditional independence[1]. We recall that $I(X,Y) = I(X,Y\,|\,\emptyset)$; it is assumed that the empty set has zero variables and thus $|\emptyset| = 1$. The above thresholds were computed with a confidence level of 95% and a sample size of $N = 280$ (this is the sample size used in the numerical simulations).

We now discuss the critical points of the Boltzmann-information curves. There are nine important critical points: the intersections of the threshold lines with

---

[1] Under the null hypothesis that conditional independence of X and Y given Z holds, the value $2NI(X,Y\,|\,Z)$ – which is called deviance against conditional independence – approximates a $\chi 2$ distribution with $|Z|(|X|-1)(|Y|-1)$ degrees of freedom, where N is the number of configurations in the sample and $|S|$ represents the number of possible values of the set of variables in S [35, Proposition 7.6.2]

the marginal and conditional Boltzmann curves determine two pairs of β values that define a marginal and a conditional dependence intervals, $[\beta^m_{min}, \beta^m_{max}]$ and $[\beta^c_{min}, \beta^c_{max}]$, respectively; the maximal values of the curves, β^m_M, β^c_M and β^G_M; the zero and minimum value of $G(1, 2, 3)$, β^G_z and β^G_m respectively.

3.2 Dissection of the Goldberg's Deceptive3 Function

In this section we investigate the separable function

$$F_{dec3} = \sum_{i=1}^{l} f_{dec3}(x_{3i-2}, x_{3i-1}, x_{3i})$$

and some other functions derived from it. As a rule we use the BFDA, but a few results are also presented for a FDA with truncation selection.

The notation used in the tables is as follows: N is the population size, n is the number of variables, $\%S$ is the success rate in 100 independent runs and G_c is the average generation where the optimum is found. For the average β values, we use β_{min} after the initial selection and β_{max} at the end of successful runs.

Deception and the Complete Bayesian Model

We start our investigation of the Deceptive3 by running the BFDA with the complete Bayesian model of the marginal distributions $p(x_{3i-2}, x_{3i-1}, x_{3i})$. In other words, it uses the factorizations

$$p(x_{3i-2}, x_{3i-1}, x_{3i}) = p(x_{3i-2})\, p(x_{3i}\,|\,x_{3i-2})\, p(x_{3i-1}\,|\,x_{3i-2}, x_{3i}) \qquad (8)$$

Equation (8) is the natural model for this function; any other model performs worse than it does. The following simulation confirms this behaviour. We run the BFDA 100 times, in a problem with 30 variables and 280 configurations. The algorithm always finds the optimum with $G_c = 12.97$. The average β at the end of the runs is 18.43, whereas the critical point β^G_z is reached as average at the generation 10. This means that for approximately 3/4 of the evolution the conditional information is stronger than the marginal information.

As can be seen from Fig. 1 (right) the variables are marginally and conditionally dependent in the range of β observed in the simulation of $[0, 18.43]$. Note that this interval is completely included in $[\beta^c_{min}, \beta^c_{max}] \subset [\beta^m_{min}, \beta^m_{max}]$. We recall that for three variables the complete model is the only one that does not have any independence relation, i.e. it is the best for the pair BFDA-Deceptive3.

We believe that deceptiveness is a direct consequence of having high values of mutual information. As we pointed out before, deception is a relative category that emerges solely in the context of the relationship problem-algorithm. In this relationship the problem contributes with high values of mutual information, whereas the algorithm's contributions are the selection and the collection of dependencies that it can deal with. The collection must be a proper sub-set of the problem's dependencies. We believe that the size and strength of the basins of attraction for any problem attractor depend on the amount of mutual information relevant to it. Without these three ingredients there can not be any deception at all. The amount of mutual information is a source of difficulty even when the right model or factorization is used.

BFDAs are perfect tools for studying the difficulty of the functions. They have everything that is needed:

- Selection is given explicitly through the parameter β.
- The collection of dependencies the algorithm can deal with are fixed by the factorization.
- The relation between mutual information and selection is given by the Boltzmann information curves.

In BFDAs, deception arises in the context of the relationship problem-factorization, i.e. a given problem may or may not be deceptive in relation to a particular factorization.

Reducing the Mutual Information

Let p_{f_{dec3},β_z^G} be the Boltzmann distribution of the Deceptive3 with β_z^G and Z_{f,β_z^G}, i.e. the distribution when the mutual and conditional information are the same (see Fig. 1).

In this section, we deal with the family of functions

$$f_{dec3}(\alpha) = \frac{log(p_\alpha)}{\beta_z^G} + \frac{log\left(Z_{f,\beta_z^G}\right)}{\beta_z^G} \qquad (9)$$

where $\alpha \in \{0, 0.05, 0.20, 0.40, 0.50\}$ and p_α is a distribution that obeys the following entropic relation

$$H(p_\alpha) = (1 - \alpha) H\left(p_{f_{dec3},\beta_z^G}\right) + 3\alpha$$

This type of entropic relation is discussed in Sect. 6.3. For the purposes of the current section it is enough to say that the mutual information in p_α decreases as α grows.

Table 2 shows the family of $f_{dec3}(\alpha)$ functions. Note that $f_{dec3}(0)$ is the Deceptive3. Besides, it is worth noting, that the symmetry of the Boltzmann information curves for the Deceptive3 is slightly broken in these functions. However, the difference is so small, that it is enough to show in Fig. 3 only the curves $I(1,2)$ and $I(1,2|3)$. The reader can easily check this by constructing the Boltzmann mutual information curves of these functions.

Table 3 presents the numerical results. The difficulty of the function decreases with increasing α, which means with increasing joint entropy and with decreasing mutual information. Note the influence of α in the convergence time: as α grows, G_c decreases. On the other hand, both β_{min} and β_{max} increase as α grows. We recall

Table 2. The family of $f_{dec3}(\alpha)$ functions

	$x_3x_2x_1$							
α	000	001	010	011	100	101	110	111
0.00	0.90	0.80	0.80	0.00	0.80	0.00	0.00	1.00
0.05	0.90	0.80	0.80	0.47	0.80	0.40	0.39	1.00
0.20	0.90	0.82	0.81	0.63	0.81	0.57	0.56	0.99
0.40	0.89	0.83	0.82	0.71	0.82	0.66	0.65	0.98
0.50	0.89	0.83	0.83	0.74	0.82	0.70	0.69	0.97

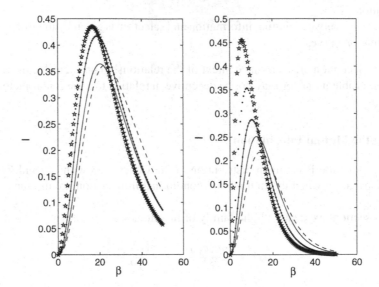

Fig. 3. Boltzmann mutual information curves for the family $f_{dec3}(\alpha)$: (*left*) marginal, (*right*) conditional. From *top* to *bottom*, f_{dec3}, $f_{dec3}(0.05)$, $f_{dec3}(0.20)$, $f_{dec3}(0.40)$ and $f_{dec3}(0.50)$

Table 3. BFDA runs with the $f_{dec3}(\alpha)$ with the complete Bayesian model. The average β values after the initial selection and at the end of successful runs are shown in columns β_{min} and β_{max}, respectively. Setting: $N = 280, n = 30$

α	%S	G_c	β_{min}	β_{max}
0.00	100	12.97	0.75	18.43
0.05	100	10.31	1.41	18.25
0.20	100	9.32	2.14	21.11
0.40	100	8.39	2.96	23.14
0.50	100	8.14	3.53	25.86

that $\beta_{min} = \beta(1) = \Delta\beta(1) = c/\sqrt{Var_f(\beta(1))}$, i.e. the standard deviation of the fitness decreases in the first generation with increasing α. Besides, for all functions we have that the interval $[\beta_{min}, \beta_{max}]$ is included in their respective $[\beta_{min}^c, \beta_{max}^c]$.

If the reader constructs the unitation representation (Fig. 1) of the functions $f_{dec3}(\alpha)$, he or she will observe that only the depth of the valley at unitation equal to two changes significantly. For example, $f_{dec3}(0.05)$ is exactly equal to the Deceptive3, except in the case when the unitation is equal to two. This is remarkable because the definition of these functions did not consider any unitation argument.

Models with a Missing Arc

We investigate the performance of the BFDA when the marginal distributions of the form $p(x_{3i-2}, x_{3i-1}, x_{3i})$ are approximated with all Bayesian models with one missing arc. Consider the following factorizations:

$$p^{12-32}(x_1, x_2, x_3) = p(x_1)p(x_3)p(x_2|x_1, x_3) \tag{10}$$

$$p^{13-32}(x_1, x_2, x_3) = p(x_1, x_3)p(x_2|x_3) \tag{11}$$

$$p^{12-13}(x_1, x_2, x_3) = p(x_1, x_3)p(x_2|x_1) \tag{12}$$

Due to the symmetry of the function with respect to the mutual information, it is enough to study these cases. For example, in the factorization 12-32 the arc 1-3 is missing and the arcs $1 \to 2$ and $3 \to 2$ are present. However, it behaves exactly as the factorizations 21-31 and 13-23.

The results are presented in the first row ($\alpha = 0$) of Table 4. The BFDA behaves much better with the factorization 12-32 than with the factorizations 12-13 and 13-32. The use of the last two factorizations leads to similar results. In what follows, we try to explain this behaviour in the context of Boltzmann information curves.

It is worth noting, that β_{max} is about 30 for all models, which is close to β_{max}^c. Furthermore, we have observed that the critical value β_z^G is reached as average in the generation 10 with the model 12-32 and in the generation 12 with the models 12-13 and 13-32. This means that a successful run occurs in range of β where both the marginal and the conditional information are above the independence thresholds, i.e. the variables are not independent. Moreover, during the first half of the evolution (before β_z^G is reached) $G(1, 2, 3) > 0$.

Table 4. BFDA runs with the $f_{dec3}(\alpha)$. The marginal distributions $p(x_{3i-2}, x_{3i-1}, x_{3i})$ are approximated with all two-arcs models. Setting: $N = 280, n = 30$

	12-32			12-13			13-32		
α	$\%S$	G_c	β_{max}	$\%S$	G_c	β_{max}	$\%S$	G_c	β_{max}
0	94	18.57	30.5	22	21.82	30.48	34	22.20	29.64
0.05	99	14.75	29.58	92	16.15	29.31	84	16.11	28.67
0.20	100	12.97	32.03	99	13.25	29.69	95	13.14	28.82
0.50	100	11.24	37.26	100	10.46	32.95	100	10.37	31.89

By comparing (10)–(12) with the chain rule, it is easy to see that each equation makes exactly one wrong assumption:

- Equation (10) assumes marginal independence of X_1 and X_3.
- Equation (11) assumes conditional independence of X_2 and X_1 given X_3.
- Equation (12) assumes conditional independence of X_2 and X_3 given X_1.

The conditional mutual information is farther away from its independence threshold than the marginal mutual information. The independence lines get closer as the sample size increases; for $N = 280$, their difference is just 0.0038. Therefore, we can assume that there is a unique threshold I_t. It is easy to see that $\int_0^{\beta_z^G} G(1, 2, 3)\, d\beta$ can be used as an estimate of the magnitude of the error of using the factorizations 12-13 or 13-32 instead of 12-32. In other words, the assumption of the model 12-32 is much less traumatic than the other assumptions when $\beta \in [\beta_{min}^c, \beta_{max}^c]$. The situation is reversed for $\beta > \beta_{max}^c$, but this happens when the first half of the evolution is already gone, thus having little impact in the outcome of the optimization.

We have also tested the above factorizations with the functions $f_{dec3}(\alpha)$. Table 4 presents the results. As was shown in Sect. 3.2, the reduction of the mutual information also implies a reduction of the difficulty of the function. Here, we can observe the effect on the convergence time as well as on the success rate. Note for example, that from $\alpha = 0$ to $\alpha = 0.05$ the success rate goes from 22% to 92% in the case of the factorization 12-13. Another interesting observation is about the difference between the performance of different factorizations as α grows. For example, the difference between the convergence time for the complete factorization (8), 12-13-32, and for the factorization 12-13 decreases as α grows: 8.85, 5.84, 3.93, 2.66 and 2.32. We believe that the last result is an evidence supporting the following statement: the reduction of the mutual information increases our choices in model selection.

Some Results with Truncation Selection

For the sake of completeness, Table 5 presents the results of running a FDA with truncation selection on the family of functions $f_{dec3}(\alpha)$. The reader can easily check the similarities of these results with those obtained with Boltzmann selection. For example, they also support the claim that the reduction of the mutual information amounts to a reduction of the functions difficulty.

4 Designing Test Functions by Maximum-Entropy

In spite of recent research advances in EDAs we still do not have a complete, sound, consistent and rigorous theory of evolutionary algorithms. In practice, this leads to the use of simulation as a fundamental tool of verification, validation and comparison of algorithms. One common simulation method is the use of test functions obtained by concatenation of elementary functions of small order. Usually, the design of such functions is focused on considerations about specific aspects of the complexity of

Table 5. FDA runs with the $f_{dec3}(\alpha)$. Setting: $N = 280$, $n = 30$ and truncation selection of 0.3

	12-13-32		12-32		12-13		13-32	
α	%S	G_c	%S	G_c	%S	G_c	%S	G_c
0	100	5.99	91	8.49	35	9.74	34	9.68
0.05	100	5.74	95	8.18	65	8.61	75	8.52
0.20	100	5.32	99	7.18	85	7.61	89	7.65
0.50	100	4.91	100	6.74	100	6.30	99	6.42

the elementary functions: multimodality, isolation of the optimum value, proximity of the function values of the good configurations, frustration of overlapped elementary functions, etc. In this scenario, it is important to know the properties of the elementary functions and how these properties are combined to define the properties of the whole function. Moreover, it would be useful to design functions that are not given as a combination of smaller elementary functions.

The design of benchmark functions for testing EDAs have to emphasize, in the first place, the complexity of the probabilistic structure of the search distributions. The fitness function, the intensity and type of selection determine for each configuration its probability of being in the selected set and consequently the probabilistic structure of the search distributions.

A successful EDA builds a probabilistic model that captures the important correlations of the search distribution, assigning high probability values to the selected configurations. Therefore, it would be convenient to design functions that enforce a given set of "important correlations", but do not enforce any other correlation constraint. In this section, we present an approach to this problem, where the designer gives a collection of probabilistic constraints that have to be fulfilled by the search distributions of the function. Our method is connected to the concept of entropy because it constructs a maximum-entropy distribution that satisfies the given constraints.

4.1 The General Framework

The corner stone of our approach to the design of benchmark functions for discrete optimization is what we have called the family of Boltzmann functions

$$f_\beta(x) = \frac{log(p_{f,\beta}(x))}{\beta} + \frac{log(Z_f(\beta))}{\beta} \tag{13}$$

Equation (13) comes from the definition of the Boltzmann probability mass $p_{f,\beta}(x)$. From the point of view of this model, (13) are members of the parametric class $\mathfrak{F}(\beta, Z, p_{f,\beta}(x))$, which could be refined by including additional parameters of the distribution $p_{f,\beta}(x)$. For example, when the distribution factorizes and no factor contains more than K variables, we are dealing with the parametric sub-class $\mathfrak{F}(\beta, Z, p_{f,\beta}(x), K)$.

Avoiding the Exponential Effort

The computation of the partition function is always problematic; it needs an exponential effort. Fortunately, in our approach this can be avoided. Note that in (13), the second term is a constant that is added in all configurations. It is a shift along the fitness dimension and has little to do with the complexity of the function. Therefore, nothing prevents us from fixing the value of the partition function. Moreover, for BFDA the following lemma holds.

Lemma 3 *The difficulty of (13) for a BFDA is completely determined by the distribution $p_{f,\beta}(\boldsymbol{x})$.*

Proof 1 *The proof follows immediately from lemma 2.*

If the distribution $p_{f,\beta}(\boldsymbol{x})$ is known and Z is set to an arbitrary value, then the function $f_\beta(\boldsymbol{x})$ is well defined for any β, i.e. for any configuration $\boldsymbol{x}$, the value $f_\beta(\boldsymbol{x})$ can be computed. This means that the computation of the function for all possible configurations is not necessary.

Usually, we use factorizations to deal with the exponential complexity of distributions. In the context of functions design, the factorizations also help to compute the optima and the central moments of the functions. This kind of information is useful to understand the functions' properties. Moreover, sometimes it is useful to have a fast procedure for computing the optima of benchmark functions when testing evolutionary algorithms. For example, when the benchmark functions are drawn from a distribution (Sect. 4.3) and the optima are needed to set the stopping criteria. The reader is referred to [27,28] for a complete description of two methods that compute the above-mentioned values for junction tree factorizations.

Whenever we have a distribution we can build a Boltzmann function. For example, there are famous Bayesian networks (like the ALARM network [2]) that can be used for this purpose. However, in this chapter we are more interested in the case when, instead of having a distribution, we have a collection of probabilistic constraints that must be satisfied by the distribution.

Dealing with Mutual Information Constraints

We have already met the family of functions (13) in Sect. 3.2. Also we have learned that the mutual information of $p_{f,\beta}(\boldsymbol{x})$ contains a lot of information about the complexity of the function $f_\beta(\boldsymbol{x})$. Therefore, when dealing with complexity issues, it makes sense to design functions that fulfill mutual information constraints like:

$$
\begin{aligned}
I\left(X_a, X_b \mid X_c\right) &\geq A \\
I\left(X_a, X_b \mid X_c\right) &\leq B \\
I\left(X_a, X_b \mid X_c\right) &\leq I\left(X_d, X_e \mid X_f\right)
\end{aligned}
\tag{14}
$$

In (14), the letters a, b, c, d, e and f denote sub-sets of indexes, and A, B are constants. Moreover, X_c and X_f may be empty, meaning that the expressions represent marginal information.

We formulate the general design problem as follows:

Given a collection of mutual information constraints $\mathfrak{C} = \{c_1, \ldots, c_L\}$, find a function $f(\boldsymbol{x})$, whose Boltzmann distribution satisfies $\mathfrak{C}$ within a given temperature interval.

Our approach to the above-mentioned problem considers structural and parametric constraints. The structural constraints are specified by Bayesian or Markov networks, which use the separation and d-separation concepts [30] to codify statements of probabilistic independence. The parametric constraints are statements about the configurations' probabilities.

In our method, the inequality

$$A_k \leq \sum_{i=0}^{M-1} a_{ik} p\left(x^{(i)}\right) \leq B_k \qquad (15)$$

denotes the k-th parametric constraint. The sum is for all configurations $x^{(i)}$ of X, i.e. M denotes the size of the space. A_k, B_k are real constants and $a_{ik} \in \{0, 1\}$.

It is worth noting, that some sub-sets of the inequalities (15) may define marginal distributions of $p(\boldsymbol{x})$ when $A_k = B_k$ for all inequalities in the sub-set. In this chapter, we deal only with this type of constraint. Therefore, the mutual information constraints (14) have to be mapped to marginal distributions. It is an interesting open question how to translate other types of constraints to probabilistic statements.

Once the collection of marginal constraints has been derived from the mutual information constraints it is necessary to compute the joint probability distribution. The next section presents the issue.

Computing the Joint Probability Distribution

Algorithm 2 presents the general scheme of the design of Boltzmann functions. In the step 2, the algorithm computes a junction tree from the given structural constraints. The computation of a junction tree out from a Bayesian or a Markov network is a well-studied problem [31]. In the step 3, is computed a maximum-entropy distribution that is compatible with the given structural and parametric constraints. There are two possibilities as it is explained below.

The classic implementation of the IPF algorithm can be used to compute the joint probability distribution when the number of variables is small. If the collection of marginals is consistent, the outcome of running the IPF is a maximum-entropy joint.

For larger number of variables, the IPF has to be combined with the junction tree technique. It is run locally on the nodes and the results are sent as messages to the neighboring nodes. It has been proved that this converges to the unique maximum-entropy solution, so it is equivalent to IPF. The interested reader is referred to [23,29] for details on the implementation of the method for computing maximum-entropy distributions on multi-connected Bayesian networks and polytrees.

Finally, in the step 4, the desired function is computed as the function that makes of $p(\boldsymbol{x})$ a Boltzmann distribution with parameters β, Z and $f(\boldsymbol{x})$.

Algorithm 2 A maximum-entropy method for designing Boltzmann functions

---

Step 1   Input β, Z, and the collection of structural and parametric constraints.
Step 2   Compute a junction tree compatible with the structural constraints.
Step 3   Compute the maximum-entropy junction tree distribution $p(\boldsymbol{x})$ that fulfill the parametric constraints.
Step 4   Output $f_\beta(\boldsymbol{x}) = \frac{log(p(\boldsymbol{x}))}{\beta} + \frac{log(Z)}{\beta}$

4.2 Designing the First-Polytree Functions

In this section, we take a closer look at our method through the design of three binary functions whose structure of the search distribution is single-connected. For obvious reasons, we say that they belong to the polytree class of functions. The functions have been called FirstPolytree3 (f_{Poly}^3), FirstPolytree5 (f_{Poly}^5) and OneEdge ($f_{OneEdge}^3$). Figure 4 presents their graph definitions, i.e. their structural constraints.

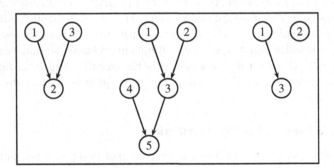

Fig. 4. Structural constraints of the first-polytree functions. From *left* to *right*: f_{Poly}^3, f_{Poly}^5 and $f_{OneEdge}^3$

The polytree functions can be specified with simple mutual information constraints. The marginal mutual information of every pair of parents of a variable should be below the marginal independence threshold I_t, for the given confidence level. Similarly, the marginal mutual information of every child-parent pair should be greater than I_t.

We first list the marginal mutual information constraints:

OneEdge: $I(1,3) > I_t$ $I(1,2) < I_t$ $I(2,3) < I_t$
FirstPolytree3: $I(1,3) < I_t$ $I(1,2) > I_t$ $I(2,3) > I_t$
FirstPolytree5: $\begin{aligned} I(1,2) < I_t \quad I(3,4) < I_t \quad I(1,3) > I_t \\ I(2,3) > I_t \quad I(3,5) > I_t \quad I(4,5) > I_t \end{aligned}$

Algorithm 3 Designing bivariate marginals

Step 1	Input $I(X,Y)$.
Step 2	Set the univariate probabilities to some random values p_x y p_y.
Step 3	**if** $I(X,Y) < I_t$, **then** set $p_{xy} = p_x p_y$.
Step 4	**if** $I(X,Y) > I_t$, **then** set p_{xy} as far as possible from $p_x p_y$.

Another type of constraints is needed to specify the orientation of the edges. The d-separation concept says that in the structure $X \to Z \gets Y$, the variables X and Y are marginally independent and conditionally dependent given Z [5]. If I_t^c denotes the conditional independence threshold, then the second list of mutual information constraints is the following:

OneEdge: $I(1,3|2) > I_t^c \; I(1,2|3) < I_t^c \; I(2,3|1) < I_t^c$

FirstPolytree3: $I(1,3|2) > I_t^c \; I(1,2|3) > I_t^c \; I(2,3|1) > I_t^c$

FirstPolytree5: $\begin{array}{l} I(1,3|2) > I_t^c \; I(1,2|3) > I_t^c \; I(2,3|1) > I_t^c \\ I(3,4|5) > I_t^c \; I(3,5|4) > I_t^c \; I(4,5|3) > I_t^c \end{array}$

Designing Bivariate Marginals with Given Mutual Information

Once the list of constraints has been given, we construct a set of bivariate marginals that satisfy the constraints. The algorithm 3 does the job.

It is known, that the sufficient statistics for the specification of any binary bivariate marginal $p(x,y)$, are the values $p_x = p(X=1)$, $p_y = p(Y=1)$ and $p = p_{xy}(X=1, Y=1)$. Moreover, either $p_{xy} \in [\max(p_x + p_y - 1, 0), p_x p_y]$ or $p_{xy} \in [p_x p_y, \min(p_x, p_y)]$. Taking the univariate probabilities p_x and p_y as input values, we proceed as follows: if $I(X,Y) < I_t$, then we just make $p_{xy} = p_x p_y$. Otherwise, we put p_{xy} as far as possible from $p_x p_y$ to maximize the mutual information. Finally, the bivariate marginal is given by

$$p_{xy}(00) = 1 - p_x - p_y + p_{xy} \quad p_{xy}(10) = p_y - p_{xy}$$
$$p_{xy}(01) = p_x - p_{xy} \qquad\qquad p_{xy}(11) = p_{xy} \tag{16}$$

After all univariate and bivariate marginals have been computed, the next step of the Algorithm 2 is the construction of the joint probability.

The classic implementation of the IPF algorithm can deal with our functions because the number of variables is small. If the IPF is run with the above marginals, a trivariate maximum-entropy joint is obtained. For larger number of variables we must resort to the junction tree implementation of the maximum-entropy algorithm.

Each node of the junction tree associated to a polytree is formed by one variable and the set of its parents. This means that the trivariate functions have only one clique and therefore, the simple IPF will be enough. The junction tree for the function FirstPolytree5 contains two cliques and therefore, the advanced implementation of the algorithm is needed. In this way, we have constructed high order marginals using only univariate and bivariate marginals. We must check that the second list of constraints are also fulfilled. Moreover, to guarantee consistency the design of these marginals must satisfy additionally the following constraint [17]:

Let d be the number of variables in a junction tree node. For all $2 \leq k \leq d$ and all possible choices $j_1, \ldots, j_k$ of k elements out of $\{1, \ldots, d\}$ the condition

$$1 \geq \sum_{i=1}^{k} p_{j_i} - \sum_{i,l=1, i \neq l}^{k} p_{j_i j_l}$$

must be fulfilled.

We use the values 12.94, 16.40 and 87.97 as input values for the partition functions of $f_{OneEdge}^3$, f_{Poly}^3 and f_{Poly}^5, respectively. The univariate probabilities also are set. For example, the values used in the function f_{Poly}^3 are 0.79, 0.46 and 0.24 for X_1, X_2 and X_3, respectively. Finally, we set $\beta = 2$.

Tables 6, 7 and 8 present the resulting functions $f_{OneEdge}^3$, f_{Poly}^3 and f_{Poly}^5, respectively. The Boltzmann distributions with parameter $\beta = 2$ are polytree distributions satisfying the structural and parametric constraints given above. The reader can easily check this by computing their Boltzmann distributions and then computing the mutual information values.

Table 6. OneEdge function

$x_3 x_2 x_1$	$f_{OneEdge}^3(\mathbf{x})$	$x_3 x_2 x_1$	$f_{OneEdge}^3(\mathbf{x})$
000	1.042	100	-0.083
001	-0.736	101	0.092
010	0.357	110	-0.768
011	-1.421	111	-0.592

Table 7. FirstPolytree3 function

$x_3 x_2 x_1$	$f_{Poly}^3(\mathbf{x})$	$x_3 x_2 x_1$	$f_{Poly}^3(\mathbf{x})$
000	-1.186	100	-4.391
001	1.074	101	-1.122
010	0.469	110	-0.083
011	0.096	111	0.553

Investigating the Polytree Functions

Figure 5 presents the Boltzmann conditional curves and the curve $G(1, 2, 3)$ for the FirstPolytree3 function. Note that the curves $I(1, 3 \mid 2)$ and $G(1, 2, 3)$ coincide

Table 8. FirstPolytree5 function ($\mathbf{x} = (x_5, x_4, x_3, x_2, x_1)$)

x	$f^5_{Poly}(\mathbf{x})$	x	$f^5_{Poly}(\mathbf{x})$	x	$f^5_{Poly}(\mathbf{x})$	x	$f^5_{Poly}(\mathbf{x})$
00000	−1.141	01000	−0.753	10000	−3.527	11000	−6.664
00001	1.334	01001	1.723	10001	−1.051	11001	4.189
00010	−5.353	01010	−4.964	10010	7.738	11010	−10.876
00011	−1.700	01011	−1.311	10011	−4.085	11011	−7.223
00100	0.063	01100	1.454	10100	1.002	11100	−1.133
00101	−0.815	01101	0.576	10101	0.124	11101	−2.011
00110	−0.952	01110	0.439	10110	−0.013	11110	−2.148
00111	−0.652	01111	0.739	10111	0.286	11111	−1.849

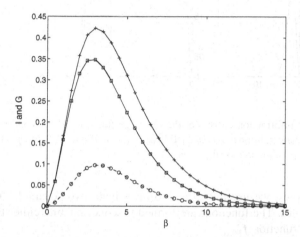

Fig. 5. Boltzmann information curves for the FirstPolytree3 function: (*plus*) $I(2,3|1)$, (*square*) $I(1,2|3)$, (*solid line*) and (*circle*) $G(1,2,3)$. Note that the last two curves coincide at the chosen scale

at the chosen scale. This means that the marginal curve $I(1,3)$ is close to zero. The actual values are below 10^{-3}, which amounts to independence for sample sizes below 2000 configurations. The other two marginal dependencies are quite strong for $\beta = 2$ (the value used in the design of the function). As far as $G(1,2,3)$ is always positive we conclude that for any selection pressure we have more evidence to decide against conditional independence than against marginal independence. Note that the conditional interval $[\beta^c_{min}, \beta^c_{max}]$ for $I(1,3|2)$ is completely included in the other two conditional intervals for any sample size. Note that in contrast with the Deceptive3, in this function the value β^G_z is not inside the interval $[\beta^c_{min}, \beta^c_{max}]$.

Figure 6 presents the conditional and marginal Boltzmann curves for the OneEdge function. For all β, the values $I(1,3)$ and $I(1,3|2)$ are very close; their difference, $G(1,2,3)$, is less than 10^{-8} and negative. The curves $I(1,2|3)$ and $I(2,3|1)$ are below 10^{-9}, which implies independence.

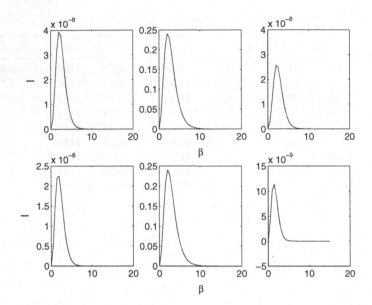

Fig. 6. Boltzmann information curves for the OneEdge function. The second row, from *left* to *right*, contains the conditional curves $I(1,2|3)$, $I(1,3|2)$ and $I(2,3|1)$. The upper row contains the corresponding marginal curves

In what follows, we use the BFDA to investigate two separable functions of 30 and 60 variables. The functions are formed by concatenating either the function $f^3_{OneEdge}$ or the function f^3_{Poly}.

By just looking at Tables 1, 6 and 7 it is difficult to draw any conclusion about what is the best factorization and which is the more difficult function for the BFDA. Following the theorem 2 the choice would be the complete model, which was shown to be the best factorization for the Deceptive3. However, the simulations of this section show that this is not the case for the other functions.

Table 9 presents the results of running the BFDA with the population size set to 120 for the FirstPolytree3. This time the factorization 12-32 is the clear winner. The convergence is almost twice as fast and its success rate is twice as high, in the factorization 12-32, as in the complete model. Similarly, the number of function evaluations is much bigger if the complete factorization is used. Therefore, we conclude

Table 9. BFDA runs with the FirstPolytree3. Setting: $N = 120$

		12-32				12-13-32		
n	$\%S$	G_c	β_{min}	β_{max}	$\%S$	G_c	β_{min}	β_{max}
30	95	13.32	0.197	4.927	55	22.29	0.139	7.954
60	85	13.49	0.196	5.526	38	21.95	0.139	8.715

that the assumption making the variables 1 and 3 marginally dependent is wrong. This is what we expected from our design decisions.

Regarding the Boltzmann curves the important observation is that the runs occur within the most inner interval $[\beta_{min}^c, \beta_{max}^c]$. Moreover, the better the conditions for the optimization are, the smaller the value of β_{max}. For example, for a fixed model, the smallest problem converges with the smallest β_{max}. Alternatively, if the size of the problem is fixed, then the best model has a smaller β_{max}. The same is observed in the simulations with the OneEdge. Table 10 presents the results.

Table 10. BFDA runs with the OneEdge. Setting: $N = 120$

	12-32			12-13-32			13		
n	$\%S$	G_c	β_{max}	$\%S$	G_c	β_{max}	$\%S$	G_c	β_{max}
30	94	11.64	4.724	98	9.36	3.917	100	9.25	3.867
60	39	20.82	7.026	75	17.067	6.151	98	17.03	6.045

For the OneEdge function three models are investigated. The model 13 – the one that is used in the design of the function – is the best. For example, compare the success rate of the complete model and the best model for 60 variables. Note that the convergence time is the same. In the model 12-32 the variables 1 and 3 are independent, which explains its poor performance.

We also have investigated the functions with the FDA. Besides the separable problem, in the simulations an overlapped additive function have been included. The overlapped case is constructed as follows: the last variable of a sub-set is also the first variable of the next sub-set in the additive decomposition. We use the letter O to denote this case. For example, contrast O-12-32 with 12-32.

Tables 11 and 12 present the numerical results. The factorizations 12-32 and O-12-32 are the best for the functions f_{Poly}^3. Similarly, the models 13 and O-13 perform better for the function $f_{OneEdge}^3$. Both the separable and the overlapped complete models do not scale well. For example, compare the success rates for the overlapped case of the OneEdge function.

Table 11. FDA runs with the FirstPolytree3

	N	$\%S$	G_c	$\%S$	G_c	$\%S$	G_c
		$n = 30$		$n = 60$		$n = 90$	
12-13-32	120	92	5.39	42	9.10	6	12.33
12-32	120	93	5.36	67	9.46	27	13.04
		$n = 31$		$n = 61$		$n = 91$	
O-12-13-32	200	83	6.81	25	11.04	4	14.00
O-12-32	200	94	6.59	63	11.28	20	14.90

Table 12. FDA and the OneEdge function

	N	$\%S$	G_c	$\%S$	G_c	$\%S$	G_c
		$n = 30$		$n = 60$		$n = 90$	
12-13-32	60	54	5.12	6	8.33	0	–
13	60	95	4.53	65	8.18	28	10.71
		$n = 31$		$n = 61$		$n = 91$	
O-12-13-32	100	71	5.57	20	10.00	2	12.50
O-13	100	100	5.20	81	9.03	57	12.52

We summarize the results as follows. The behaviour of the polytree functions investigated in this section, agrees with our design expectations. On the other hand, a clear correspondence between what happened in the simulations and the Boltzmann curves was observed. We take this as a sort of validation of both the usefulness of the analysis and design method introduced in this chapter.

4.3 Designing Random Class of Functions

In the previous section, we followed the common practice of concatenating low order functions to form larger additively decomposable functions. However, it would be useful if we could design a complete additive function with a given structure without resorting to the trick of concatenating small sub-functions. Moreover, it would be even more useful to design random class of functions, instead of isolated functions. To accomplish this task our method has to be extended.

In this section, we restrict ourselves to the design of the random class of binary polytree functions. This will provide the reader with general ideas and guidelines that might be helpful to undertake other design efforts.

Sampling the Structural Constraints

The first step is the generation of a random polytree graph. As was explained in Sect. 4.2, it is the structural constraint.

There exist simple methods for generating random graphs. Any of these algorithms together with a rejection sampling technique to reject graphs with directed cycles and undirected cycles, will do the job. At this stage the method outputs the graph, its junction tree and two lists, L_1 and L_2. If a pair (i, j) belongs to the first list, both i and j are parents of the same node and therefore, $I(X_i, X_j) < I_t$. On the other hand, the second list contains a pair (i, j), if and only if, j is the parent of i. In this case, $I(X_i, X_j) > I_t$. For each pair (i, j) in the lists, we sample a bivariate marginal distribution $p(x_i, x_j)$, that obeys the corresponding mutual information constraint. This non-trivial task is discussed in what follows.

Sampling Bivariate Marginals Under Independence

The problem is related to the evaluation of the exact sampling distributions of the cell counts in applied multivariate analysis [35]. Therefore, we set $n_i = Np(x_i)$, $n_j = Np(x_j)$ and $n_{ij} = Np(x_i, x_j)$, where N is the sample size.

Let assume Poisson, multinomial or independent multinomial sampling. Under the null hypothesis of independence, the conditional distribution of n_{ij} given the observed marginal counts n_i and n_j is the central hyper-geometric distribution, which is known exactly. The random scalar variable N_{ij} is given by

$$N_{ij} \sim \frac{\binom{n_i}{n_{ij}} \binom{N - n_i}{n_j - n_{ij}}}{\binom{N}{n_j}} \tag{17}$$

Let n_i and n_j be given. Then, for any pair (i, j) in the list L_1 we generate the bivariate marginal $p(x_i, x_j)$ by sampling n_{ij} from (17), and then substituting $p_{ij} = n_{ij}/N$, p_i and p_j in (16).

It is worth noting, that the method can be extended to deal with variables of cardinality greater than two [35].

Sampling Correlated Bivariate Marginals

For the computation of the marginals associated to the list L_2, the solution comes from the exact non-null distribution theory [1].

Let assume multinomial sampling and let θ be the odds ratio [35]. Conditional on n_i and n_j, the distribution of n_{ij} depend only on θ, and is given by

$$N_{ij} \sim \frac{\binom{n_i}{n_{ij}} \binom{N - n_i}{n_j - n_{ij}} \theta^{n_{ij}}}{\sum\limits_{u=m}^{M} \binom{n_i}{u} \binom{N - n_i}{n_j - u} \theta^u} \tag{18}$$

where $m = max(0, n_i + n_j - n)$ and $M = min(n_i, n_j)$.

As far as the constraints are specified using the mutual information, one could try a reparameterization of (18). However, we use directly the odds ratio, which obeys $0 \leq \theta < \infty$. Values of θ farther from 1.0 in a given directions represent higher values of mutual information. Moreover, if $\theta_1 = 1/\theta_2$, then both θ_1 and θ_2 represent the same level of dependence.

Let n_i and n_j be given. For any pair (i, j) in the list L_2, we compute θ according to the mutual information $I(X_i, X_j)$. Then, n_{ij} is sampled from (18) and $p(x_i, x_j)$ is obtained from (16).

Once all the bivariate marginals have been computed we are ready to build the maximum-entropy junction tree. Afterwards, we obtain an instance of the random class by substituting in (13) the distribution and the given β.

How to Test EDA Algorithms

The procedure introduced in the previous sections allows us to define a large class of functions: the class of random Boltzmann polytree functions (RBPF). We denote the class by $RBPF(n, K, \beta)$, where K is the maximum number of parents in the polytree. Note that Z is not included as a parameter because it is chosen automatically in such a way to make the function non-negative for any configuration x.

Testing evolutionary algorithms have been recognized as a major problem in current EDA research [23]. We believe that the approach presented in this chapter will improve the ability of the research community to test and compare EDA algorithms. Moreover, the design of random classes of Boltzmann functions should help to understand the complex mechanisms involved in EDA optimization, because now we have an explicit control of the dependencies presented in the functions. We are confident that others random classes can be designed using similar ideas to the ones presented in this chapter.

Within our framework, any optimization algorithm should be tested in samples of carefully designed random classes of functions. In other words, instead of using a single function and running the algorithm 100 times, we prefer to use once 100 different functions sampled from the same random class.

5 Learning Low Cost Max-Entropy Distributions

A critical problem of learning search distributions in EDAs is the sample complexity, which is related with the number of functions evaluations. One important challenge of an evolutionary algorithm is the reduction of the number of evaluations, while the effectiveness and efficiency of the search is preserved. In this section we will use the concept of entropy to achieve this goal. Our idea is simple: the true search distribution is substituted by an approximation, which can be reliably computed with less population size.

The following definitions will help to clarify our ideas.

Algorithm 4 Maximum-entropy EDA

Step 1   Set $t \leftarrow 1$. Generate $N \gg 0$ points randomly.
Step 2   Select M points according to a selection method.
Step 3   Find a suitable R and learn a $\mathbb{R}_{p^s(x), R}$ from the selected set.
Step 4   Compute the maximum entropy distribution $p_R^s(X)$.
Step 5   Sample N new points according to the distribution

$$p(x, t+1) = p_R^s(x_1, \ldots, x_n)$$

Step 6   Set $t \leftarrow t + 1$. If termination criteria are not met, go to step 2.

Definition 5 *Let $p(X_1, \ldots, X_n)$ be the factorization of the selected set. We say that $p(X_1, \ldots, X_n)$ is a true search distribution if it was computed from a data set, whose size allows reliable estimates of the factors' probabilities.*

Definition 6 *Let X be a random vector of dimension n and $R = \{r_1, \ldots, r_m\}$ be a set of index-sets. A restriction of a joint distribution $p(x)$ is a set of marginal distributions*

$$\mathbb{R}_{p(\mathbf{X}),R} = \{p(X_{r_1}), \ldots, p(X_{r_m})\}$$

of $p(x)$, such that the following holds:

1. $\forall i, 1 \leq i \leq m, r_i \subset \{1, \ldots, n\}$ and $r_i \neq \emptyset$
2. $\forall i, j, 1 \leq i, j \leq m, r_i \nsubseteq r_j$

Definition 7 *Let $\mathbb{R}_{p(\mathbf{X}),R}$ be a restriction of $p(x)$, then $p_R(x)$ is defined as the maximum-entropy distribution that fulfills the constraints $\mathbb{R}_{p(\mathbf{X}),R}$.*

Using the above definitions, we introduce an EDA that uses the MEP (see algorithm 4). We have called it maximum-entropy EDA (meEDA).

Step 2 is a critical point of the meEDA algorithm because the algorithm has to choose a suitable restriction set. It is an open problem how to identify good restrictions of the search distributions. For example, besides the primary goal of getting a sampling distribution with less cost than the true distribution, there could be other reasons that determine a good choice of the restriction set. On the other hand, an efficient procedure for the computation of the maximum-entropy distribution exists only if the structure of the restriction set satisfies certain constraints. The next section presents an algorithm EDA where the maximum-entropy distribution can be computed efficiently.

5.1 Extending PADA2 with Maximum-Entropy

The polytree functions designed in Sect. 4 have a common property: their search distributions are single-connected. In this section we modify PADA2 – an algorithm specially designed to deal with single connected Bayesian networks – to transform it into a meEDA.

The polytree approximation distribution algorithm (PADA) [33,34] was designed to deal with the whole class of single-connected Bayesian networks; also called the polytree class. It uses first, second and third order marginals to recover polytrees from data. In this work we will use PADA2 [33] – variant of PADA, which learns only first and second order marginals distributions. PADA2 is inspired by an algorithm proposed by Rebane and Pearl [30]. We shortly outline the basic ideas behind the algorithm.

A polytree with n variables has a maximum of $n - 1$ arcs, otherwise it would not be single connected. PADA2 chooses the edges that have the largest values of the magnitude $H(X) + H(Y) - H(X, Y)$, which is also called mutual information [3]. The selection of the edges is done by a greedy maximum weight spanning tree algorithm. These edges form the so-called skeleton (the underlying undirected graph).

After the construction of the skeleton is done, a procedure tries to orient the edges by using the following scheme: if $X - Z - Y \in skeleton$, then whenever $H(X) + H(Y) = H(X, Y)$ holds statistically it orients the edges to Z. In this case it is said that Z is a head to head connection. The edges that were not oriented after the above test are directed at random without introducing new head to head connections.

Both during learning and sampling, EDAs that learn general Bayesian networks need a population size, which is exponential in the number of parents. This is important to get reliable estimates of the conditional probabilities. However, although PADA2 only learns first and second order marginals, it has to deal with the same exponential problem in the sampling step, i.e. what is gained in learning is lost in the sampling.

To transform PADA2 into mePADA2 we must define the polytree's restriction set, i.e. all bivariate marginals that belong to the skeleton and the bivariate marginals defined for each pair parent-child. Note that this restriction set was used as a parametric constraint in Sect. 4.2. The next step consists in computing the higher order marginals as the maximum-entropy distributions that obey the given second order marginals. Consistency is guaranteed by propagating across the junction tree associated to the polytree as was explained in Sect. 2.4.

Now we present some numerical results to support the theoretical claims. We use two separable ADF functions, which are based on the Deceptive3 and FirstPolytree5. Although the structure of the Deceptive3 function is not single-connected, PADA2 tries to build the better single-connected approximation it can. It is remarkable that the method still produces very good results. We recall that the basic claim of our research is that the maximum-entropy distribution, which can be computed with a smaller population size than the true search distribution, is suitable for sampling. Moreover, sometimes it gives better results than the true distribution.

The algorithms are run until a maximum of 20 generations with a truncation selection of 0.3 and without elitism. Each experiment is repeated 100 times. The problem sizes were set to 21 variables for the Deceptive3 and 20 variables for the FirstPolytree5.

As can be seen from Table 13 the improvement of mePADA2 is enormous as compared to PADA2. For the f^5_{Poly}, the superiority of mePADA is more evident; not only it scales much better than PADA2, but the convergence time is drastically reduced. It is also remarkable that the number of generations until success always stays the same or even improves. It has also stabilized as can be seen from the decrease in the standard deviation.

The idea of improving the performance of EDAs by constructing maximum-entropy approximations of the search distributions was first introduced in [29]. Later it was further developed in [23] for multi-connected networks.

Table 13. PADA2 vs. mePADA2 with f_{dec3} and f_{Poly}^5

| | f_{dec3} | | | | f_{Poly}^5 | | | |
| | PADA2 | | mePADA2 | | PADA2 | | mePADA2 | |
N	$\%S$	G_c	$\%S$	G_c	$\%S$	G_c	$\%S$	G_c
200	0	–	2	8.5 ± 0.7	25	10.1 ± 2.1	59	5.1 ± 1.1
600	8	9.7 ± 1.5	69	7.4 ± 1.1	50	10.4 ± 2.6	100	3.9 ± 0.7
800	10	8.7 ± 3.2	90	7.0 ± 1.2	54	10.6 ± 2.3	100	3.7 ± 0.6
5000	92	7.2 ± 1.2	100	5.8 ± 0.9	55	10.8 ± 1.5	100	2.9 ± 0.4

6 Entropy and Mutation

The last section of this chapter relates the concept of entropy toa powerful operator of evolutionary algorithms: mutation.

The mutation operator did not receive much attention during the early years of research in EDAs. It was believed to play no important role due the dramatic improvement in search efficiency achieved by EDAs, with regard to GAs. People profoundly believed that the success of EDAs is determined by the amount of knowledge it has about the search distributions, i.e. the best informed models were considered – and still are considered – the best models. Within this way of thinking there was little space for mutations. However, after some years of hard work, researchers have come to the conclusion that mutation is also a powerful operator within EDAs. Therefore, new and original developments are needed in the field to deal with this issue.

To begin with, we must draw the reader attention to the fundamental shift in the interpretation of mutation: EDAs have to approach mutation from a distribution perspective, in contrast with the genotype perspective of GAs. While a GA mutates single individuals, an EDA must mutate distributions. We have developed an approach to fulfill this requirement.

6.1 How do we Measure the Effect of the Mutation?

A major problem with the mutation operator in evolutionary algorithms, is the lack of a comprehensible, uniform and standard mechanism for measuring its impact in the evolution. There are almost as many mutation operators as problems, and only few of them are problem-independent. The common way of assessing the amount of mutation considers the probability or frequency of application of the operator, i.e. there are no measurements units for mutation. The obvious drawback of this approach is that it is difficult to compare the impact of different operators or the effect of the same operator in different situations.

Our approach to mutation solves the above-mentioned problems. It has a distribution perspective, is problem-independent, has measurements units, and its impact in different scenarios can be easily compared. It is based on the concept of entropy.

The relation between entropy and mutation is quite intuitive: when a random variable is mutated a certain degree of randomness is added to it. In others words, mutation increases the level of uncertainty we have about the exact value of a random variable. Therefore, it seems reasonable to measure the amount of mutation applied to a variable as the increase of its entropy. This connection was first made in [33].

Linking the concepts of mutation and entropy has some important advantages:

- Entropy is a well understood information-theoretic concept, which encapsulates the notion of randomness and uncertainty.
- It connects the mutation to other fundamental concepts like mutual information and relative entropy.
- Mutation also gets from entropy measurements units: bits or nats, instead of using the popular, but less clear notion of probability of application of the mutation operator.

6.2 From Bit-flip to Entropic Mutation

In this section, we shortly discuss two important mutation schemes that precede our proposal. One was introduced in GAs, and the other was recently introduced in EDAs. The observation of the entropic variations produced by these schemes was a major motivation for our work.

Bit-flip Mutation

The classical GA mutation operator for binary problems is a bit-flip (BF) operation that is applied to each gene with a certain given probability μ [8]. The next lemma relates BF-mutation with the univariate probabilities.

Lemma 4 *For binary variables, BF mutation changes the probability according to*

$$p_f - p_i = \mu \left(1 - 2p_i\right)$$

where p_f is the probability after mutation and p_i is the probability before mutation.

Proof 2 *Let the probability of a bit flip be μ, and p_i be the probability of a gene being 1 before mutation. As these events are independent, we can write for the probability p_f of the gene being 1 after mutation*

$$p_f = p_i \left(1 - \mu\right) + \left(1 - p_i\right)\mu = p_i \left(1 - 2\mu\right) + \mu \tag{19}$$

and from this we get

$$p_f - p_i = \mu \left(1 - 2p_i\right) \quad \square \tag{20}$$

If we compute the entropy of a variable before and after the BF-mutation, $H\left(p_i\right)$ and $H\left(p_f\right)$ respectively, then we can measure the increase of entropy produced by this operation

$$\delta H = H\left(p_f\right) - H\left(p_i\right)$$

Figure 7 shows δH curves for six different values of the probability of mutation μ. Note that δH is nonlinear for small values of the initial entropy, $H\left(p_i\right)$, and small μ. However, for large values of $H\left(p_i\right)$ the curves approach a linear function. Moreover, for large μ the curves approach lines. The limit case, $\mu = 0.5$, defines a random walk: for any p_i the probability after mutation is 0.5.

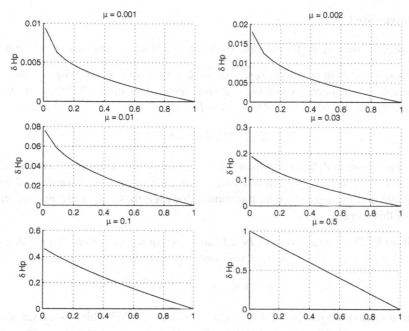

Fig. 7. Entropic curves δH vs H for bit-flip mutation

Prior Mutation

Prior mutation was introduced in [20]. It uses the concept of Bayesian prior, which assumes that the probability of an event has an a priori known distribution. Usually, for binomial variables, the family of Dirichlet distributions plays the role of prior distributions.

In an EDA with prior mutation, the univariate probabilities are not approximated by the maximum likelihood estimates m/N (m is the number of 1 in N cases). Instead the approximation $(m+r)/(N+2r)$ is used, where r is the hyper-parameter of the Dirichlet distribution. Prior mutation is linked to bit-flip mutation. The following theorem was proved in [20].

Theorem 5 *For binary variables, a Bayesian prior with parameter r corresponds to mutation rate $\mu = r/(N+2r)$*

Therefore, for the univariate case bit-flip mutation amounts to prior mutation, and as a consequence, they have the same entropic curves.

6.3 Entropic Mutation

The linear properties of both the bit-flip and prior entropic curves, have suggested that we consider a mutation scheme where δH changes linearly. As a result we have come out with a novel mutation scheme that we have called linear entropic mutation (LEM). In this chapter, we just outline the general ideas.

The Univariate Case

In this section, we discuss the entropic mutation of a binary scalar random variable.

Definition 8 *Let X be a random scalar variable with entropy $H(X)$. We say that to the variable X has been applied the univariate entropic mutation $\delta H(X)$, if after mutation the entropy of the variable is given by*

$$H_m(X) = H(X) + \delta H(X)$$

This is a general definition, which can be applied as well to discrete and continuous random vector variables. Besides the univariate mutation, we have defined the conditional and the joint entropic mutations. However, these cases are beyond the scope of this work.

Definition 9 (Full mutation) *Let X be a binary random scalar variable with entropy $H(X)$. We say that $\delta H(X)$ is a full (or complete) mutation of the variable X if*

$$\delta H(X) = 1 - H(X)$$

Full mutation amounts to bit-flip mutation with $\mu = 0.5$. In this case, a variable gets an increase of entropy equal to what it needs to reach its maximum entropy. This kind of mutation has little use in an optimization context. At this point it is natural to ask ourselves when and how much the entropy of a given variable should be changed. A simple answer based on common sense says that one would like to change a variable if it has low entropy. Indeed, it does not make any sense to mutate a variable with probability $p = 0.5$ ($H(p) = 1$).

Figure 8 shows the line of full mutation as a function of the initial entropy, together with two others linear functions of H. The slopes of the lines are the mutation intensities, α. The following definition formalizes this idea.

Definition 10 *Let X be a random scalar variable with entropy $H(X)$. We say that to the variable X has been applied the linear entropic mutation $\delta H(X)$ with parameter α if after mutation it has entropy $H_\alpha(X)$ and the following holds*

$$\delta H(X) = (1 - H(X))\alpha \;\;\Leftrightarrow\;\; H_\alpha(X) = (1 - \alpha)H(X) + \alpha \qquad (21)$$

Note in Fig. 8, that α is the ordinate for $H(X) = 0$. So, it is bounded by $\alpha = 1$ (full mutation) and $\alpha = 0$ (no mutation).

The mutation intensity α controls the strength of the mutation, i.e. how much the entropy of a variable is changed. In an optimization scenario α might change across time; thus, the general form of the mutation intensity is $\alpha(t)$.

The computation of a LEM-mutation of $p(X)$ is accomplished in two steps. Firstly, $H_\alpha(X)$ is computed according to (21), and then the new probability distribution $p_\alpha(X)$ is obtained from $H_\alpha(X)$. However, as the entropy of binary variables is symmetric – each entropy value is mapped to exactly two probability values – we introduce the following definition to resolve the ambiguity.

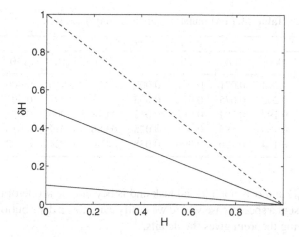

Fig. 8. LEM-mutation of a random binary variable. From *top* to *bottom* the slopes (α) are equal to 1, 0.5 and 0.1

Definition 11 *(Inverse function of $H(X)$). Let $H^{(-1)} : [0,1] \times [0,1] \to [0,1]$ be a function such that for any real numbers p and q, with $0 \leq p, q \leq 1$,*

$$p = H^{(-1)}(H(p), q) \Rightarrow (2p - 1)(2q - 1) \geq 0$$

Definition 11 says that for a given pair $\langle h, q \rangle$ (with $h = H(p)$), the function $H^{(-1)}(h, q)$ returns a probability p, such that both p and q lie together in the interval $[0, 0.5)$ or in $[0.5, 1]$. This definition is useful because for any p, p_α lie in the same half of $[0, 1]$ as p. Finally we can write the expression for p_α as follows:

$$p_\alpha = H^{(-1)}((1 - \alpha)H(X) + \alpha, p) \tag{22}$$

A Note on the Multivariate Case

The multivariate LEM is more difficult than the univariate case, even for binary variables. Here we just give a necessary condition. Other results for multidimensional distributions will be published elsewhere soon.

Definition 12 *Let $p(x_1, x_2, \ldots, x_n)$ and $p_\alpha(x_1, x_2, \ldots, x_n)$ denote a binary joint probability mass and its LEM-mutation with mutation intensity α. If $H(X)$ and $H_\alpha(X)$ are their respective entropy values, then the following holds:*

$$\delta H(X) = (n - H(X))\alpha \quad and \quad H_\alpha(X) = (1 - \alpha)H(X) + n\alpha \tag{23}$$

Table 14 shows the set of joint probability distributions $p_\alpha(x_1, x_2, \ldots, x_n)$ that were used to compute the family of functions $f_{dec3}(\alpha)$ in Sect. 3.2. Note in the second column that the entropy values obey the relation (23), where $H(X)$ is the entropy of the first row and $n = 3$. However, computing $p_\alpha(x_1, x_2, \ldots, x_n)$ from

Table 14. LEM mutation and the family $f_{dec3}(\alpha)$.

					$x_3x_2x_1$				
α	H_α	000	001	010	011	100	101	110	111
0.00	1.74	0.204	0.071	0.071	0.000	0.071	0.000	0.000	0.582
0.05	1.80	0.202	0.075	0.074	0.002	0.073	0.001	0.001	0.572
0.20	1.99	0.197	0.084	0.082	0.012	0.078	0.006	0.006	0.536
0.40	2.24	0.189	0.095	0.091	0.028	0.086	0.017	0.015	0.479
0.50	2.37	0.184	0.100	0.096	0.038	0.090	0.024	0.022	0.446

$p(x_1, x_2, \ldots, x_n)$ and α is not a trivial task and is beyond the scope of this chapter. Here we just present a special case where we easily can show a distribution that fulfill (23). The following theorem gives the details.

Theorem 6 *Let $p(\boldsymbol{x})$ be the joint probability mass of a set of independent random variables $\boldsymbol{X} = (X_1, X_2, \ldots, X_n)$. If*

$$p_\alpha(x_1, x_2, \ldots, x_n) = \prod_{i=1}^{n} H^{(-1)}((1-\alpha)H(X_i) + \alpha, p_i) \tag{24}$$

then

$$H(p_\alpha(X)) = (1-\alpha)H(X) + n\alpha \tag{25}$$

Proof 3 *The lemma follows from theorem 3 and the linearity of the LEM-mutation. We rewrite the right term of (25)*

$$(1-\alpha)H(X) + n\alpha = (1-\alpha)\sum_{i=1}^{n}H(X_i) + n\alpha$$

$$= \sum_{i=1}^{n}((1-\alpha)H(X_i) + \alpha)$$

$$= \sum_{i=1}^{n}H_\alpha(X_i)$$

From (22) and (24) follows that $p_\alpha(x_1, x_2, \ldots, x_n)$ is the distribution of independence with univariate probabilities $p_\alpha(x_i)$. Therefore, the left term of (25) is given by

$$H_\alpha(X) = \sum_{i=1}^{n}H_\alpha(X_i)$$

This completes the proof. □

Closely related to the above theorem is the following general result.

Theorem 7 *Let $p(\boldsymbol{x})$ be any joint probability mass of a set of random variables $\boldsymbol{X} = (X_1, X_2, \ldots, X_n)$, then*

$$H_\alpha(\boldsymbol{X}) \leq \sum_{i=1}^{n} H(X_i)_\alpha$$

with equality if and only if the variables are independent.

Proof 4 *The proof follows immediately from theorem 3 and the linearity of LEM.*

6.4 Testing the UMDA with LEM

On the basis of theorem 6 we can add LEM-mutation to the UMDA [22], which is a FDA with full factorization. The mutation operation is inserted before the sampling step, i.e. the distribution of the selected set is mutated.

Mutation is a powerful mechanism that does not only makes the optimization algorithm more robust and effective, but also might reduce its population size requirements. The search using mutation takes more time and less population size than without it. With regard to the number of function evaluations these are conflicting factors. We just illustrate this issue with an example.

We run the UMDA with the OneMax function, which outputs the number of variables set to one in its input. The UMDA solves this function (with high probability) if the population size is close to the problem size [22]. For the experiment we have chosen a population size that is half the problem size ($N = 30$, $n = 60$), which implies a dramatic reduction of the success rate. Figure 9 shows the success rate and the number of function evaluations as a function of α. Note that for $\alpha = 0$ (no mutation), the success rate is $\approx 18\%$ (out from 100 runs). However, for $\alpha \in [0.06, 0.2]$ the success rate is above 90%.

Note that for $\alpha \in [0.08, 0.12]$, the number of functions evaluations reaches the minimum. This value is less than the minimum population size ($N \approx 55$) that is needed to have a success rate above 90% without mutation. This value is shown as a threshold dot line in the figure. We conclude that the gain due to the population size is not eliminated by the increment in the convergence time.

In summary, with small populations and low or high mutation rates the algorithm performs badly. However, there exists a window $[\alpha_{min}, \alpha_{max}]$ where the success rate is high, that might contain another window where the algorithm reaches the minimum possible number of functions evaluations.

7 Conclusions

This chapter has highlighted several important issues regarding the relation between the concept of entropy and EDAs.

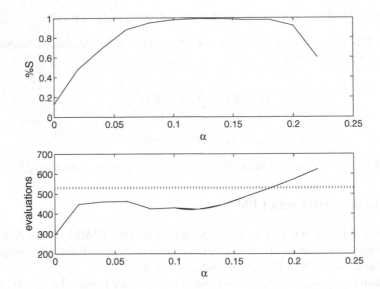

Fig. 9. UMDA and the Onemax function. Success rate and number of function evaluations vs. the mutation intensity α. Setting $N = 30$, $n = 60$

We have introduced a tool to investigate the levels of interactions of the variables under Boltzmann selection: the Boltzmann mutual information curves. It constitutes the corner stone of a method for analysing the complexity of functions for EDAs.

Closely related to the analysis method, is our approach to the design of single and random classes of benchmark functions. We are confident that the use of random classes of Boltzmann functions improves our ability to test EDA algorithms in a more scientific way giving to the benchmark approach a sound theoretical basis. The point is that our method offers an explicit control of the dependencies presented in the functions.

We have used the maximum entropy principle as a key element of the design method and also to build low cost approximations of search distributions that obey a given collection of constraints. We believe that the building of low cost distributions may have tremendous impact on real-world applications of EDAs, so it deserves the special attention of the research community.

Finally, a short introduction to a new scheme of mutation, which is based on the concept of entropy was presented. The linear entropic mutation is a natural operator for EDAs because it mutates distributions instead of single individuals. From a theoretical point of view it opens new exciting directions of research toward a better understanding of the complex dynamics describing the golden equilibrium between exploration and exploitation.

Acknowledgments

We would like to thank Rolando Biscay, Roberto Santana, Omar Ochoa, the anonymous referees and the editors, for their useful comments, fruitful discussions, and support. We specially thank Heinz Mühlenbein for his oustanding contribution to our research work.

References

1. A. Agresti. *Categorical Data Analysis*. John Wiley and Sons, 1990.
2. I. Beinlich, H. R. Chavez, and G. Cooper. The ALARM monitoring system: A case study with two probabilistic inference techniques for belief networks. In *Artificial Intelligence in Medical Care*, pp. 247–256, 1989.
3. T. M. Cover and J. A. Thomas. *Elements of Information Theory*. Jonh Wiley and Sons, New York, 1991.
4. I. Csiszar. I-Divergence geometry of probability distributions and minimization problems. *Annals of Probability*, 3:146–158, 1975.
5. L. M. de Campos. Independency relationship and learning algorithms for singly connected networks. *Experimental and Theoretical Artificial Intelligence*, 10:511–549, 1998.
6. K. Deb, J. Horn, and D. E. Goldberg. Multimodal deceptive function. *Complex Systems*, 7:131–153, 1993.
7. D. E. Goldberg. Simple genetic algorithms and the minimal deceptive problem. In Lawerance Davis, editor, *Genetic Algorithms and Simulated Annealing*, pp. 74–88. Pitman, 1987.
8. D. E. Goldberg. *Genetic Algorithms in Search, Optimization and Machine Learning*. Addison-Wesley, Reading, MA, 1989.
9. D. E. Goldberg, K. Deb, and J. Horns. Massive multimodality, deception, and genetic algorithms. *Lecture Notes in Computer Sciences, Parallel Problem Solving from Nature PPSN II*, pp. 37–46, 1992.
10. C. Huang and A. Darwiche. Inference in belief networks: A procedural guide. *Journal of Approximate Reasoning*, 15(3):225–263, 1996.
11. C. T. Ireland and S. Kullback. Contingency tables with given marginals. *Biometrika*, 55:179–188, 1968.
12. E. T. Jaynes. Information theory and statistical mechanics. *Physics Review*, 6:620–643, 1957.
13. E. T. Jaynes. Where do we stand on maximum entropy? In R. D. Levine and M. Tribus, editors, *The Maximum Entropy Formalism*. MIT Press, 1978.
14. F.V. Jensen and F. Jensen. Optimal junction trees. In *10th Conference on Uncertainty in Artificial Intelligence*, pp. 360–366, Seattle, 1994.
15. R. Jiroušek and S. Přeučil. On the effective implementation of the iterative proportional fitting procedure. *Computational Statistics and Data Analysis*, 19:177–189, 1995.
16. S. L. Lauritzen. *Graphical Models*. Oxford Press, 1996.
17. F. Leisch, A. Weingessel, and K. Hornik. On the Generation of Correlated Artificial Binary Data. Technical Report 13, Viena University of Economics and Bussines Administration, Viena, 1998.
18. P. M. Lewis. Approximating probability distributions to reduce storage requirements. *Information and Control*, 2:214–225, 1959.

19. T. Mahnig and H. Mühlenbein. Comparing the adaptive Boltzmann selection schedule SDS to truncation selection. In *Third International Symposium on Adaptive Systems ISAS 2001, Evolutionary Computation and Probabilistic Graphical Models*, pp. 121–128, La Habana, 2001.

20. T. Mahnig and H. Mühlenbein. Optimal mutation rate using Bayesian priors for estimation of distribution algorithms. *Lecture Notes in Computer Sciences*, 2264:33–48, 2001.

21. C. H. Meyer. *Korrektes Schließen bei Unvollständiger Information*. PhD thesis, Fernuniversität Hagen, 1998. In German.

22. H. Mühlenbein. The equation for the response to selection and its use for prediction. *Evolutionary Computation*, 5(3):303–346, 1998.

23. H. Mühlenbein and R. Höns. The estimation of distributions and the maximum entropy principle. *Evolutionary Computation*, 2004. To appear.

24. H. Mühlenbein and T. Mahnig. Evolutionary optimization and the estimation of search distributions. *Journal of Approximate Reasoning*, 31(3):157–192, 2002.

25. H. Mühlenbein, T. Mahnig, and A. Ochoa. Schemata, distributions and graphical models in evolutionary optimization. *Journal of Heuristics*, 5(2):213–247, 1999.

26. H. Mühlenbein and G. Paas. From recombination of genes to the estimation of distributions I. Binary parameters. *Lecture Notes in Computer Sciences, Parallel Problem Solving from Nature PPSN IV*, 1141:178–187, 1996.

27. D. Nilsson. An efficient algorithm for finding the M most probable configuration in Bayesian networks. *Statistics and Computing*, 2:159–173, 1998.

28. D. Nilsson. The computation of moments of decomposable functions in probabilistics expert systems. In *Third International Symposium on Adaptive Systems ISAS 2001, Evolutionary Computation and Probabilistic Graphical Models*, pp. 116–120, La Habana, 2001.

29. A. Ochoa, R. Höns, M. Soto, and H. Müehlenbein. A maximum entropy approach to sampling in EDA – the single connected case. *Lecture Notes in Computer Sciences, 8th Iberoamerican Congress on Pattern Recognition CIARP 2003*, 2905:683–690, 2003.

30. J. Pearl. *Probabilistic Reasoning in Intelligent Systems: Networks of Plausible Inference*. Morgan Kaufmann, 1988.

31. J. Pearl. *Causality: Models, Reasoning and Inference*. Cambridge University Press, 2000.

32. C. E. Shannon. A mathematical theory of communication. *Bell System Technical Journal*, 27:379–423, 1948.

33. M. Soto. *Un Estudio sobre los Algoritmos Evolutivos Basados en Redes Bayesianas Simplemente Conectadas y su Costo de Evaluación*. PhD thesis, Instituto de Cibernética, Matemática y Física, La Habana, 2003. In Spanish.

34. M. Soto and A. Ochoa. A factorized distribution algorithm based on polytrees. In *Congress on Evolutionary Computation CEC 2000*, pp. 232–237, California, 2000.

35. J. Whittaker. *Graphical Models in Applied Multivariate Statistics*. John Wiley and Sons, 1989.

Entropy-based Convergence Measurement in Discrete Estimation of Distribution Algorithms

Jiri Ocenasek

Computational Laboratory (CoLab), Swiss Federal Institute of Technology ETH,
Hirschengraben 84, 8092 Zürich, Switzerland
jiri@ocenasek.com

Summary. This chapter presents an entropy-based convergence measurement applicable to Estimation of Distribution Algorithms. Based on the measured entropy, the time point when the generation of new solutions becomes ineffective, can be detected. The proposed termination criterion is inherent to the complexity of used probabilistic models and automatically postpones the termination if inappropriate models are used.

1 Introduction

In most Estimation of Distribution Algorithms (EDAs) [6, 8, 11] the probabilistic model learned from the population of candidate solutions is used mainly for generating new solutions. In this chapter we propose an additional usage of the probabilistic model in an information-theoretical way. The entropy of the probabilistic model provides a measure of the amount of information contained in the population. This can be used for controlling the EDA, for example for detecting the proper termination point.

The proposed termination criterion has been integrated with the Mixed Bayesian Optimization Algorithm (MBOA) [9]. The following sections focus mainly on its mathematical formulation, implementation aspects, and demonstration of its behavior on the 2D Ising spin glass optimization problem.

2 Main Principles of EDAs

EDAs explore the search space by sampling a probability distribution that is developed during the optimization. They work with a population of candidate solutions. Each generation, the fittest solutions are used for the model building or model updating and new solutions are generated from the model. These new solutions are evaluated and incorporated into the original population, replacing some or all of the old ones. This process is repeated until the termination criterion is met.

J. Ocenasek: *Entropy-based Convergence Measurement in Discrete Estimation of Distribution Algorithms*, StudFuzz
192, 39–50 (2006)
www.springerlink.com

We will focus on discrete domains. A Bayesian network (BN) is one of the general models to express discrete probability distributions. The underlying probability distribution $p(\mathbf{X})$ is estimated as the product of conditional probability distributions of each parameter X_i given $\mathbf{\Pi}_i$ – the parameters that influence X_i.

$$p(X_0, ..., X_{n-1}) = \prod_{i=0}^{n-1} p(X_i|\mathbf{\Pi}_i) \tag{1}$$

We use upper case symbol X_i to denote the ith design parameter (or the ith gene in Evolutionary Algorithms terminology or the ith random variable in mathematical terminology) whereas lower-case symbols x_i denote a realization of this parameter. Boldface symbols distinguish vectors from scalars. The symbol N denotes the population size whereas n denotes the problem size.

The construction of an optimal Bayesian network from the population of candidate solutions is itself an NP-hard problem [1], and EDAs usually use either an incremental or a greedy version of the learning algorithm to accelerate the BN construction. An example of an algorithm to learn a Bayesian network with implementation details can be found in [5].

Well known EDAs using Bayesian networks are for example the Bayesian Optimization Algorithm (BOA) [10], the Estimation of Bayesian Network Algorithm (EBNA) [3] and the Learning Factorized Distribution Algorithm (LFDA) [7]. The Bayesian network can be also considered as a generalization of models with restricted cardinality of interactions used in early EDAs.

3 Entropy Computation

3.1 Bayesian Networks with Tabulated Conditional Probabilities

The entropy $H(\mathbf{X})$ can be computed as the sum of local conditional entropies according to the factorization of the probability distribution $p(\mathbf{X})$ in (1):

$$H(X_0, ..., X_{n-1}) = \sum_{i=0}^{n-1} H(X_i|\mathbf{\Pi}_i) = -\sum_{i=0}^{n-1} \sum_{\pi_i \in \mathcal{P}_i} \sum_{x_i \in \mathcal{X}_i} p(x_i, \pi_i) \log_2 p(x_i|\pi_i)$$

$$\tag{2}$$

where the outer sum loops over all design parameters X_i, the middle sum loops over $\mathcal{P}_i$ – the set of possible vectors that can be assigned to $\mathbf{\Pi}_i$ – and the inner sum loops over $\mathcal{X}_i$ – the set of possible values of X_i. For example the two inner sums go over all rows of the local conditional probability table if $p(X_i|\mathbf{\Pi}_i)$ is given in tabular form.

The probabilities $p(x_i, \pi_i)$ and $p(x_i|\pi_i)$ can be estimated from the population D as $p(x_i, \pi_i) = m(x_i, \pi_i)/N$ and $p(x_i|\pi_i) = m(x_i, \pi_i)/m(\pi_i)$, where $m(x_i, \pi_i)$ is the number of solutions in D having parameter X_i set to x_i and parameters $\mathbf{\Pi}_i$ set to π_i; $m(\pi_i)$ counts solutions in D with $\mathbf{\Pi}_i$ set to π_i.

3.2 Bayesian Networks with Local Structures

In the previous section we focused on the Bayesian network with tabular form of the conditional probability distributions. Unfortunately, the size of the necessary tables grows exponentially with respect to the order of captured dependencies. For example in the binary case one needs a table with 2^k rows to describe the probability of $X_i = 1$ given all combinations of its k parents. This exponential complexity emerged as a major problem in learning Bayesian network models with tabular representation.

More compact representations can be achieved by using local structures in the form of decision trees or decision graphs[1]. For each variable one decision tree is constructed. The variables that determine X_i are used as decision nodes in that tree (no dependency graph has to be explicitly maintained) and concrete values of $p(x_i|\pi_i)$ are stated in the leaves. Usually the number of leaves of the decision tree is smaller than the number of rows of the corresponding table, thus the frequency estimation is more precise.

The entropy of a Bayesian network with local structures represented in the form of decision trees can be computed as

$$H(X_0, ..., X_{n-1}) = -\sum_{i=0}^{n-1} \sum_{j \in L_i} \sum_{x_i \in \mathcal{X}_i} p(x_i, j) \log_2 p(x_i|j) \qquad (3)$$

where L_i denotes the set of leaf nodes in the ith decision tree corresponding to variable X_i; $\mathcal{X}_i$ denotes the set of possible values of X_i; $p(x_i, j)$ denotes the probability that the solution traverses to the jth leaf of the ith tree and has the parameter X_i set to x_i; $p(x_i|j)$ denotes the probability of $X_i = x_i$ only within the jth leaf of the ith tree. These probabilities are estimated using the frequencies computed from population D as $p(x_i, j) = m(x_i, j)/N$ and $p(x_i|j) = m(x_i, j)/m(j)$, where $m(x_i, j)$ is the number of solutions in D having parameter X_i set to x_i and traversing to the jth leaf of the ith tree; $m(j)$ counts for solutions in D traversing to the jth leaf of the ith tree.

The first EDA with decision graphs was the hierarchical Bayesian Optimization Algorithm (hBOA) [12]. Another example is the Mixed Bayesian Optimization Algorithm (MBOA)[2] [9] which uses various types of graph nodes to optimize problems with both continuous and discrete parameters.

4 Entropy-based Convergence Measurement

Every unique discrete solution which can be sampled from the model can be encoded by a unique string. Let us assume that such encoding is chosen optimally –

---

[1] In accordance with [2], we use the term *decision trees* to denote the tree representation of probability distributions. Some references distinguish between deterministic *decision trees* and stochastic *probability trees*.

[2] MBOA can be downloaded from *http://jiri.ocenasek.com/*.

less frequently sampled solutions receive longer representations and more frequently sampled solutions receive shorter representations – such that the average encoding per sample is minimal. This average length for optimal encoding is given by the entropy. Since we used logarithms with base 2 in (2) and (3), the entropy is given in bits, regardless of the cardinality of the original alphabet. In other words, the model entropy denotes the average number of binary decisions per individual that the random generator has to take to generate the new population. Intuitively, the models built during the initial few generations are of high entropy, since the sampling process in the initial stages is close to uniform sampling. As the evolutionary process starts converging, the sampling preferably generates solutions that are mutually similar and the entropy drops. Therefore, we propose that the entropy can be utilized as convergence measurement in those Evolutionary Algorithms that operate via probabilistic model building and sampling.

Figure 1 shows an example of entropy changes during the optimization. The run shows how MBOA solves a random 2D Ising spin glass benchmark of size 10×10. The population size was $N = 2000$ and the problem size was $n = 100$. The exact definition of Ising spin glass model is presented in Sect. 8.1. In each generation the entropy was computed from the decision trees using (3). We see that the entropy decreases during the run, which means that the population gets less stochastic as longer partial solutions occur and multiply. As the evolution continues, the information about the dependencies between parameters becomes evident and sizeable by the model, thus decreasing the entropy of the search distribution. In another words, the

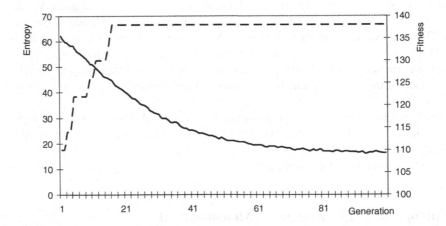

Fig. 1. The change of entropy during an optimization of a 10×10 random 2D Ising spin glass instance (see Sect. 8.1) with MBOA (*solid line*, *left* value axis) and the corresponding change of fitness value (*dashed line*, *right* value axis). Population size is $N = 2000$ and the problem size is $n = 100$; tournament selection and Restricted Tournament Replacement are used. The employed model is a Bayesian network with local probabilities in form of decision trees. Each line represents the median values of 10 runs. On average, the globally optimal fitness value 138 is reached in generation 17

original n-dimensional spin glass problem is being transformed into lower dimensional problem as the dependencies are captured. However, note that MBOA uses a Restricted Tournament Replacement (RTR) [4] as a niching technique to protect diversity in the genotype space. This phenomenon is also evident in Fig. 1 – from the generation number 45 the entropy decrease is decelerated.

5 Entropy-based Termination Criterion

Our goal is to identify the optimal time point for terminating an EDA. In the ideal case the termination should be allowed only if the entropy drops to zero, because then there is a guarantee that no improved solution can be discovered by sampling the model. However, in our approach we are interested in the efficiency. We would like to detect the moment when the sampling process becomes ineffective. Precisely speaking, we would like to stop when the probability that the newly generated solution was already observed in the past, reaches some confidence level α close to 1.

To make this development tractable, we use the assumption that sampling the original n-dimensional distribution over the search space $\mathbf{X}$ is equivalent to sampling the uniform distribution with $H(\mathbf{X})$ independent binary parameters. The problem of effective stopping can then be formulated as "How many observations of uniformly generated vectors of $H(\mathbf{X})$ binary variables are sufficient to be sure (with confidence level α) that we observed each of $2^{H(\mathbf{X})}$ vectors at least once?" The probability that a concrete sample of probability $1/2^{H(\mathbf{X})}$ appears at least once during k trials is given by

$$1 - \left(1 - \frac{1}{2^{H(\mathbf{X})}}\right)^k \tag{4}$$

The probability that all $2^{H(\mathbf{X})}$ different samples appear should be greater than α:

$$\left(1 - \left(1 - \frac{1}{2^{H(\mathbf{X})}}\right)^k\right)^{2^{H(\mathbf{X})}} \geq \alpha \tag{5}$$

The k for which this inequality holds can be approximated using power expansion for α close to 1 as:

$$k \geq 2^{H(\mathbf{X})} H(\mathbf{X}) \ln(2) - 2^{H(\mathbf{X})} \ln \ln \frac{1}{\alpha} \tag{6}$$

This gives us the number of solutions that have to be generated from the given model before we can be sure with probability α that the solution sampled afterwards is just a duplicate of some formerly observed solution, thus reaching inefficiency.

Note that the simplification used to formulate the stopping problem ignores the true distribution of solutions in the population and assumes that all solutions are equally likely to be sampled, thus focusing on the average-case samples. This is an optimistic assumption – in real situations usually the good solutions are harder to get and reside in the tail of the distribution. Therefore, the proposed termination criterion should be considered as a necessary condition, but not a sufficient condition for termination.

6 Implementation Details

Let us denote by $H^{(t)}(\mathbf{X})$ the entropy of the model built in generation t and the corresponding number of samples suggested according to (6) by $k^{(t)}$. The number of solutions sampled each generation from the model will be denoted by N'. If $k^{(t)}$ is less or equal to N', then the algorithm stops in the current generation after evaluating the new solutions. If $k^{(t)}$ is greater than the size of the newly generated population N', the algorithm continues until generation number $t + k^{(t)}/N'$, assuming that the models in generations $t+1$ to $t+k^{(t)}/N'$ are not divergent with respect to the model in generation t and that the generality of constructed models does not change (see discussion in Sect. 7).

The whole pseudocode of an EDA driven by the proposed termination criterion can be written as in Fig. 2.

```
t = 0
Uniformly_generate(Population(0)); Evaluate(Population(0));
termination = infinity; while (termination > t) do
   Parents(t) = Select(Population(t));
   M(t) = Build_Bayesian_network(Parents(t));
   Offspring(t) = Sample(M(t));
   H(t) = Compute_entropy(M(t));
   k(t) = Compute_required_samples(H(t));
   if (termination > (t + k(t)/Size_of(Offspring(t))))
      termination = t + k(t)/Size_of(Offspring(t));
   end if
   Evaluate(Offspring(t));
   Population(t+1) = Replace(Population(t),Offspring(t));
   t = t + 1;
end while
```

Fig. 2. Pseudocode of an EDA driven by the proposed termination criterion

7 Model Generality Issues

7.1 Inappropriate Model Class

In the previous sections we assumed that the allowed model complexity is appropriate to capture the nonlinear interaction between the variables of the optimized problem. Now we will discuss what happens if the used probabilistic model is inappropriate to the optimized problem.

As an example, let us consider a Bayesian network without dependencies. The probabilities are just captured in the form of a vector of marginal probabilities. The evolution of the entropy of this model used for optimizing the 2D spin glass problem of size 10×10 is shown in Fig. 3. One can see that the computed entropy is higher than that in Fig. 1.

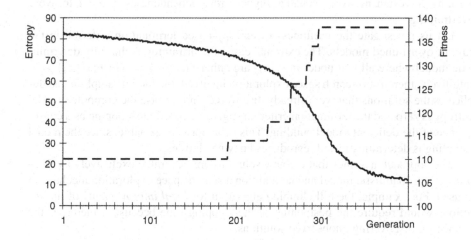

Fig. 3. The change of entropy during an optimization of a 10×10 random 2D Ising spin glass instance with MBOA (*solid line, left* value axis) and the corresponding change of fitness value (*dashed line, right* value axis). Population size is $N = 2000$ and the problem size is $n = 100$; tournament selection and Restricted Tournament Replacement are used. In contrast to Fig. 1, the employed model is a univariate vector of marginal probabilities. Each line represents the median values of 10 runs. On average, the globally optimal fitness value 138 is reached in generation 297

From the Gibbs' theorem it follows that the entropy computed using the approximative probability distribution q cannot be lower than the entropy computed using the true probability distribution p:

$$-\sum_{\mathbf{x}} p(\mathbf{x}) \log p(\mathbf{x}) \leq -\sum_{\mathbf{x}} p(\mathbf{x}) \log q(\mathbf{x}) \tag{7}$$

where $\mathbf{x}$ goes for all possible instances of $\mathbf{X}$ and the equality holds only if both distributions are equal.

In other words, the model is unable to capture the underlying probability distribution, which makes the population appear to be more stochastic. Compared to the case with an appropriate model, the termination will be postponed. This behavior is desirable, since it reflects the inability of the evolutionary algorithm to effectively converge using an insufficient model.

7.2 Overtrained Model

The goal of building a Bayesian network is to capture the general dependencies between parameters of the problem being optimized, but to avoid the spurious dependencies that are specific to the concrete population instance. Most algorithms for learning Bayesian networks penalize higher order dependencies in order to avoid overtraining.

Let us investigate the usefulness of entropy-based termination criterion in the case of overtrained models. The extremal case of overtraining is the fully deterministic model where all leaf nodes $p(x_i, \pi_i)$ are either $p(x_i, \pi_i) = 0$ or $p(x_i, \pi_i) = 1$. Intuitively, there is no search space exploration involved, the model sampling just duplicates the solutions that were already discovered. In this case the computed model entropy is zero and the termination criterion suggests immediate stopping as a consequence of the deficient model building. This behavior is reasonable, since the model sampling is deterministic and reproduces known solutions.

The approach assumes that the new solutions are generated exclusively by sampling the probabilistic model and no additional search space exploration mechanism is used. For example, the hill climbing algorithm for *local improvement*[3] of the solutions would require the postponing of the stopping time because it increases the chance for discovering unobserved solutions.

8 Experiments

The simplification used to derive the termination criterion in Sect. 5 ignores the true distribution of solutions in the population and assumes that all solutions are equally likely to be sampled, thus focusing on the average-case sampling complexity. To investigate the behavior of the proposed termination criterion on real problems, we experimented with the Ising spin glass benchmark.

8.1 Ising Spin Glass Benchmark

Finding the lowest energy configuration of spin glass system is an important task in modern quantum physics. We choose the spin glass optimization problem as a typical example of problem which does not fulfill the said average-case assumption. In [13] it was shown that the computational complexity of studied spin glass systems was dominated by rare events of extremely hard spin glass samples.

Each configuration of spin glass system is defined by the set of spins

$$S = \{s_i | \forall i \in \{1, \dots, s^d\} : s_i \in \{+1, -1\}\}, \tag{8}$$

---

[3] Also known in the literature as *local optimization*.

where d is the dimension of spin glass grid and s is the size of spin glass grid. For the optimization by MBOA the size of spin glass problem s^d is equal to the length of chromosome n, thus

$$n = s^d \qquad (9)$$

Each spin glass benchmark instance is defined by the set of interactions $\{J_{i,j}\}$ between neighboring spins s_i and s_j in the grid. The energy of given spin glass configuration S can be computed as

$$E(S) = \sum_{i,j \in \{1,\ldots,n\}} J_{i,j} s_i s_j \qquad (10)$$

where the sum runs over all neighboring positions i and j in the grid. For general spin glass systems the interaction is a continuous value $J_{i,j} \in [-1, 1]$, but we focus only on Ising model with either ferromagnetic bond $J_{i,j} = -1$ or antiferromagnetic bond $J_{i,j} = +1$, thus $J_{i,j} \in \{-1, +1\}$. Obviously, in the case of ferromagnetic bond the lower (negative) contribution to the total energy is achieved if both spin are oriented in the same direction, whereas in the case of antiferromagnetic bond the lower (negative) contribution to the total energy is achieved if both spins are oriented in opposite directions.

8.2 Empirical Results

Figures 4 and 5 indicate how realistic the proposed termination criterion for the real spin glass problem is. We measure how the efficiency of MBOA changes during the optimization. Each spin glass configuration sampled from the model is archived. In each generation the archive is searched to compute how many of the newly generated configurations were already seen at least once in the previous generations. In Fig. 4 we see that in the final stages of optimization MBOA becomes ineffective because most configurations are seen more than once. The termination criterion proposed in Sect. 5 suggested stopping in 350th generation. On the one hand, we see that the suggested stopping point (350) is more than 20 times larger than the average number of generations needed to discover the global optimum (17). This indicates that from the empirical point of view the proposed termination criterion is reliable for terminating the optimization of 2D Ising spin glasses. On the other hand, we see that the observed portion of resampled solutions in 350th generation is approximately 78%, which is lower than the chosen confidence level $\alpha = 0.95$. This indicates that for real problems the confidence level α appears to be too optimistic. To make the criterion more realistic, the existence of extremal solutions would have to be taken into account.

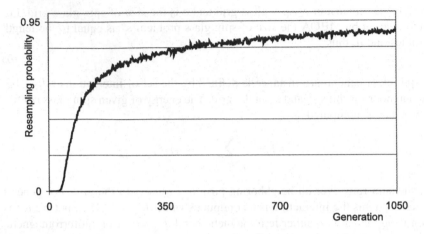

Fig. 4. The empirical efficiency of MBOA during an optimization of a 10×10 random 2D Ising spin glass instance. In each generation the portion of generated solutions that were already visited at least once in the previous generations is shown. Population size is $N = 2000$ and the problem size is $n = 100$; tournament selection and Restricted Tournament Replacement are used. The line represents the average values of 10 runs. On average, the globally optimal fitness value 138 is reached in generation 17

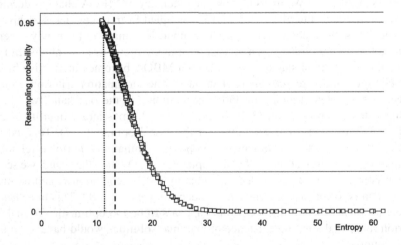

Fig. 5. The relation between the model entropy and the empirically measured efficiency of MBOA. This experiment is identical to the experiment from Fig. 4, but the horizontal axis displays $H(\mathbf{X})$ for each generation. The entropy in the first generation was $H(\mathbf{X}) = 61.75$. The *dashed line* indicates the proposed termination point in 350th generation (with $H(\mathbf{X}) = 13.52$), whereas the *dotted line* indicates the 1650th generation where the measured resampling probability reached the desired 95% level (with $H(\mathbf{X}) = 11.67$)

9 Conclusions

This chapter presents an entropy-based convergence measurement for EDAs.

Based on the measured entropy, we propose a method for detecting the time point when the sampling process becomes ineffective. The proposed termination criterion is inherent to the complexity of used probabilistic models and automatically postpones the termination if an inappropriate class of models is used.

We focus mainly on the mathematical and implementation aspects of the proposed termination criterion. On several instances of spin glass problems we also demonstrate the practical usefulness of this approach.

Future work will be oriented on the usage of the proposed approach in EDAs working in continuous domains and on the investigation of the limits of this approach in different scenarios, for example in the case of problems with a great number of local optima.

Acknowledgement

The author would like to thank to Petros Koumoutsakos, Martin Pelikan, Joonas Asikainen, Nikolaus Hansen and Stefan Kern for constructive comments.

References

1. D. M. Chickering, D. Geiger, and D. Heckerman. Learning Bayesian networks is NP–hard. Technical report, Microsoft Research, Redmond, WA, 1994.
2. D. M. Chickering, D. Heckerman, and C. Meek. A Bayesian approach to learning Bayesian networks with local structure. In *Proceedings of Thirteenth Conference on Uncertainty in Artificial Intelligence*, pp. 80–89. Morgan Kaufmann, 1997.
3. R. Etxeberria and P. Larrañaga. Global optimization with Bayesian networks. In *II Symposium on Artificial Intelligence. CIMAF99. Special Session on Distributions and Evolutionary Optimization*, pp. 332–339, 1999.
4. G. Harik. Finding multimodal solutions using restricted tournament selection. In *Proceedings of the Sixth International Conference on Genetic Algorithms (ICGA 6)*, pp. 24–31. Morgan Kaufmann, 1995.
5. D. Heckerman, D. Geiger, and D. M. Chickering. Learning Bayesian networks: The combination of knowledge and statistical data. Technical report, Microsoft Research, Redmond, WA, 1994.
6. P. Larrañaga. A review on estimation of distribution algorithms. In P. Larrañaga and J. A. Lozano, editors, *Estimation of Distribution Algorithms. A New Tool for Evolutionary Computation*, pp. 80–90. Kluwer Academic Publishers, 2002.
7. H. Mühlenbein and T. Mahnig. FDA – a scalable evolutionary algorithm for the optimization of additively decomposed functions. *Evolutionary Computation*, 7(4):353–376, 1999.
8. H. Mühlenbein and G. Paaß. From recombination of genes to the estimation of distributions I. Binary parameters. In *Lecture Notes in Computer Science 1411: Parallel Problem Solving from Nature - PPSN IV*, pp. 178–187, 1996.

9. J. Ocenasek and J. Schwarz. Estimation of distribution algorithm for mixed continuous-discrete optimization problems. In IOS Press, editor, *2nd Euro-International Symposium on Computational Intelligence*, pp. 227–232, 2002.

10. M. Pelikan, D. E. Goldberg, and E. Cantú-Paz. BOA: The Bayesian optimization algorithm. In W. Banzhaf, J. Daida, A. E. Eiben, M. H. Garzon, V. Honavar, M. Jakiela, and R. E. Smith, editors, *Proceedings of the Genetic and Evolutionary Computation Conference GECCO-99*, volume 1, pp. 525–532. Morgan Kaufmann Publishers, San Francisco, CA, 1999. Orlando, FL.

11. M. Pelikan, D. E. Goldberg, and F. Lobo. A survey of optimization by building and using probabilistic models. *Computational Optimization and Applications*, 21(1):5–20, 2002.

12. M. Pelikan, D. E. Goldberg, and K. Sastry. Bayesian optimization algorithm, decision graphs, and Occam's razor. In *Proceedings of the Genetic and Evolutionary Computation Conference GECCO-2001*, pp. 519–526. Morgan Kaufman, 2000.

13. M. Pelikan, J. Ocenasek, S. Trebst, M. Troyer, and F. Alet. Computational complexity and simulation of rare events of ising spin glasses. In *Proceedings of the Genetic and Evolutionary Computation Conference GECCO-2004*, pp. 36–47. Springer-Verlag, 2004.

Real-coded Bayesian Optimization Algorithm

Chang Wook Ahn[1], R.S. Ramakrishna[1] and David E. Goldberg[2]

[1] Department of Information and Communications
Gwangju Institute of Science and Technology (GIST)
Oryong-dong, Buk-gu, Gwangju 500-712, Korea
cwan@evolution.re.kr
rsr@gist.ac.kr

[2] Department of General Engineering
University of Illinois at Urbana-Champaign (UIUC)
Urbana, IL 61801, USA
deg@uiuc.edu

Summary. This chapter describes a real-coded (i.e., continuous) Estimation of Distribution Algorithm (EDA) that solves real-valued (i.e., numerical) optimization problems of bounded difficulty quickly, accurately, and reliably. This is the *real-coded Bayesian Optimization Algorithm* (rBOA). The objective is to bring the power of (discrete) BOA to bear upon the area of real-valued optimization. That is, the rBOA must properly decompose a problem and effectively perform *Probabilistic Building-Block Crossover* (PBBC) for real-valued multivariate data. In other words, a unique feature of rBOA is to learn complex dependencies of variables and make use of mixture models at the level of substructures. To begin with, a Bayesian factorization is performed. The resulting factorization that contains linkage information is then utilized for finding implicit subproblems (i.e., substructures). Mixture models are employed for independently fitting each of these substructures. Subsequently, an independent substructure-wise sampling draws the offspring. Experimental studies show that the rBOA finds, with a sub-quadratic scale-up behavior for (additively) decomposable problems, a solution that is superior in quality to that found by advanced real-coded EDAs regardless of inherent problem characteristics. Moreover, comparable or better performance is achieved for nondecomposable problems.

1 Introduction

Estimation of Distribution Algorithms (EDAs), also known as *Probabilistic Model Building Genetic Algorithms* (PMBGAs), signal a paradigm shift in genetic and evolutionary computation research [13, 23]. Incorporating (automated) linkage learning techniques into a graphical probabilistic model, EDAs exploit a feasible probabilistic model built around superior solutions found thus far while efficiently traversing the search space [23]. EDAs iterate the three steps listed below, until some termination criterion is satisfied:

1. Select good candidates (i.e., solutions) from a (initially randomly generated) population of solutions.

C.W. Ahn et al.: *Real-coded Bayesian Optimization Algorithm*, StudFuzz **192**, 51–73 (2006)
www.springerlink.com   © Springer-Verlag Berlin Heidelberg 2006

2. Estimate the probability distribution from the selected individuals.
3. Generate new candidates (i.e., offspring) from the estimated distribution.

It must be noted that the third step uniquely characterizes EDAs as it replaces traditional recombination and mutation operators employed by simple Genetic and Evolutionary Algorithms (sGEAs). Although the sGEAs (with well-designed mixing operator) and EDAs deal with solutions (i.e., individuals) in quite different ways, it has been theoretically shown (and empirically observed) that their performances are quite close (to each other) [13, 23]. Moreover, EDAs ensure an effective mixing and reproduction of Building Blocks (BBs) due to their ability to accurately capture the (BB) structure of a given problem, thereby solving GA-hard problems with a linear or sub-quadratic performance in terms of (fitness) function evaluations [1, 3, 19, 21, 23]. However, there is a trade-off between the accuracy of the estimated distribution and the efficiency of computation. For instance, a close and complicated model is recommended if the fitness function to be evaluated is computationally expensive.

A large number of EDAs have been proposed for discrete and real-valued (i.e., continuous) variables in this regard. Depending on how intricate and involved the probabilistic models are, they are divided into three categories: *no dependencies*, *pairwise dependencies*, and *multivariate dependencies* [23]. Among them, the category of multivariate dependencies endeavors to use general probabilistic models, thereby solving many difficult problems quickly, accurately, and reliably [13, 23]. The more complex the probabilistic model the harder as well is the task of finding the best structure. At the expense of some computational efficiency (with regard to learning the probabilistic model), they can significantly improve the overall time complexity for large (additively) decomposable problems due to their innate ability to reduce the number of (computationally expensive) fitness function evaluations. *Extended compact Genetic Algorithm* (EcGA) [7], *Factorized Distribution Algorithm* (FDA) [18], *Estimation of Bayesian Networks Algorithm* (EBNA) [11], and (*hierarchical*) *Bayesian Optimization Algorithm* ((h)BOA) [21, 22] are some leading examples for discrete variables.

Note that the BOA is perceived to be an important effort that employs general probabilistic models for discrete variables [1, 23]. It utilizes techniques for modeling multivariate data by Bayesian networks so as to estimate the (joint) probability distribution of promising solutions. The BOA is very effective even on large decomposable (discrete) problems with loose and tight linkage of BBs. It is important to note that the power of BOA arises from realizing *Probabilistic Building-Block Crossover* (PBBC) that approximates *population-wise building-block crossover* by a probability distribution estimated on the basis of proper (problem) decomposition [21, 22]. The underlying decomposition can be performed regardless of types of dependency between variables because it is capable of accurately modeling any type of dependency due to the inherent characteristic (i.e., finite cardinality) of the discrete world. The PBBC may shuffle as many superior partial solutions (i.e., BBs) as possible in order to bring about an efficient and reliable search for the optimum. Therefore, it is only natural that the principles of BOA be tried on real-valued variables. In this regard, *real-coded* EDAs, also known as *continuous* EDAs, have been developed. *Estimation of Gaussian Networks Algorithm* (EGNA) [12,14], *Iterative Density-estimation Evolutionary Algorithms* (IDEAs) [3, 4], and *Mixed Bayesian Optimization Algorithm* (MBOA) [19, 20] are representative schemes. A brief review of them is presented in the sequel.

In the EGNA, the Gaussian network is induced in each generation by means of a chosen (scoring) metric and the offspring is created by simulating the learned network. However, the EGNA is not suitable for solving complicated problems because it only constructs a single-peak (Gaussian) model.

The IDEAs exploit Bayesian factorizations and mixture distributions for learning probabilistic models. There is a general, but simple factorization mixture selection to be named 'mixed IDEA' (mIDEA) in this chapter. It clusters the selected individuals and subsequently estimates a factorized probability distribution in each cluster separately. It is evident that the mIDEA can learn various types of dependency. However, it cannot realize the PBBC because different clusters (that may create important BBs) do not share all the common features.

The MBOA learns a Bayesian network with local structures in the form of decision trees coming with univariate normal-kernel leaves. One decision tree is built for each target variable, and the split nodes of the decision tree are used to linearly split the domain of parent variables into parts. This results in a decomposition of the conditional distribution's domain into axis-parallel partitions, thereby efficiently approximating the variables by univariate (kernel) distributions [19, 20]. Although the MBOA can be very effective for problems involving variables with simple interactions (i.e., linearity), it is inefficient for nonlinear, symmetric problems because finding the (linear) split boundaries for detecting the inherent characteristics is very difficult and quite often even impossible.

In this chapter, we propose a real-coded BOA (rBOA) along the lines of (discrete) BOA. The rBOA can solve various types of decomposable problem in an efficient and scalable manner, and also find a high quality solution to traditional real-valued benchmarks that represent a variety of difficulties beyond decomposability.

The rest of the chapter is organized as follows. Section 2 outlines rBOA. Section 3 suggests a learning strategy for probabilistic models. Section 4 presents a popular technique for model sampling. Real-valued test problems are cited in Sect. 5. Experimental results are presented in Sect. 6. We conclude with a summary in Sect. 7.

2 Description of Real-coded BOA

This section describes the rBOA as an efficient tool for solving real-valued problems of bounded difficulty with a sub-quadratic scale-up behavior. The purpose is to transplant the strong points of BOA into the continuous world.

Generously drawing on generic procedures of EDAs (Sect. 1), the following pseudo-code summarizes the rBOA:

STEP 1. INITIALIZATION
    Randomly generate initial population $\mathcal{P}$
STEP 2. SELECTION
    Select a set of promising candidates $\mathcal{S}$ from $\mathcal{P}$
STEP 3. LEARNING
    Learn a probabilistic model $\mathcal{M}$ from $\mathcal{S}$ using a metric (and constraints)
STEP 4. SAMPLING
    Generate a set of offspring $\mathcal{O}$ from the estimated probability distribution
STEP 5. REPLACEMENT
    Create a new population $\mathcal{P}$ by replacing some individuals of $\mathcal{P}$ with $\mathcal{O}$
STEP 6. TERMINATION
    If the termination criteria are not satisfied, go to STEP 2

In spite of similar behavior patterns, EDAs can be characterized by the method of learning a probabilistic model (in the STEP 3). That is, the performance of EDAs depends rather directly on the efficiency of probabilistic model learning. In general, the learning of probabilistic models consists of two tasks: learning the structure and learning the parameters [21], also known as *model selection* and *model fitting*, respectively [3]. The former determines the structure of a probabilistic model. The structure defines conditional dependencies (and independencies). Model fitting estimates the (conditional) probability distributions with regard to the found structure.

It is noted that model selection is closely related to model fitting. In the model selection phase, the best structure is searched by investigating the values of a chosen metric for all possible structures. However, the results of model fitting are directly or indirectly needed for computing the metric. Due to the large number of possible structures, the outcome may be unacceptably high computational complexity unless model fitting is performed in some simple way. (A detailed investigation is described in Sect. 3.1)

On the other hand, there is a significant difference between discrete and real-coded EDAs from the viewpoint of probabilistic model learning. Discrete EDAs can easily estimate a probability distribution for a given/observed data by simply counting the number of instances for possible combinations. Moreover, the estimated distribution converges to its true distribution as the data size increases. Thus, discrete EDAs can quickly and accurately carry out model selection and model fitting at the same time.

A typical attempt to bring the merit of discrete EDAs to bear on real-valued variables is to use histogram methods. This follows from the observation that constructing the histogram for a discrete distribution (from population statistics) and approximating it for a continuous distribution are analogous tasks [27]. Of course, the problem is tricky in higher dimensions, but nonetheless, it is theoretically possible. Indeed, it converges as the population size tends to infinity.

On the other hand, real-coded EDAs cannot use this simple (counting) method to estimate a probability distribution for real-valued data due to (uncountably) infinite cardinality. There is an efficient method of reliably approximating the true probability distribution. The method relies on (*finite*) *mixture models* [16]. Some recent methods for unsupervised learning of mixture models are capable of automatically selecting the exact number of mixture components and overcoming some drawbacks of the *Expectation-Maximization* (EM) algorithm [6, 16]. Due to its iterative nature, however, reconciling the unsupervised mixture learning techniques with the EDA framework is obviously hopeless (regardless of the frequency of its use). In this regard, faster mixture models are believed to be useful for efficiently estimating the probability distribution, in spite of sacrificing the accuracy. Although the fast alternatives can significantly reduce the computational cost, they are still not suitable candidates as model fitting is required for every considered structure.

It is, therefore, impossible to directly employ the learning procedure of discrete EDAs (such as BOA) in order to learn a probabilistic model for real-valued variables. An alternative technique for learning probabilistic models in real space is needed. Such a technique can draw on the power of EDAs in the discrete domain. By incorporating the solution with offspring generation procedure (i.e., model sampling), the proper decomposition and the PBBC that are important characteristics of BOA can be realized. The solution is explained in Sect. 3.

3 Learning of Probabilistic Models

This section presents an efficient technique for learning probabilistic models. Two tasks stand out in this regard: model selection and model fitting.

3.1 Model Selection

Factorizations (or *factorized probability distributions*) discover dependencies (and independencies) among random variables. A factorization is a probability distribution that can be described as a product of Generalized Probability Density Functions (GPDFs) which are themselves Probability Density Functions (PDFs) involving real-valued random variables [3,5]. *Bayesian factorizations*, also known as *Bayesian factorized probability distributions* come under a general class of factorizations [3, 15]. A Bayesian factorization estimates a (joint) GPDF for multivariate (dependent) variables by a product of univariate conditional GPDF of each random variable. The Bayesian factorization is represented by a directed acyclic graph, called a Bayesian factorization graph, in which nodes (or vertices) and arcs identify the corresponding variables (in the data set) and the conditional dependencies between variables, respectively [3, 15].

An n-dimensional real-valued optimization problem is considered for discussion. We denote the random variables in the problem by $\mathbf{Y} = (Y_1, \ldots, Y_n)$ and their instantiations by $\mathbf{y} = (y_1, \ldots, y_n)$. The PDF of $\mathbf{Y}$ is represented by $f(\mathbf{Y})(\mathbf{y})$. The second parenthesis of probability distribution can be omitted for convenience. (This causes no ambiguity.)

In general, a PDF is represented by a probabilistic model $\mathcal{M}$ that consists of a structure ζ and an associated vector of parameters $\boldsymbol{\theta}$ (i.e., $\mathcal{M} = (\zeta, \boldsymbol{\theta})$) [3, 4]. As the rBOA employs the Bayesian factorization, the PDF $f(\mathbf{Y})$ for the problem can be encoded as

$$f(\mathbf{Y}) = f_{(\zeta,\boldsymbol{\theta})}(\mathbf{Y}) = \prod_{i=1}^{n} f_{\dot{\boldsymbol{\theta}}^{Y_i}}(Y_i|\boldsymbol{\Pi}_{Y_i}) \tag{1}$$

where $\mathbf{Y} = (Y_1, \ldots, Y_n)$ presents a vector of real-valued random variables, $\boldsymbol{\Pi}_{Y_i}$ is the set of parents of Y_i (i.e., the set of nodes from which there exists an arc to Y_i), and $f_{\dot{\boldsymbol{\theta}}^{Y_i}}(Y_i|\boldsymbol{\Pi}_{Y_i})$ is the univariate conditional PDF of Y_i conditioned on $\boldsymbol{\Pi}_{Y_i}$ with its parameters $\dot{\boldsymbol{\theta}}^{Y_i}$.

Although there are various methods for learning the structure of a probabilistic model (i.e., model selection), a widely used approach has two basic factors: a scoring metric and a search procedure [3,21]. The scoring metric measures the quality of the structures of probabilistic models (i.e., Bayesian factorization graphs) and the search procedure efficiently traverses the space of feasible structures for finding the best one with regard to a given scoring metric.

Scoring Metric

A penalized maximum likelihood criterion known as the *Bayesian Information Criterion* (BIC) is employed as the scoring metric. Although any metric can be used, the reason for choosing the BIC is its empirically observed effectiveness in greedy estimation of factorized probability distributions [4,13,18]. Let $\mathcal{S}$ be the set of selected individuals, viz. $\mathcal{S} = (\mathbf{y}^1, \ldots, \mathbf{y}^{|\mathcal{S}|})$, where $|\mathcal{S}|$ is the number of the individuals. The BIC that assigns the structure ζ a score is formulated as follows [3,4]:

$$BIC(\zeta) = \ln\left(\prod_{j=1}^{|\mathcal{S}|} f_{(\zeta,\boldsymbol{\theta})}(\mathbf{Y})(\mathbf{y}^j)\right) - \lambda \ln(|\mathcal{S}|)\,|\boldsymbol{\theta}|$$

$$= \sum_{j=1}^{|\mathcal{S}|} \ln f_{(\zeta,\boldsymbol{\theta})}(\mathbf{Y})(\mathbf{y}^j) - \lambda \ln(|\mathcal{S}|)\,|\boldsymbol{\theta}|. \tag{2}$$

Here, λ regularizes the extent of penalty and $|\boldsymbol{\theta}|$ is the number of parameters of $f_{(\zeta,\boldsymbol{\theta})}(\mathbf{Y})$. Physically, the first and second terms represent the model accuracy and the model complexity, respectively.

Computing the BIC score for the structure ζ requires its parameters $\boldsymbol{\theta}$ which fit the structure. However, the relations of cause and effect among them lead to unacceptably high computational complexity. This is because the number of possible structures to be tested/traversed increases exponentially with the problem size and the parameter fitting for the data set in real space is by no means a simple undertaking.

In short, the impracticality arises from the close relationship between model selection and model fitting. One way to cross the hurdle is to break the connection without obscuring their intrinsic objectives. An important feature of model selection is to acquire out a priori knowledge of the variables which are (conditionally) dependent regardless of linearity, nonlinearity, or symmetry. The reason is that the dependent type itself is learned (with probability distributions) by model fitting (in Sect. 3.2). Decoupling the connection can be achieved by computing the needed probability distributions for possible structures from a reference distribution. This is so because computing a marginal distribution (with regard to an interesting structure) from a (reference) probability distribution fitted on the whole (problem) space is much simpler than directly estimating the exact probability distribution corresponding to the real-valued data set. EGNA and IDEAs are widely known in this respect. However, this can be hazardous in that it may fail to discover specific dependencies such as nonlinearity or symmetry.

In order to overcome the difficulty, multiple (probability) distributions are employed instead of one, with a view to capture the specific dependencies by a combination of piecewise linear interactions. In other words, the probability distributions used should lead to correct structures by capturing the dependency itself. We define the correct structure as the Bayesian factorization graph that encodes only the true or false interactions of the variables, regardless of the types of dependencies. Moreover, we learn one structure because it has been shown empirically that using one suitably constructed structure is sufficient to solve difficult problems [1, 20, 22].

We employ mixture models for efficiently modeling the selected individuals by a mixture of probability distributions. With this in view, the BIC in (2) must be modified further.

As the PDF $f_{(\zeta,\theta)}(\mathbf{Y})$ can be described by a linear combination of a number of mixture components, (2) can be extended to

$$BIC\,(\zeta) = \sum_{j=1}^{|\mathcal{S}|} \ln\left(\sum_{i=1}^{K} \alpha_i f_{(\zeta,\theta^i)}(\mathbf{Y})(\mathbf{y}^j)\right) - \lambda \ln\left(|\mathcal{S}|\right) \sum_{i=1}^{K} |\theta^i| \qquad (3)$$

where K is the number of mixture components, $\alpha_1, \ldots, \alpha_K$ are the mixing probabilities satisfying $\alpha_i \geq 0$, $\forall i$, and $\sum_{i=1}^{K} \alpha_i = 1$, and θ^i is the set of parameters defined on the ith mixture component.

The observed-data vector (i.e., the selected individuals $\mathcal{S}$) can be viewed as being incomplete due to the unavailability of the associated component-label vectors, $\mathbf{w}^1, \ldots, \mathbf{w}^{|\mathcal{S}|}$ [6, 16]. Each label $\mathbf{w}^i$ is a K-dimensional binary vector and each element w_j^i is defined to be 0 or 1, depending on whether $\mathbf{y}^j$ did or did not arise from the ith mixture component. The component-label vectors are taken to be the realized values of the random vectors, $\mathbf{W}^1, \ldots, \mathbf{W}^{|\mathcal{S}|}$, in which it is assumed that they agree with an unconditional multinomial distribution [16]. That is, the probability distribution of the complete-data vector carries an appropriate distribution for the incomplete-data vector. Hence, (3) can be rewritten as

$$BIC\left(\zeta\right) = \sum_{i=1}^{K}\sum_{j=1}^{|\boldsymbol{S}|} w_j^i \left\{\ln\alpha_i + \ln f_{(\varsigma,\boldsymbol{\theta}^i)}(\mathbf{Y})(\mathbf{y}^j)\right\} - \lambda\ln\left(|\boldsymbol{S}|\right)\sum_{i=1}^{K}\left|\boldsymbol{\theta}^i\right|$$

$$= \sum_{i=1}^{K}\ln\alpha_i\sum_{j=1}^{|\boldsymbol{S}|} w_j^i + \sum_{i=1}^{K}\sum_{j=1}^{|\boldsymbol{S}|} w_j^i\ln f_{(\varsigma,\boldsymbol{\theta}^i)}(\mathbf{Y})(\mathbf{y}^j) - \lambda\ln\left(|\boldsymbol{S}|\right)\sum_{i=1}^{K}\left|\boldsymbol{\theta}^i\right|.$$

$$\tag{4}$$

As the vectors $\mathbf{w}^1, \ldots, \mathbf{w}^{|\boldsymbol{S}|}$ can be simulated by the resulting mixture distribution, it is natural that $\sum_{j=1}^{|\boldsymbol{S}|} w_j^i$ coincides with the expected number of selected individuals drawn from the probability distribution $f_{(\varsigma,\boldsymbol{\theta}^i)}(\mathbf{Y})$, denoted by $|\boldsymbol{S}_i|$, and the maximal log-likelihood is equivalent to the maximal negative entropy, i.e.:

$$\sum_{j=1}^{|\boldsymbol{S}|} w_j^i\ln f_{(\varsigma,\boldsymbol{\theta}^i)}(\mathbf{Y})(\mathbf{y}^j) = -\left|\boldsymbol{S}_i\right| h\left(f_{(\varsigma,\boldsymbol{\theta}^i)}(\mathbf{Y})\right)$$

where $h\left(f_{(\varsigma,\boldsymbol{\theta}^i)}(\mathbf{Y})\right)$ represents the differential entropy of $f_{(\varsigma,\boldsymbol{\theta}^i)}(\mathbf{Y})$. Moreover, the number of parameters for each distribution is the same (i.e., $|\boldsymbol{\theta}'| = |\boldsymbol{\theta}^1| = \ldots = |\boldsymbol{\theta}^K|$) because the structure ζ is fixed for every distribution to be mixed. Thus, (4) is rewritten as

$$BIC\left(\zeta\right) = \sum_{i=1}^{K}\left|\boldsymbol{S}_i\right|\left\{\ln\alpha_i - h\left(f_{(\varsigma,\boldsymbol{\theta}^i)}(\mathbf{Y})\right)\right\} - K\lambda\ln\left(|\boldsymbol{S}|\right)\left|\boldsymbol{\theta}'\right|. \tag{5}$$

Since the terms $|\boldsymbol{S}_i|$ and $\ln\alpha_i$ are not affected by the structure ζ, (5) can be further reduced to

$$BIC\left(\zeta\right) = -\sum_{i=1}^{K}\left|\boldsymbol{S}_i\right| h\left(f_{(\varsigma,\boldsymbol{\theta}^i)}(\mathbf{Y})\right) - K\lambda\ln\left(|\boldsymbol{S}|\right)\left|\boldsymbol{\theta}'\right|. \tag{6}$$

Thus, the BIC in (6) leads to a correct factorization even if there is some kind of nonlinearity and/or symmetry between variables.

Search Procedure

Learning the structure of a probabilistic model given a scoring metric is NP-complete [3, 9, 21]. However, most EDAs have successfully employed a greedy approach for searching a promising structure with a chosen metric. We employ the *incremental greedy algorithm*, a kind of greedy search algorithm. Being one among many variants, this greedy algorithm starts with an empty graph with no arcs, and proceeds by (incrementally) adding an arc (such that no cycles are introduced) that maximally improves the metric until no more improvement is possible [9]. The greedy algorithm is not guaranteed to discover an optimal structure in general because searching for the structure is an NP-complete problem. However, the resulting structure is good enough for encoding most important interactions between variables of the problem [21, 22].

3.2 Model Fitting

Note that the BOA models any type of dependency because it maintains all the conditional probabilities according to the learned structure, without losing any information due to the finite cardinality (of discrete variables). Moreover, the BOA naturally performs the PBBC with regard to the proper decomposition as it treats all the subproblems independently through the model selection, model fitting, and model sampling (i.e., offspring generation) phases. Hence, the BOA can solve difficult problems quickly, accurately, and reliably [21, 22].

With this in view, the model fitting (of the rBOA) must realise the probability distribution of a problem as a product of conditionally independent distributions accurately estimated on the basis of subproblems. In other words, the PBBC can be prepared by subspace-based model fitting. Unlike discrete EDAs, however, a preprocessing step for explicitly discovering subproblems (i.e., problem decomposition) is essential in real-coded EDAs, before performing the subspace-based model fitting. This is because discrete EDAs can implicitly carry out the problem decomposition in the course of (probabilistic) model learning while real-coded EDAs cannot do so (see Sect. 2).

Problem Decomposition

Problem decomposition can be easily accomplished because a set consisting of a node and its parents in the Bayesian factorization graph represents a component subproblem of decomposable problems. Here, the sets of variables of component subproblems may or may not be disjoint, but they cannot properly contain each other. In Fig. 1, the Bayesian factorization graph consists of five component subproblems, viz., $\{Y_2, Y_3\}$, $\{Y_3, Y_1\}$, $\{Y_2, Y_3, Y_6\}$, $\{Y_2, Y_6, Y_5\}$, $\{Y_4, Y_7\}$. However, it is not proper to directly use the component subproblems for model fitting. The reason is explained below.

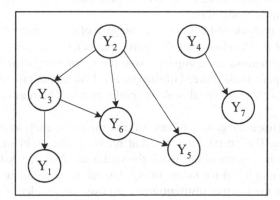

Fig. 1. Bayesian factorization graph involving component subproblems

The probability distribution of a problem can be constructed as a product of univariate conditional distributions which are computed from the probability distributions of component subproblems. Hence, the fitting process must be applied to every component subproblem. Since the fitting process itself is relatively complex (even with a simple technique), it follows that fitting the model on the basis of component subproblems is not adequate, especially as the problem size increases.

Thus, an alternative decomposition is required for quickly and accurately performing the model fitting on the basis of subproblems. In this regard, there is an observation that the set of a parent and its child nodes can be grouped as a kind of subproblem because the child nodes share a common feature even though they do not directly interact with each other. The set is called the *dual component* subproblem. It follows that the conditional distributions can be accurately computed from the probability distributions over the dual component subproblems. At this juncture, *minimal compound* subproblems are defined as the largest component or dual component subproblems that are not proper subsets of each other. In this way, a large number of fitting processes can be avoided (in proportion to the problem size) without losing fitting accuracy. For the problem in Fig. 1, the five component subproblems reduce to three minimal compound subproblems, viz., $\{Y_2, Y_3, Y_6, Y_5\}, \{Y_3, Y_1\}, \{Y_4, Y_7\}$ shown in Fig. 2(a).

There is another decomposition that is simple and also quite efficient for large problems. Consider the maximal connected subgraphs of a Bayesian factorization graph. Nodes in a maximally connected subgraph are looked on as a family; they have a common feature of being bound with common ancestors or descendants. Thus, the nodes can be thought of as interacting with each other in some sense. The conditional distributions can then be obtained from the probability distributions fitted over the maximally connected subgraphs without unduly compromising on the fitting accuracy. Here, the maximal connected subgraph is called the *maximal compound* subproblem. In Fig. 2(b), three minimal compound subproblems of Fig. 2(a) can be reduced to two maximal compound subproblems, viz., $\{Y_2, Y_3, Y_6, Y_5, Y_1\}, \{Y_4, Y_7\}$. Since this decomposition is a special case of decomposing the problem by minimal compound subproblems, minimal compound subproblems are employed for explaining the subspace-based model fitting.

Note that most real-coded EDAs (in the category of multivariate dependencies) such as EGNA and IDEAs choose an alternative that is far from being perfect. That is, conditional distributions are computed from the referencing distributions fitted over the problem space itself (instead of subspaces). This cannot provide the PBBC, thereby resulting in an exponential scale-up performance. The reason is explained below.

BBs can be defined by groups of real-valued variables, each having values in some neighborhood (i.e., small interval), that break up the problem into smaller chunks which can be intermixed to reach the optimum. Assume that the mixture models have been employed for model fitting. Univariate conditional distributions are computed from the mixture distributions fitted over the problem space itself. In the model sampling phase, an entire individual is drawn from a proportionately chosen mixture component. Regardless of the result of Bayesian factorization, it does not

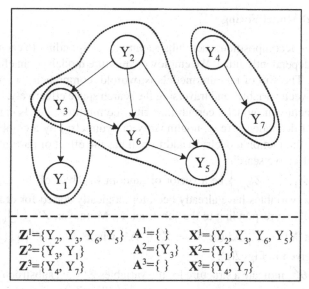

(a) Minimal compound subproblems.

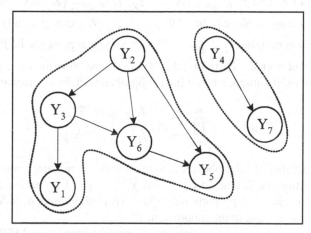

(b) Maximal compound subproblems.

Fig. 2. Examples of the problem decomposition

perform the PBBC as any mutual information of different regions cannot be shared. Instead, at least one mixture component must contain almost all the (superior) BBs of the problems for the sake of finding an optimal solution. In order to construct the mixture distribution that contain such mixture components, however, a huge population and a very large number of mixture components are required. It may result in an exponential scale-up behavior, even if the problem can be decomposable into subproblems of bounded order.

Subspace-based Model Fitting

Following proper decomposition, each substructure (corresponding to each subproblem) must be independently fitted. We employ the mixture models as an efficient tool for the purpose. The aim of mixture models is twofold: comprehending the type of dependency between variables and traversing the search space effectively. Since each mixture component can model a certain linearity, the mixture models can approximate any type of dependency (e.g., nonlinearity or symmetry) by a combination of piecewise linear interaction models. In addition, it has the effect of partitioning each (sub)space for effective search.

Let $\mathbf{Z}^i = \left\{ Z_1^i, \ldots, Z_{|\mathbf{Z}^i|}^i \right\}$ be a vector of random variables of the ith subproblem in which the variables have already been topologically sorted for drawing new partial-individuals corresponding to the substructure. Moreover, $\mathbf{Z}^i \nsubseteq \bigcup_{k=1}^{i-1} \mathbf{Z}^k$ and $\bigcup_i \mathbf{Z}^i = \mathbf{Y}$. Let $\mathbf{X}^i = \mathbf{Z}^i / \mathbf{A}^i$ (or $\mathbf{X}^i = \mathbf{Z}^i - \mathbf{A}^i$) where $\mathbf{A}^i = \mathbf{Z}^i \bigcap \left(\bigcup_{k=1}^{i-1} \mathbf{Z}^k \right)$. An example is given in Fig. 2(a).

Let $\zeta^{\mathbf{Z}^i}$ and $\theta^{\mathbf{Z}^i}$ indicate a structure for the variables $\mathbf{Z}^i$ (i.e., substructure) and its associated parameters, respectively (viz., $\mathcal{M}^{\mathbf{Z}^i} = \left(\zeta^{\mathbf{Z}^i}, \theta^{\mathbf{Z}^i} \right)$). Let $f_{\left(\zeta^{\mathbf{Z}^i}, \theta^{\mathbf{Z}^i} \right)} \left(\mathbf{Z}^i \right)$ represent a PDF of $\mathbf{Z}^i$ and $f_{\left(\zeta^{\mathbf{A}^i}, \theta^{\mathbf{A}^i} \right)} \left(\mathbf{A}^i \right) = \int_{\mathbf{X}_i} f_{\left(\zeta^{\mathbf{Z}^i}, \theta^{\mathbf{Z}^i} \right)} \left(\mathbf{Z}^i \right) d\mathbf{X}_i$. As the mixture models are being employed, the PDF $f_{\left(\zeta^{\mathbf{Z}^i}, \theta^{\mathbf{Z}^i} \right)} \left(\mathbf{Z}^i \right)$ can generally be represented by linearly combining $f_{\left(\zeta^{\mathbf{Z}^i}, \theta_j^{\mathbf{Z}^i} \right)} \left(\mathbf{Z}^i \right)$ (for all j) that presents the PDF of jth mixture component over $\mathbf{Z}^i$. Therefore, the PDF of $\mathbf{Y}$ can be written as a product of linear combinations of subspace-based (i.e., subproblem) PDFs as given by

$$f_{(\zeta, \theta)}(\mathbf{Y}) = \prod_{i=1}^{m} \sum_{j=1}^{c_i} \beta_{ij} \frac{f_{\left(\zeta^{\mathbf{Z}^i}, \theta_j^{\mathbf{Z}^i} \right)} \left(\mathbf{Z}^i \right)}{f_{\left(\zeta^{\mathbf{A}^i}, \theta_j^{\mathbf{A}^i} \right)} \left(\mathbf{A}^i \right)} \tag{7}$$

where m is the number of subproblems, c_i is the number of mixture components for $\mathbf{Z}^i$, β_{ij} is the mixture coefficients, $\beta_{ij} \geq 0$, and $\sum_{j=1}^{c_i} \beta_{ij} = 1$ for each i. In general, the mixture coefficient β_{ij} is proportional to the (expected) number of individuals of the jth mixture component of the subproblem $\mathbf{Z}^i$.

Any PDF can be rewritten as the product of univariate conditional PDFs according to its probabilistic model structure. Therefore, (7) can be rewritten as

$$f_{(\zeta, \theta)}(\mathbf{Y}) = \prod_{i=1}^{m} \sum_{j=1}^{c_i} \beta_{ij} \frac{\prod_{k=1}^{|\mathbf{Z}^i|} f_{\dot{\theta}_j^{Z_k^i}} \left(Z_k^i | \mathbf{\Pi}_{Z_k^i} \right)}{\prod_{l=1}^{|\mathbf{A}^i|} f_{\dot{\theta}_j^{A_l^i}} \left(A_l^i | \mathbf{\Pi}_{A_l^i} \right)} . \tag{8}$$

With a view to generating the offspring (i.e., model sampling), (8) can be simplified to

$$f_{(\zeta, \theta)}(\mathbf{Y}) = \prod_{i=1}^{m} \sum_{j=1}^{c_i} \beta_{ij} \prod_{k=1}^{|\mathbf{X}^i|} f_{\ddot{\theta}_j^{X_k^i}} \left(X_k^i | \mathbf{\Pi}_{X_k^i} \right) . \tag{9}$$

Therefore, the structure learned is efficiently fitted by the subspace-based mixture distributions even in the presence of nonlinearly and/or symmetrically dependent variables.

4 Sampling of Probabilistic Models

After model fitting, new individuals (i.e., offspring) are generated from sampling the resulting factorization of (9). Due to its simplicity and efficiency, Probabilistic Logic Sampling (PLS) is employed [10]. Model sampling is performed in a straight-forward manner. At first, the PDF of the jth mixture component for the ith sub-problem is selected with a probability β_{ij}. Subsequently, a multivariate string (i.e., partial-individual) corresponding to $\mathbf{Z}^i$ can be drawn by simulating the univariate conditional PDFs of the chosen PDF which models one of the promising partitions (i.e., a superior BB) of a subspace (i.e., subproblem). By repeating this for all the subproblems, superior BBs can be mixed and bred for subsequent search.

To sum up, model selection amounts to a proper decomposition. The PBBC is realized successfully by model fitting and sampling on the basis of the proper decomposition.

5 Real-valued Test Problems

This section presents real-valued test problems: (additively) decomposable problems and traditional real-valued optimization problems.

Decomposable problems are created by concatenating basis functions of a certain order. The overall fitness is equal to the sum of all the basis functions. Two types of real-valued decomposable problem are presented.

The first problem is a (Real-valued) Deceptive Problem (RDP) composed of trap functions. The RDP to be maximized is defined by

$$f_{RDP}(\mathbf{y}) = \sum_{i=1}^{m} f_{trap}\left(y_{2i-1}, y_{2i}\right) \tag{10}$$

where $y_j \in [0, 1]$, $\forall j$, m are the number of subproblems, and $f_{trap}(\bullet, \bullet)$ is defined as follows:

$$f_{trap}(y_j, y_{j+1}) = \begin{cases} \alpha, & \text{if } y_j, y_{j+1} \geq \delta \,, \\ \frac{\beta}{\delta}\left(\delta - \sqrt{\frac{y_j^2 + y_{j+1}^2}{2}}\right), & \text{otherwise} \,. \end{cases} \tag{11}$$

Here, α and β are the global and the local (i.e, deceptive) optimum, respectively, so that α/β indicates the signal to noise ratio (SNR), and δ is the border of attractors.

Note that the trap function is not only flexible but also quite simple because δ controls the degree of BB supply and the SNR is adjusted by α/β. As an interesting characteristic, it retains 2^m optimal plateaus, out of which there is only one global optimum. The optimum is isolated and there is no attractor around the region, thereby not being amenable to hill climbing strategies (such as mutation) only. It is clear that recombination is essential to efficiently solve the RDP. In other words, linkage friendly recombination operation should be included for preventing disruption of (superior) partial solutions (i.e., BBs).

The second problem is a (Real-valued) Nonlinear, Symmetric Problem (RNSP) that is constructed by concatenating nonlinear, symmetric functions. The RNSP to be maximized is

$$f_{RNSP}(\mathbf{y}) = \sum_{i=1}^{m} f_{non\text{-}sym}(y_{2i-1}, y_{2i}) \tag{12}$$

where $y_j \in [-5.12, 5.12]$, $\forall j$, and $f_{non\text{-}sym}(\bullet, \bullet)$ is defined by

$$f_{non\text{-}sym}(y_j, y_{j+1}) = \begin{cases} 0.0, & \text{if } 1 - \delta \leq y_j, y_{j+1} \leq 1 + \delta, \\ -100(y_{j+1} - y_j^2)^2 - (1 - y_j)^2, & \text{otherwise}. \end{cases} \tag{13}$$

Here, δ adjusts the degree of BB supply, and the nonlinear, symmetric function retains the traits of Rosenbrock function presented in Table 1.

It is important to note that linkage friendly recombination which is also capable of capturing nonlinear, symmetric interactions is required for effectively solving the RNSP. It is seen that the RNSP provides a real challenge for real-coded optimization algorithms. Moreover, incorporating the mutation operation further helps find the global optimum as the nonlinear, symmetric function (i.e., basis function) is unimodal so that the hill-climbing strategy at any point eventually leads toward its optimum.

On the other hand, four well-known real-valued optimization problems shown in Table 1 are investigated. The task is to minimized the problems. They have some intriguing characteristics beyond decomposability which most optimization algorithms find hard to negotiate.

Griewangk function [26] consists of many local optima that prevent optimization algorithms from converging to the global optimum if (fine-grained) gradient information is incorporated.

Table 1. Traditional problems for numerical optimization.

Problem	Definition	Range
Griewangk	$\frac{1}{4000} \sum_{j=1}^{n} (y_j - 100)^2 - \prod_{j=1}^{n} cos\left(\frac{y_j - 100}{\sqrt{j}}\right) + 1$	$y_j \in [-600, 600]$
Michalewicz	$-\sum_{j=1}^{n} sin(y_j) sin^{20}\left(\frac{j \cdot y_j^2}{\pi}\right)$	$y_j \in [0, \pi]$
Cancellation	$\left(10^{-5} + \sum_{j=1}^{n} \left| y_j + \sum_{i=1}^{j-1} y_i \right|\right) / 100$	$y_j \in [-3, 3]$
Rosenbrock	$\sum_{j=2}^{n} \left\{ 100 \cdot (y_j - y_{j-1}^2)^2 + (1 - y_{j-1})^2 \right\}$	$y_j \in [-5.12, 5.12]$

Michalewicz function [17] also has many suboptimal solutions (albeit to a lesser degree than the Griewangk) and some long valleys along which the minimum value is the same. Thus, gradient information does not lead to better local optima found at the intersections of the channels.

Summation cancellation function [2] has (multivariate) linear interactions between variables. Moreover, the optimum is located on a very narrow peak. Thus, it is hard to find the optimal solution without some information on dependencies (of the variables) and dense-searching in the vicinity of the optimum.

Rosenbrock function [24, 25] is highly nonlinear and symmetric around quite a flat curved valley. Due to the very small gradient and the strong signal (to solution quality) along the bottom of the valley, it is very hard to find the (global) optimum. Oscillations from one side of the valley to the other is likely unless a starting point is selected in the vicinity of the optimum. No algorithm finds it easy to discover the global optimum of Rosenbrock function.

6 Experimental Results

This section investigates the ability of rBOA to benefit from the strengths of BOA (i.e., the proper decomposition and the PBBC) in real space.

6.1 Experiment Setup

The performance of rBOA is measured by the average number of (function) evaluations until convergence to the optimum. A comparative study is performed by comparing the solution quality (returned by the fixed number of evaluations) of rBOA with that of EGNA [14], mIDEA [4], and MBOA [20] (these are advanced real-coded EDAs). The references are appropriately tuned in the interest of fair comparison. For instance, the references employ selection and replacement strategies which are identical to those of rBOA.

Among various (un)supervised learning algorithms for accomplishing mixture models, clustering is perceived to be a suitable candidate in terms of computational efficiency [1, 3]. In this respect, k-means algorithm [8] is employed for model selection and BEND (random) leader algorithm [3, 8] (with a threshold value of 0.3) is used for model fitting. A promising number of clusters (i.e., mixture components K) empirically obtained for each problem is used for model selection. Model fitting and model sampling are carried out on the basis of maximal compound subproblems in view of their efficiency for large decomposable problems. Moreover, normal probability distribution has been employed due to its inherent advantages – close approximation and simple analytic properties. Truncation selection that picks the top half of the population and the BIC of (6) whose regularization parameter λ is 0.5 have been invoked for learning a probabilistic model. The renewal policy replaces the worst half of the population with the newly generated offspring (i.e., elitism-preserving replacement). Since no prior information about the problem structure is available in practice, we set $|\mathbf{Y}| - 1$ for the number of allowable parents (i.e., no constraint in the model selection). Each experiment is terminated when the optimum is found or the number of generations reaches 200. All the results were averaged over 100 runs.

6.2 Results and Discussion

Figure 3 shows the average number of evaluations that rBOA performs to find the optimum of RDP with $\alpha = 1.0$, $\beta = 0.8$, $\delta = 0.8$, and n ranging from 10 to 100. The figure also shows results for RNSP with $\delta = 0.2$ and $n = 10$ to 60. The population size supplied is empirically determined by a bisection method so that the optimum is found. In Fig. 3, it is seen that the results for the RDP and the RNSP are closely approximated (fitted) by $\Theta(n^{1.9})$ and $\Theta(n^{1.8})$, respectively. Thus, rBOA can solve (additively) decomposable problem of bounded difficulty with a sub-quadratic complexity (in terms of fitness function evaluations).

Figure 4 provides a comparative study of the performance of rBOA and references (i.e., EGNA, mIDEA, and MBOA) as applied to the decomposable problems (i.e., RDP and RNSP). Since a decomposable problem consists of m subproblems, the effective problem difficulty tends to be proportional to m. Hence, the population is supplied by a linear model, viz., $N = r \cdot m$, for simplicity.

Figure 4(a) compares the proportion of correct BBs as applied to the RDP with $\alpha = 1.0$, $\beta = 0.8$, $\delta = 0.8$, and varying m. The rBOA employs one mixture component, viz., $K = 1$, for model selection. The population is supplied by $N = 100m$. The results show that the solutions found by rBOA and MBOA are much better than those computed by mIDEA and EGNA. Although the MBOA seems to be somewhat superior to the rBOA, it has no statistical significance. Table 2 supports this assertion. It is also seen that the rBOA and the MBOA achieve stable quality of solutions while the performance of mIDEA and EGNA rapidly deteriorates as the problem size increases. From Figs. 3(a) and 4(a), it is clear that the scale-up behavior of rBOA and MBOA is sub-quadratic; while the mIDEA and the EGNA have an exponential scalability.

Figure 4(b) depicts the BB-wise objective function values returned by the algorithms when applied to the RNSP with $\delta = 0.2$ and varying m. Mixture models for model selection use three mixture components ($K = 3$). A linear model, viz., $N = 200m$, is used for supplying the population. As in the RDP, it is seen that the performance of rBOA and MBOA remains uniform irrespective of the problem size. It can mean that they have a sub-quadratic scalability for the RNSP. However, the results show that the rBOA outperforms the MBOA quite substantially with regard to the quality of solution. This consequence is clearly seen in the statistical test in Table 2. It is also observed that the mIDEA and the EGNA find solutions of unacceptable quality as the problem size increases and their scalabilities obviously become exponential.

From the results, we may conclude that the rBOA finds a better solution with a sub-quadratic scale-up behavior for decomposable problems than does the MBOA, the mIDEA, and the EGNA, especially as the size and difficulty of problems increase.

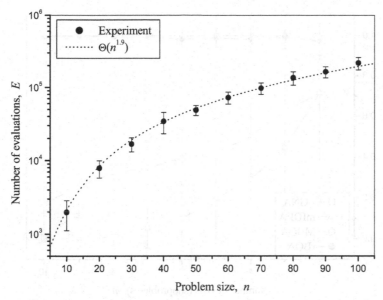

(a) Performance of rBOA on f_{RDP} with $\alpha = 1.0$, $\beta = 0.8$, and $\delta = 0.8$.

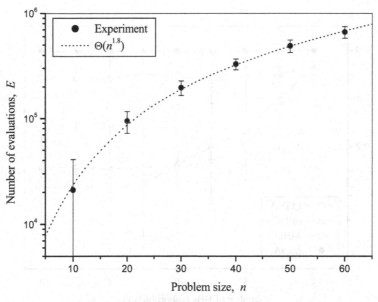

(b) Performance of rBOA on f_{RNSP} with $\delta = 0.2$.

Fig. 3. Performance of rBOA on decomposable problems

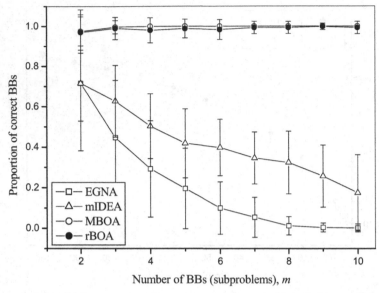

(a) Performance of algorithms on f_{RDP} with $\alpha = 1.0, \beta = 0.8, \delta = 0.8$, and varying m.

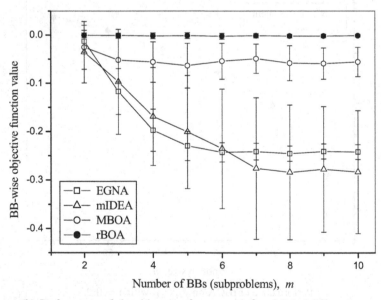

(b) Performance of algorithms on f_{RNSP} with $\delta = 0.2$ and different m.

Fig. 4. Comparison results of algorithms on decomposable problems

Table 2. Performance comparison of algorithms on f_{RDP} and f_{RNSP}

Problem	Measure	EGNA	mIDEA	MBOA	rBOA
RDP	μ_{QoS}	0.196000	0.418000	1.0	0.988000
$(m = 5)$	σ_{QoS}	0.197949	0.169900	0.0	0.047497
RDP	μ_{QoS}	0.002000	0.175000	1.0	0.992000
$(m = 10)$	σ_{QoS}	0.019900	0.187283	0.0	0.030590
RNSP	μ_{QoS}	-0.229916	-0.200973	-0.063843	-0.001384
$(m = 5)$	σ_{QoS}	0.030276	0.136850	0.056469	0.005965
RNSP	μ_{QoS}	-0.238623	-0.299768	-0.056143	-0.001456
$(m = 10)$	σ_{QoS}	0.017609	0.111364	0.030395	0.002651

Statistical t-test

Test case	RDP		RNSP	
	$m = 5$	$m = 10$	$m = 5$	$m = 10$
rBOA − EGNA	$38.30^{\dagger}$	$273.20^{\dagger}$	$71.72^{\dagger}$	$110.78^{\dagger}$
rBOA − mIDEA	$32.80^{\dagger}$	$41.92^{\dagger}$	$14.45^{\dagger}$	$13.59^{\dagger}$
rBOA − MBOA	-2.51	-1.99	$11.10^{\dagger}$	$13.34^{\dagger}$
MBOA − EGNA	$40.41^{\dagger}$	$499.00^{\dagger}$	$27.18^{\dagger}$	$33.51^{\dagger}$
MBOA − mIDEA	$34.08^{\dagger}$	$43.83^{\dagger}$	$14.71^{\dagger}$	$10.43^{\dagger}$
mIDEA − EGNA	$8.10^{\dagger}$	$9.05^{\dagger}$	2.19	-1.97

† The value of t is *significant* at $\alpha = 0.01$ by a paired, two-tailed test.

Table 3 compares the solutions found by the algorithms as applied to the well-known real-valued optimization problems depicted in Table 1. Three mixture components are employed for all the benchmarks. However, any number of components is acceptable for Griewangk and Michalewicz functions as there is no interaction between variables. The results show that the MBOA is superior to the rBOA, the mIDEA, and the EGNA (they find acceptable solutions, however) for the Griewangk function because it can capture some knowledge about independence as well as overcome numerous traps (i.e., local optima) due to the kernel distributions. In the Michalewicz function, the performances of MBOA and rBOA are comparable, and both algorithms outperform the EGNA and the mIDEA. It means that the EGNA and the mIDEA fail to discover independent interactions between variables. It is also seen that the EGNA and the rBOA are quite superior to the mIDEA and the MBOA in the Cancellation function. Although all the algorithms can successfully capture the information about linear interactions, the EGNA achieves the best performance due to its inherent efficiency when it comes to single-peak functions. Even though the rBOA traverses multiple regions of the unimodal function, its performance is acceptably high. It is important to note that the rBOA outperforms the MBOA, the mIDEA, and the EGNA in the case of the Rosenbrock function whose optimum is hard to find. Further, the performance of MBOA and EGNA is very poor. This is explained below.

Table 3. Performance of algorithms on real-valued benchmarks ($n = 5$)

Problem	Measure	EGNA	mIDEA	MBOA	rBOA
Griewangk	μ_{QoS}	0.061968	0.067873	0.003258	0.065993
($N = 2000$)	σ_{QoS}	0.016287	0.018634	0.005205	0.017604
Michalewicz	μ_{QoS}	−4.637647	−4.613430	−4.687653	−4.687640
($N = 500$)	σ_{QoS}	0.013388	0.076301	0.005857	0.000044
Cancellation	μ_{QoS}	0.000034	0.014854	0.001654	0.000557
($N = 100$)	σ_{QoS}	0.000122	0.006420	0.001663	0.000740
Rosenbrock	μ_{QoS}	2.141721	0.003518	0.664121	0.000177
($N = 3000$)	σ_{QoS}	0.182596	0.017894	0.521631	0.001283

Statistical t-test				
Test case	Griewangk	Michalewicz	Cancellation	Rosenbrock
EGNA − rBOA	−1.70	37.17[†]	−6.69[†]	116.64[†]
mIDEA − rBOA	0.74	9.68[†]	21.97[†]	1.83
MBOA − rBOA	−33.77[†]	0.00	5.72[†]	12.67[†]
EGNA − MBOA	32.76[†]	37.16[†]	−9.58[†]	25.49[†]
mIDEA − MBOA	33.07[†]	9.68[†]	19.83[†]	−12.65[†]
EGNA − mIDEA	−2.35	−3.30[†]	−22.91[†]	115.46[†]

[†] The value of t is *significant* at $\alpha = 0.01$ by a paired, two-tailed test.

The variables of the Rosenbrock function strongly interact around a curved valley. Also, the function is symmetric. It is clear that incorrect factorizations (i.e., no dependencies between variables) are encountered at an early stage of the algorithms. Due to the incorrect structure, they try to solve the problems by treating the variables in isolation. Of course, finding an optimum in this way is difficult because any given algorithm does not cross the intrinsic barrier. After a few generations, however, individuals start to collect around the curved valley. At this time, the rBOA can easily capture such a nonlinear, symmetric dependency due to mixture models. On the other hand, the mIDEA can cope with the cancellation effect (arising from symmetry) to some extent by clustering in the overall problem space. However, the MBOA does not deal successfully with the situation because finding a promising set of split boundaries so as to cross the barrier is very difficult. In addition, the EGNA finds it impossible to overcome the hurdles by a (simple) single-peak model.

From Table 3, it can be concluded that the rBOA finds good solutions to complicated problems in terms of dependencies (of decision variables) while achieving comparable or acceptable solutions to others.

As a result, the rBOA achieves the optimal solution with a sub-quadratic scale-up behavior for decomposable problems. Note that the sub-quadratic scalability is solely due to proper decomposition brought about by correct factorization and the PBBC realized by the subspace-based model fitting and model sampling.

Moreover, the rBOA finds better solutions for decomposable problems and acceptable (or even better) solutions to traditional real-valued optimization benchmarks, than those found by the state-of-the-art real-coded EDAs.

7 Conclusion

In this chapter, we have presented a real-coded BOA in the form of (advanced) real-coded EDAs. Decomposable problems were the prime targets and sub-quadratic scale-up behavior (of rBOA) was a major objective. This was achieved by proper decomposition (i.e., linkage learning) and probabilistic building-block crossover on real-valued variables. As a step in this direction, Bayesian factorization was performed by means of mixture models, the substructures were extracted from the resulting Bayesian factorization graph (i.e., problem decomposition), and each substructure was fitted by mixture distributions whose parameters were extracted (by estimation) from the subspaces (i.e., subproblems). In the model sampling phase, the offspring was generated by an independent subproblem-wise sampling procedure.

Experimental studies demonstrated that that the rBOA finds the optimal solution with a sub-quadratic scale-up behavior. The comparative studies exhibited that the rBOA outperforms the up-to-date real-coded EDAs (EGNA, mIDEA, and MBOA) when faced with decomposable problems regardless of inherent problem characteristics such as deception, nonlinearity, and symmetry. Moreover, the solutions computed by rBOA are acceptable in the case of traditional real-valued optimization problems while they are generally better than those found by EGNA, mIDEA, and MBOA. Further, the quality of solutions improves with the degree of problem difficulty.

It is noted that the rBOA learns complex dependencies of variables by means of mixture distributions and estimate the distribution of population by exploiting mixture models at the level of substructures. This allows us to keep options open at the right level of attention throughout the run. In the past, most (advanced) real-coded EDAs used single normal models or mixtures at the level of the problem, but these are unable to capture the critical detail.

More work on the proper number of mixture components and fast mixture models needs to be done. However, rBOA's strategy of decomposing problems, modeling the resulting building blocks, and then searching for better solutions appears to have certain advantages over existing advanced probabilistic model building methods that have been suggested and used elsewhere. Certainly, there can be many alternatives with regard to exploring the method of decomposition, the types of probabilistic models utilized, as well as their computational efficiency, but this avenue appears to lead to a class of practical procedures that should find widespread use in many engineering and scientific applications.

References

1. C. W. Ahn, D. E. Goldberg, and R. S. Ramakrishna. Real-coded bayesian optimization algorithm: Bringing the strength of boa into the continuous world. In *enetic and Evolutionary Computation Conference – GECCO 2004*, pp. 840–851, 2004.

2. S. Baluja and S. Davies. Using optimal dependency-trees for combinatorial optimization: Learning the structure of the search space. Technical Report CMU-CS-97-107, Carnegie Mellon University, 1997.

3. P. A. N. Bosman. Design and application of iterated density-estimation evolutionary algorithms. Doctoral Dissertation, Utrecht University, Utrecht, The Netherlands, 2003.

4. P. A. N. Bosman and D. Thierens. Optimization in continuous domains by learning and simulation of Gaussian networks. In *Proceedings of OBUPM workshop at the Genectic and Evolutionary Computation Conference*, pp. 208–212, 2001.

5. M. DeGroot. *Optimal Statistical Decisions*. McGraw–Hill, New York, 1970.

6. M. A. T. Figueiredo and A. K. Jain. Unsupervised learning of finite mixture models. *IEEE Transactions on Pattern Analysis and Machine Intelligence*, 24(3):381–385, 2002.

7. G. Harik. Linkage learning in via probabilistic modeling in the ECGA. Technical Report IlliGAL-99010. University of Illinois at Urbana-Champaign, Urbana, IL, 1999.

8. J. Hartigan. *Clustering algorithms*. John Wiley & Sons, Inc., New York, 1975.

9. D. Heckerman, D. Geiger, and D. M. Chickering. Bayesian networks: The combination of knowledge and statistical data. Technical Report MSR-TR-94-09, Microsoft Advanced Technology Division, Microsoft Corporation, Seattle, Washington, USA, 2004.

10. M. Henrion. Propagating uncertainty in bayesian networks by probabilistic logic sampling. In *Proceedings of the Second Conference on Uncertainty in Artificial Intelligence*, pp. 149–163, 1988. Amsterdam.

11. P. Larrañaga, R. Etxeberria, J. A. Lozano, and J. M. Peña. Combinatorial optimization by learning and simulation of Bayesian networks. In *Proceedings of the Sixteenth Conference on Uncertainty in Artificial Intelligence*, pp. 343–352, 2000. Stanford.

12. P. Larrañaga, R. Etxeberria, J. A. Lozano, and J. M. Peña. Optimization in continuous domains by learning and simulation of Gaussian networks. In A. S. Wu, editor, *Proceedings of the 2000 Genetic and Evolutionary Computation Conference Workshop Program*, pp. 201–204, 2000.

13. P. Larrañaga and J. A. Lozano. *Estimation of Distribution Algorithms. A New Tool for Evolutionary Computation*. Kluwer Academic Publishers, 2002.

14. P. Larrañaga, J. A. Lozano, and E. Bengoetxea. Estimation of distribution algorithms based on multivariate normal and gaussian networks. Technical Report KZZA-1K-1-01, Department of Computer Science and Artificial Intelligence, University of the Basque Country, Donostia, Spain, 2001.

15. S. L. Lauritzen. *Graphical Models*. Oxford University Press, 1996.

16. G. McLachlan and D. Peel. *Finite mixture models*. John Wiley & Sons, Inc, 2000.

17. Z. Michalewicz. Genetic Algorithms + Data Structures = Evolution Programs. Springer–Verlag, 1992.

18. H. Mühlenbein and T. Mahnig. FDA - a scalable evolutionary algorithm for the optimization of additively decomposed functions. *Evolutionary Computation*, 7(4):353–376, 1999.

19. J. Ocenasek. Parallel estimation of distribution algorithms. Doctoral Dissertation, Brno University of Technology, Brno, Czech, 2002.

20. J. Ocenasek and J. Schwarz. Estimation of distribution algorithm for mixed continuous-discrete optimization problems. In IOS Press, editor, *2nd Euro-International Symposium on Computational Intelligence*, pp. 227–232, 2002.

21. M. Pelikan. Bayesian optimization algorithm: From single level to hierarchy. Doctoral Dissertation, University of Illinois at Urbana-Champaign, Urbana, IL, USA, 2002.

22. M. Pelikan, D. E. Goldberg, and E. Cantú-Paz. BOA: The Bayesian optimization algorithm. In W. Banzhaf, J. Daida, A. E. Eiben, M. H. Garzon, V. Honavar, M. Jakiela, and R. E. Smith, editors, *Proceedings of the Genetic and Evolutionary Computation Conference GECCO-99*, volume 1, pp. 525–532. Morgan Kaufmann Publishers, San Francisco, CA, 1999. Orlando, FL.

23. M. Pelikan, D.E. Goldberg, and F.G. Lobo. A survey of optimization by building and using probabilistic models. *Computational Optimization and Applications*, 21(1):5–20, 2002.

24. H. H. Rosenbrock. An automatic method for finding the greatest or least value of a function. *The Computer Journal*, 3(3):175–184, 1960.

25. R. Salomon. Evolutionary algorithms and gradient search: Simiarities and differences. *IEEE Transactions on Evolutionary Computation*, 2(2):45–55, 1998.

26. A. Törn and A. Žilinskas. *Global optimization*. Springer–Verlag, 1989.

27. S. Tsutsui, M. Pelikan, and D. E. Goldberg. Evolutionary algorithm using marginal histogram models in continuous domain. Technical Report 2001019, University of Illinois at Urbana-Champaign, Illinois Genetic Algorithms Laboratory, Urbana, IL, USA, 2001.

The CMA Evolution Strategy: A Comparing Review

Nikolaus Hansen

CoLab Computational Laboratory, ETH Zürich
ICoS Institute of Computational Science, ETH Zürich
nikolaus.hansen@inf.ethz.ch

Summary. Derived from the concept of self-adaptation in evolution strategies, the CMA (Covariance Matrix Adaptation) adapts the covariance matrix of a multi-variate normal search distribution. The CMA was originally designed to perform well with small populations. In this review, the argument starts out with large population sizes, reflecting recent extensions of the CMA algorithm. Commonalities and differences to continuous Estimation of Distribution Algorithms are analyzed. The aspects of reliability of the estimation, overall step size control, and independence from the coordinate system (invariance) become particularly important in small populations sizes. Consequently, performing the adaptation task with small populations is more intricate.

Nomenclature

Abbreviations

CMA Covariance Matrix Adaptation

EDA Estimation of Distribution Algorithm

EMNA Estimation of Multivariate Normal Algorithm

ES Evolution Strategy

$(\mu/\mu_{\{I,W\}}, \lambda)$-ES, evolution strategy with μ parents, with recombination of all μ parents, either Intermediate or Weighted, and λ offspring.

OP : $\mathbb{R}^n \to \mathbb{R}^{n \times n}, \mathbf{x} \mapsto \mathbf{x}\mathbf{x}^T$, denotes the outer product of a vector with itself, which is a matrix of rank one with eigenvector $\mathbf{x}$ and eigenvalue $\|\mathbf{x}\|^2$.

RHS Right Hand Side.

Greek symbols

$\lambda \geq 2$, population size, sample size, number of offspring.

$\mu \leq \lambda$ parent number, number of selected search points in the population.

N. Hansen: *The CMA Evolution Strategy: A Comparing Review*, StudFuzz **192**, 75–102 (2006)
www.springerlink.com

μ_{cov}, parameter for weighting between rank-one and rank-μ update, see (22).

$\mu_{\text{eff}} = \left(\sum_{i=1}^{\mu} w_i^2 \right)^{-1}$, the variance effective selection mass, see (5).

$\sigma^{(g)} \in \mathbb{R}_+$, step size.

Latin symbols

$\mathbf{B} \in \mathbb{R}^n$, an orthogonal matrix. Columns of $\mathbf{B}$ are eigenvectors of $\mathbf{C}$ with unit length and correspond to the diagonal elements of $\mathbf{D}$.

$\mathbf{C}^{(g)} \in \mathbb{R}^{n \times n}$, covariance matrix at generation g.

c_{ii}, diagonal elements of $\mathbf{C}$.

$c_{\text{c}} \leq 1$, learning rate for cumulation for the rank-one update of the covariance matrix, see (17) and (33).

$c_{\text{cov}} \leq 1$, learning rate for the covariance matrix update, see (11), (21), (22), and (34).

$c_{\sigma} < 1$, learning rate for the cumulation for the step size control, see (23) and (31).

$\mathbf{D} \in \mathbb{R}^n$, a diagonal matrix. The diagonal elements of $\mathbf{D}$ are square roots of eigenvalues of $\mathbf{C}$ and correspond to the respective columns of $\mathbf{B}$.

d_{ii}, diagonal elements of $\mathbf{D}$.

$d_{\sigma} \approx 1$, damping parameter for step size update, see (24), (28), and (32).

E Expectation value

$f : \mathbb{R}^n \to \mathbb{R}, \mathbf{x} \mapsto f(\mathbf{x})$, objective function (fitness function) to be minimized.

$f_{\text{sphere}} : \mathbb{R}^n \to \mathbb{R}, \mathbf{x} \mapsto f_{\text{sphere}}(\mathbf{x}) = \|\mathbf{x}\|^2 = \sum_{i=1}^{n} x_i^2$.

$g \in \mathbb{N}$, generation counter, iteration number.

$\mathbf{I} \in \mathbb{R}^{n \times n}$, Identity matrix, unity matrix.

$\mathbf{m}^{(g)} \in \mathbb{R}^n$, mean value of the search distribution at generation g.

$n \in \mathbb{N}_{>0}$, search space dimension, see f.

$\mathcal{N}(\mathbf{0}, \mathbf{I})$, multi-variate normal distribution with zero mean and unity covariance matrix. A vector distributed according to $\mathcal{N}(\mathbf{0}, \mathbf{I})$ has independent, $(0, 1)$-normally distributed components.

$\mathcal{N}(\mathbf{m}, \mathbf{C}) \sim \mathbf{m} + \mathcal{N}(\mathbf{0}, \mathbf{C})$, multi-variate normal distribution with mean $\mathbf{m} \in \mathbb{R}^n$ and covariance matrix $\mathbf{C} \in \mathbb{R}^{n \times n}$. The matrix $\mathbf{C}$ is symmetric and positive definite.

$\mathbf{p} \in \mathbb{R}^n$, evolution path, a sequence of successive (normalized) steps, the strategy takes over a number of generations.

w_i, where $i = 1, \ldots, \mu$, recombination weights, see also (3).

$\mathbf{x}_k^{(g+1)} \in \mathbb{R}^n$, k-th offspring from generation $g + 1$. We refer to $\mathbf{x}^{(g+1)}$, as search point, or object parameters/variables, commonly used synonyms are candidate solution, or design variables.

$\mathbf{x}_{i:\lambda}^{(g+1)}$, i-th best individual out of $\mathbf{x}_1^{(g+1)}, \ldots, \mathbf{x}_\lambda^{(g+1)}$.

1 Introduction

We assume a search scenario, where we want to minimize an objective function $f : \mathbb{R}^n \to \mathbb{R}, \mathbf{x} \mapsto f(\mathbf{x})$.[1] The only accessible information on f are function values of evaluated search points. Our performance measure is the number of function evaluations needed to reach a certain function value. Many continuous domain evolutionary algorithms use a normal distribution to sample new search points. In this chapter, we focus on algorithms with a multi-variate normal search distribution, where the covariance matrix of the distribution *is not restricted to a priori*, e.g., not a diagonal matrix. Estimation of Distribution Algorithms (EDAs) falling into this class, include the Estimation of Multi-variate Normal Algorithm (EMNA), the Estimation of Gaussian Network Algorithm (EGNA) [15, 16], and the Iterated Density Estimation Evolutionary Algorithm (IDEA) [4]. Evolution Strategies (ESs) falling into this class include a $(\mu/\mu_I, \lambda)$-ES[2] with self-adaptation of correlated mutations [19], and the ES with Covariance Matrix Adaptation (CMA) [10]. Originally, the CMA was interpreted as *derandomized self-adaptation* [12]: in contrast to the original self-adaptation, where changes of the distribution parameters obey their own stochastics, in the CMA, changes of the distribution parameters are *deterministically* linked to the object parameter variations. In this chapter, we will review the CMA from a different perspective revealing the close relationship to EDAs like the EMNA.

The Multi-variate Normal Distribution

Any normal distribution, $\mathcal{N}(\mathbf{m}, \mathbf{C})$, is uniquely determined by its mean $\mathbf{m} \in \mathbb{R}^n$ and its symmetric and positive definite covariance matrix $\mathbf{C} \in \mathbb{R}^{n \times n}$. Covariance matrices have an appealing geometrical interpretation: they can be uniquely identified with the (hyper-)ellipsoid $\{\mathbf{x} \in \mathbb{R}^n \mid \mathbf{x}^T \mathbf{C}^{-1} \mathbf{x} = 1\}$, as shown in Fig. 1. The ellipsoid is a surface of equal density of the distribution. The principal axes of the ellipsoid correspond to the eigenvectors of $\mathbf{C}$, the squared axes lengths correspond to the eigenvalues. The eigendecomposition is denoted by $\mathbf{C} = \mathbf{B} (\mathbf{D})^2 \mathbf{B}^T$, where columns of $\mathbf{B}$ are eigenvectors of $\mathbf{C}$ with unit length ($\mathbf{B}$ is orthogonal), and the squared diagonal elements of the diagonal matrix $\mathbf{D}$ are the corresponding eigenvalues.

The normal distribution $\mathcal{N}(\mathbf{m}, \mathbf{C})$ can be written in different forms.

$$\mathcal{N}(\mathbf{m}, \mathbf{C}) \sim \mathbf{m} + \mathcal{N}(\mathbf{0}, \mathbf{C}) \sim \mathbf{m} + \mathbf{BDB}^T \mathcal{N}(\mathbf{0}, \mathbf{I}) \sim \mathbf{m} + \mathbf{BD}\mathcal{N}(\mathbf{0}, \mathbf{I}) \quad (1)$$

---

[1] In fact, the image needs not to be $\mathbb{R}$. Any totally ordered set is sufficient.

[2] $(\mu/\mu_I, \lambda)$ refers to the non-elitist selection scheme with μ parents, Intermediate recombination of all μ parents, and λ offspring.

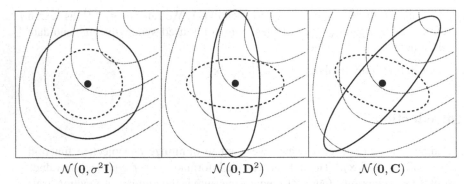

$$\mathcal{N}(\mathbf{0},\sigma^2\mathbf{I}) \qquad \mathcal{N}(\mathbf{0},\mathbf{D}^2) \qquad \mathcal{N}(\mathbf{0},\mathbf{C})$$

Fig. 1. Six ellipsoids, depicting one-σ lines of equal density of six different normal distributions, where $\sigma \in \mathbb{R}_+$, $\mathbf{D}$ is a diagonal matrix, and $\mathbf{C}$ is a positive definite full covariance matrix. Thin lines depict exemplary objective function contour lines

where "$\sim$" denotes equality in distribution and $\mathbf{I}$ denotes the identity matrix. If $\mathbf{D} = \sigma\mathbf{I}$, where $\sigma \in \mathbb{R}_+$, $\mathbf{C} = \sigma^2\mathbf{I}$ and the ellipsoid is isotropic (Fig. 1, left). If $\mathbf{B} = \mathbf{I}$, the ellipsoid is axis parallel oriented (middle). In the coordinate system given by $\mathbf{B}$, the distribution $\mathcal{N}(\mathbf{0},\mathbf{C})$ is uncorrelated.

Objective

The objective of covariance matrix adaptation is, loosely speaking, to fit the search distribution to the contour lines of the objective function f to be minimized. In Fig. 1 the solid-line distribution in the right figure follows the objective function contour most suitably, and it is easy to foresee that it will help to approach the optimum the most. On convex-quadratic objective functions, setting the covariance matrix of the search distribution to the inverse Hessian matrix is equivalent to rescaling the ellipsoid function into a spherical one. We assume that the optimal covariance matrix equals the inverse Hessian matrix, up to a constant factor.[3] Consequently, the adaptation mechanism should aim to *approximate the inverse Hessian matrix*. Choosing a covariance matrix or choosing a respective affine linear transformation of the search space is equivalent [7].

Basic Equation

In the CMA evolution strategy, a population of new search points is generated by sampling a multi-variate normal distribution. The basic equation for sampling the search points, for generation number $g = 0, 1, 2, \ldots$, reads[4]

$$\mathbf{x}_k^{(g+1)} \sim \mathcal{N}\left(\mathbf{m}^{(g)}, \left(\sigma^{(g)}\right)^2 \mathbf{C}^{(g)}\right) \qquad \text{for } k = 1, \ldots, \lambda \qquad (2)$$

---

[3] Even though there is good intuition and strong empirical evidence for this statement, a rigorous proof is missing.

[4] Framed equations belong to the final algorithm of a CMA evolution strategy.

where

$\sim$ denotes the same distribution on the left and right side.

$\mathcal{N}(\mathbf{m}^{(g)}, (\sigma^{(g)})^2 \mathbf{C}^{(g)}) \sim \mathbf{m}^{(g)} + \sigma^{(g)} \mathcal{N}(\mathbf{0}, \mathbf{C}^{(g)}) \sim \mathbf{m}^{(g)} + \sigma^{(g)} \mathbf{B}^{(g)} \mathbf{D}^{(g)} \mathcal{N}(\mathbf{0}, \mathbf{I})$
   is the multi-variate normal search distribution.

$\mathbf{x}_k^{(g+1)} \in \mathbb{R}^n$, k-th offspring (search point) from generation $g + 1$.

$\mathbf{m}^{(g)} \in \mathbb{R}^n$, mean value of the search distribution at generation g.

$\sigma^{(g)} \in \mathbb{R}_+$, "overall" standard deviation, step size, at generation g.

$\mathbf{C}^{(g)} \in \mathbb{R}^{n \times n}$, covariance matrix at generation g.

$\lambda \geq 2$, population size, sample size, number of offspring.

To define the complete iteration step, the remaining question is, how to calculate $\mathbf{m}^{(g+1)}$, $\mathbf{C}^{(g+1)}$, and $\sigma^{(g+1)}$ for the next generation $g + 1$. The next three sections will answer these questions, respectively.

2 Selection and Recombination: Choosing the Mean

The new mean $\mathbf{m}^{(g+1)}$ of the search distribution is a *weighted average of μ selected points* from the sample $\mathbf{x}_1^{(g+1)}, \ldots, \mathbf{x}_\lambda^{(g+1)}$:

$$\mathbf{m}^{(g+1)} = \sum_{i=1}^{\mu} w_i \, \mathbf{x}_{i:\lambda}^{(g+1)} \tag{3}$$

$$\sum_{i=1}^{\mu} w_i = 1, \qquad w_i > 0 \ \text{ for } i = 1, \ldots, \mu \tag{4}$$

where

$\mu \leq \lambda$ is the parent population size, i.e. the number of selected points.

$w_{i=1\ldots\mu} \in \mathbb{R}_+$, positive weight coefficients for recombination, where $w_1 \geq w_2 \geq \cdots \geq w_\mu > 0$. Setting $w_i = 1/\mu$, (3) calculates the mean value of μ selected points.

$\mathbf{x}_{i:\lambda}^{(g+1)}$, i-th best individual out of $\mathbf{x}_1^{(g+1)}, \ldots, \mathbf{x}_\lambda^{(g+1)}$ from (2). The index $i : \lambda$ denotes the index of the i-th ranked individual and $f(\mathbf{x}_{1:\lambda}^{(g+1)}) \leq f(\mathbf{x}_{2:\lambda}^{(g+1)}) \leq \cdots \leq f(\mathbf{x}_{\lambda:\lambda}^{(g+1)})$, where f is the objective function to be minimized.

Equation (3) implements *recombination* by taking a weighted sum of μ individuals, *and selection* by choosing $\mu < \lambda$ and/or assigning different weights w_i.

The measure

$$\mu_{\text{eff}} = \left(\sum_{i=1}^{\mu} w_i^2 \right)^{-1} \tag{5}$$

can be paraphrased as *variance effective selection mass*. From the definition of w_i we derive $1 \le \mu_{\text{eff}} \le \mu$, and $\mu_{\text{eff}} = \mu$ for equal recombination weights, i.e. $w_i = 1/\mu$ for all $i = 1 \ldots \mu$. Usually, $\mu_{\text{eff}} \approx \lambda/4$ indicates a reasonable setting of w_i. A typical setting would be $w_i \propto \mu - i + 1$, and $\mu \approx \lambda/2$.

3 Adapting the Covariance Matrix

In this section, the update of the covariance matrix, $\mathbf{C}$, is derived. We will start out estimating the covariance matrix from a single population of one generation (Sect. 3.1). For small populations this estimation is unreliable and an adaptation procedure has to be invented (Sect. 3.2). The adaptation procedure takes into account more than one generation and can be further enhanced by exploiting dependencies between successive steps (Sect. 3.3).

3.1 Estimating the Covariance Matrix

For the moment we assume that the population contains enough information to reliably estimate a covariance matrix from the population.[5] For the sake of convenience we assume $\sigma^{(g)} = 1$ in this section. For $\sigma^{(g)} \neq 1$ the discussion holds except for a constant factor.

Referring to (2), we can (re-)estimate the original covariance matrix $\mathbf{C}^{(g)}$ from the sample population, $\mathbf{x}_1^{(g+1)} \ldots \mathbf{x}_\lambda^{(g+1)}$, by

$$\mathbf{C}_{\text{emp}}^{(g+1)} = \frac{1}{\lambda - 1} \sum_{i=1}^{\lambda} \left(\mathbf{x}_i^{(g+1)} - \frac{1}{\lambda} \sum_{j=1}^{\lambda} \mathbf{x}_j^{(g+1)} \right) \left(\mathbf{x}_i^{(g+1)} - \frac{1}{\lambda} \sum_{j=1}^{\lambda} \mathbf{x}_j^{(g+1)} \right)^{\mathrm{T}} . \tag{6}$$

The empirical covariance matrix $\mathbf{C}_{\text{emp}}^{(g+1)}$ is an unbiased estimator of $\mathbf{C}^{(g)}$: assuming the $\mathbf{x}_i^{(g+1)}, i = 1 \ldots \lambda$, to be random variables (rather than a realized sample), we have that $\mathrm{E}\left[\mathbf{C}_{\text{emp}}^{(g+1)} \mid \mathbf{C}^{(g)} \right] = \mathbf{C}^{(g)}$. Consider now a slightly different approach to get an estimator for $\mathbf{C}^{(g)}$.

---

[5] To re-estimate the covariance matrix, $\mathbf{C}$, from a $\mathcal{N}(\mathbf{0}, \mathbf{I})$ distributed sample such that $\text{cond}(\mathbf{C}) < 10$ a sample size $\lambda \ge 4n$ is needed. The condition number of the matrix $\mathbf{C}$ is defined via the Euclidean norm: $\text{cond}(\mathbf{C}) \stackrel{\text{def}}{=} \|\mathbf{C}\| \times \|\mathbf{C}^{-1}\|$, where $\|\mathbf{C}\| = \sup_{\|\mathbf{x}\|=1} \|\mathbf{C}\mathbf{x}\|$. For the covariance matrix $\mathbf{C}$ holds $\text{cond}(\mathbf{C}) = \frac{\lambda_{\max}}{\lambda_{\min}} \ge 1$, where $\lambda_{\max}$ and $\lambda_{\min}$ are the largest and smallest eigenvalue of $\mathbf{C}$.

$$\mathbf{C}_\lambda^{(g+1)} = \frac{1}{\lambda} \sum_{i=1}^{\lambda} \left(\mathbf{x}_i^{(g+1)} - \mathbf{m}^{(g)} \right) \left(\mathbf{x}_i^{(g+1)} - \mathbf{m}^{(g)} \right)^{\mathrm{T}} \qquad (7)$$

The matrix $\mathbf{C}_\lambda^{(g+1)}$ is an unbiased maximum likelihood estimator of $\mathbf{C}^{(g)}$. The remarkable difference between (6) and (7) is the reference mean value. For $\mathbf{C}_{\mathrm{emp}}^{(g+1)}$ it is the mean of the *actually realized* sample. For $\mathbf{C}_\lambda^{(g+1)}$ it is the true mean value of the distribution, $\mathbf{m}^{(g)}$ (see (2)). Therefore, the estimators $\mathbf{C}_{\mathrm{emp}}^{(g+1)}$ and $\mathbf{C}_\lambda^{(g+1)}$ can be interpreted differently: while $\mathbf{C}_{\mathrm{emp}}^{(g+1)}$ estimates the distribution variance *within the sampled points*, $\mathbf{C}_\lambda^{(g+1)}$ estimates variances of sampled *steps*, $\mathbf{x}_i^{(g+1)} - \mathbf{m}^{(g)}$. For the CMA the second approach is chosen.

Equation (7) re-estimates *the original* covariance matrix. To "estimate" a "better" covariance matrix (7) is modified and the same, *weighted selection* mechanism as in (3) is used [8].

$$\mathbf{C}_\mu^{(g+1)} = \sum_{i=1}^{\mu} w_i \left(\mathbf{x}_{i:\lambda}^{(g+1)} - \mathbf{m}^{(g)} \right) \left(\mathbf{x}_{i:\lambda}^{(g+1)} - \mathbf{m}^{(g)} \right)^{\mathrm{T}} \qquad (8)$$

The matrix $\mathbf{C}_\mu^{(g+1)}$ is an estimator for the distribution of *selected steps*, just as $\mathbf{C}_\lambda^{(g+1)}$ is an estimator of the original distribution of steps before selection. Sampling from $\mathbf{C}_\mu^{(g+1)}$ tends to reproduce selected, 1.e. *successful* steps, giving a justification for what a "better" covariance matrix means.

We compare (8) with the EMNA$_{global}$ approach [15, 16], where

$$\mathbf{C}_{\mathrm{EMNA}_{global}}^{(g+1)} = \frac{1}{\mu} \sum_{i=1}^{\mu} \left(\mathbf{x}_{i:\lambda}^{(g+1)} - \mathbf{m}^{(g+1)} \right) \left(\mathbf{x}_{i:\lambda}^{(g+1)} - \mathbf{m}^{(g+1)} \right)^{\mathrm{T}} , \qquad (9)$$

and $\mathbf{m}^{(g+1)} = \frac{1}{\mu} \sum_{i=1}^{\mu} \mathbf{x}_{i:\lambda}^{(g+1)}$. The subtle difference is, again, the choice of the reference mean value.[6] Equation (8) estimates selected steps while in (9) the variance within the selected population is estimated. Equation (8) always reveals larger variances than (9), because the reference mean value in (9) is the minimizer for the variances. Moreover, in most conceivable selection situations (9) decreases the variances.

Figure 2 demonstrates the estimation results on *a linear* objective function for $\lambda = 150$, $\mu = 50$, and $w_i = 1/\mu$. While (8) increases the expected variance in direction of the gradient (where the selection takes place, here the diagonal), given ordinary settings for parent number μ and recombination weights $w_1, \ldots, w_n$, (9) decreases this variance! Therefore, (9) is highly susceptible to premature convergence, in particular with small parent populations, where the population cannot be expected to bracket the optimum at any time. However, for large values of μ in large populations with large initial variances, the impact of the different reference mean value can be marginal.

---

[6] Taking a weighted sum, $\sum_{i=1}^{\mu} w_i \ldots$, instead of the mean, $\frac{1}{\mu} \sum_{i=1}^{\mu} \ldots$, is an appealing, but less important, difference.

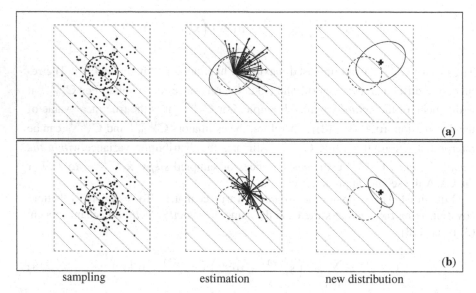

sampling   estimation   new distribution

Fig. 2. Estimation of the covariance matrix on $f_{\text{linear}}(\mathbf{x}) = -\sum_{i=1}^{2} x_i$ to be minimized. Contour lines (*dotted*) indicate that the strategy should move toward the upper right corner. (a) Estimation of $\mathbf{C}_{\mu}^{(g+1)}$ according to (8), where $w_i = 1/\mu$; (b) estimation of $\mathbf{C}_{\text{EMNA}_{global}}^{(g+1)}$ according to (9). *Left*: sample of $\lambda = 150$ $\mathcal{N}(\mathbf{0}, \mathbf{I})$ distributed points. *Middle*: the $\mu = 50$ selected points (*dots*) determining the entries for the estimation equation (*solid straight lines*), and the estimated covariance matrix (*ellipsoid*). *Right*: search distribution of the next generation. Given $w_i = 1/\mu$, (a) *increases* the expected variance in gradient direction for all $\mu < \lambda/2$, while (b) *decreases* this variance for any $\mu < \lambda$

To ensure $\mathbf{C}_{\mu}^{(g+1)}$ is a *reliable* estimator implementing (2), (3), and (8), the variance effective selection mass μ_{eff} (cf. (5)) must be large enough: to get condition numbers smaller than ten for $\mathbf{C}_{\mu}^{(g)}$ on $f_{\text{sphere}}(\mathbf{x}) = \sum_{i=1}^{n} x_i^2$, to our experience, $\mu_{\text{eff}} \approx 10n$ is needed. The next step is to circumvent this restriction on μ_{eff}.

3.2 Rank-μ-Update

To achieve *fast* search (opposite to more *robust* or more *global* search), e.g. competitive performance on f_{sphere}, the population size λ must be small. Because $\mu_{\text{eff}} \approx \lambda/4$ also μ_{eff} must be small and we may assume, e.g., $\mu_{\text{eff}} \leq 1 + \ln n$. Then, it is not possible to get a *reliable* estimator for a good covariance matrix from (8) alone. As a remedy, information from previous generations is added. For example, after a sufficient number of generations, the mean of the estimated covariance matrices from all generations,

$$\mathbf{C}^{(g+1)} = \frac{1}{g+1} \sum_{i=0}^{g} \frac{1}{\sigma^{(i)2}} \mathbf{C}_{\mu}^{(i+1)} \tag{10}$$

becomes a reliable estimator for the selected steps. To make $\mathbf{C}_\mu^{(g)}$ from different generations comparable, the different $\sigma^{(i)}$ are incorporated. (Assuming $\sigma^{(i)} = 1$, (10) resembles the covariance matrix from EMNA$_i$ [16].)

In (10), all generation steps have the same weight. To assign recent generations a higher weight, exponential smoothing is introduced. Choosing $\mathbf{C}^{(0)} = \mathbf{I}$ to be the unity matrix and a learning rate $0 < c_{\text{cov}} \leq 1$, then $\mathbf{C}^{(g+1)}$ reads

$$\mathbf{C}^{(g+1)} = (1 - c_{\text{cov}})\mathbf{C}^{(g)} + c_{\text{cov}}\frac{1}{\sigma^{(g)^2}}\mathbf{C}_\mu^{(g+1)}$$

$$= (1 - c_{\text{cov}})\mathbf{C}^{(g)} + c_{\text{cov}}\sum_{i=1}^{\mu} w_i \operatorname{OP}\left(\frac{\mathbf{x}_{i:\lambda}^{(g+1)} - \mathbf{m}^{(g)}}{\sigma^{(g)}}\right) \quad (11)$$

where

$c_{\text{cov}} \leq 1$ learning rate for updating the covariance matrix. For $c_{\text{cov}} = 1$, no prior information is retained and $\mathbf{C}^{(g+1)} = \frac{1}{\sigma^{(g)2}}\mathbf{C}_\mu^{(g+1)}$. For $c_{\text{cov}} = 0$, no learning takes place and $\mathbf{C}^{(g+1)} = \mathbf{C}^{(0)}$.

$\operatorname{OP} : \mathbb{R}^n \to \mathbb{R}^{n \times n}, \mathbf{x} \mapsto \mathbf{x}\mathbf{x}^{\mathrm{T}}$, denotes the outer product of a vector with itself.

This covariance matrix update is called rank-μ-update [9], because the sum of outer products in (11) is of rank $\min(\mu, n)$ (with probability one). Note that this sum can even consist of a single term, if $\mu = 1$.

The factor $1/c_{\text{cov}}$ can be interpreted as the *backward time horizon*. Because (11) expands to the weighted sum

$$\mathbf{C}^{(g+1)} = (1 - c_{\text{cov}})^{g+1}\mathbf{C}^{(0)} + c_{\text{cov}}\sum_{i=0}^{g}(1 - c_{\text{cov}})^{g-i}\frac{1}{\sigma^{(i)^2}}\mathbf{C}_\mu^{(i+1)} \ , \quad (12)$$

the backward time horizon, Δg, where about 63% of the overall weight is summed up, is defined by

$$c_{\text{cov}}\sum_{i=g+1-\Delta g}^{g}(1 - c_{\text{cov}})^{g-i} \approx 0.63 \approx 1 - \frac{1}{e} \ . \quad (13)$$

Resolving the sum yields

$$(1 - c_{\text{cov}})^{\Delta g} \approx \frac{1}{e} \ , \quad (14)$$

and resolving for Δg, using the Taylor approximation for ln, yields

$$\Delta g \approx \frac{1}{c_{\text{cov}}} \ . \quad (15)$$

That is, approximately 37% of the information in $\mathbf{C}^{(g+1)}$ is older than $1/c_{\text{cov}}$ generations, and, according to (14), the original weight is reduced by a factor of 0.37 after approximately $1/c_{\text{cov}}$ generations.

The choice of c_{cov} is crucial. Small values lead to slow learning, too large values lead to a failure, because the covariance matrix degenerates. Fortunately, a good setting seems to be largely independent of the function to be optimized.[7] A first order approximation for a good choice is $c_{cov} \approx \mu_{eff}/n^2$. Therefore, the characteristic time horizon for (11) is roughly n^2/μ_{eff}.

Even for the learning rate $c_{cov} = 1$, adapting the covariance matrix cannot be accomplished within one generation. The effect of the original sample distribution does not vanish until a sufficient number of generations. Assuming fixed search costs (number of function evaluations), a small population size λ allows a larger number of generations and therefore usually leads to a faster adaptation of the covariance matrix.

3.3 Cumulation: Utilizing the Evolution Path

We have used the selected steps, $(\mathbf{x}_{i:\lambda}^{(g+1)} - \mathbf{m}^{(g)})/\sigma^{(g)}$, to update the covariance matrix in (11). Because $\mathrm{OP}(\mathbf{x}) = \mathbf{x}\mathbf{x}^{\mathrm{T}} = \mathrm{OP}(-\mathbf{x})$, *the sign of the steps in (11) is irrelevant* – that is, the sign information is not used for calculating $\mathbf{C}^{(g+1)}$. To exploit this information, the so-called *evolution path* is introduced [10, 12].

We call a sequence of successive steps, the strategy takes over a number of generations, an evolution path. An evolution path can be expressed by a sum of consecutive steps. This summation is referred to as cumulation. To construct an evolution path, the step size σ is disregarded. For example, an evolution path of three steps can be constructed by the sum

$$\frac{\mathbf{m}^{(g+1)} - \mathbf{m}^{(g)}}{\sigma^{(g)}} + \frac{\mathbf{m}^{(g)} - \mathbf{m}^{(g-1)}}{\sigma^{(g-1)}} + \frac{\mathbf{m}^{(g-1)} - \mathbf{m}^{(g-2)}}{\sigma^{(g-2)}} \ . \tag{16}$$

Again, we use exponential smoothing as in (11), to construct the evolution path, $\mathbf{p}_c \in \mathbb{R}^n$, starting with $\mathbf{p}_c^{(0)} = \mathbf{0}$.

$$\mathbf{p}_c^{(g+1)} = (1 - c_c)\mathbf{p}_c^{(g)} + \sqrt{c_c(2 - c_c)\mu_{eff}} \ \frac{\mathbf{m}^{(g+1)} - \mathbf{m}^{(g)}}{\sigma^{(g)}} \tag{17}$$

where

$\mathbf{p}_c^{(g)} \in \mathbb{R}^n$, evolution path at generation g.

$c_c \leq 1$. Again, $1/c_c$ is the backward time horizon of the evolution path $\mathbf{p}_c$ (compare (15)). A time horizon between $\sqrt{n}$ and n is reasonable.

The factor $\sqrt{c_c(2 - c_c)\mu_{eff}}$ is a normalization constant for $\mathbf{p}_c^{(g)}$. For $c_c = 1$ and $\mu_{eff} = 1$, the factor reduces to one, and $\mathbf{p}_c^{(g+1)} = (\mathbf{x}_{1:\lambda}^{(g+1)} - \mathbf{m}^{(g)})/\sigma^{(g)}$. The factor is chosen, such that

---

[7] We use the sphere model $f_{sphere}(\mathbf{x}) = \sum_i x_i^2$ to empirically find a good setting for the parameter c_{cov}, dependent on n and μ_{eff}. The setting found was applicable to any non-noisy objective function we had tried so far.

$$\mathbf{p}_c^{(g+1)} \sim \mathcal{N}(\mathbf{0}, \mathbf{C}) \tag{18}$$

if

$$\mathbf{p}_c^{(g)} \sim \frac{\mathbf{x}_{i:\lambda}^{(g+1)} - \mathbf{m}^{(g)}}{\sigma^{(g)}} \sim \mathcal{N}(\mathbf{0}, \mathbf{C}) \quad \text{for all } i = 1, \ldots, \mu \ . \tag{19}$$

To derive (18) from (19) and (17) remark that

$$(1 - c_c)^2 + \sqrt{c_c(2 - c_c)}^2 = 1 \quad \text{and} \quad \sum_{i=1}^{\mu} w_i \mathcal{N}_i(\mathbf{0}, \mathbf{C}) \sim \frac{1}{\sqrt{\mu_{\text{eff}}}} \mathcal{N}(\mathbf{0}, \mathbf{C}) \ . \tag{20}$$

The (rank-one) update of the covariance matrix $\mathbf{C}^{(g)}$ via the evolution path $\mathbf{p}_c^{(g+1)}$ reads [10]

$$\mathbf{C}^{(g+1)} = (1 - c_{\text{cov}})\mathbf{C}^{(g)} + c_{\text{cov}} \mathbf{p}_c^{(g+1)} \mathbf{p}_c^{(g+1)\,\mathrm{T}} \ . \tag{21}$$

An empirically validated choice for the learning rate in (21) is $c_{\text{cov}} \approx 2/n^2$. For $c_c = 1$ and $\mu = 1$, (21) and (11) are identical.

Using the evolution path for the update of $\mathbf{C}$ is a significant improvement of (11) for small μ_{eff}, because correlations between consecutive steps are exploited. The leading signs of steps, and the dependencies between consecutive steps, play a significant role in the resulting evolution path $\mathbf{p}_c^{(g+1)}$. For $c_c \approx 3/n$ the number of function evaluations needed to adapt a nearly optimal covariance matrix on cigar-like objective functions becomes $\mathcal{O}(n)$.

As a last step, we combine (11) and (21).

3.4 Combining Rank-μ-Update and Cumulation

The final CMA update of the covariance matrix combines (11) and (21), where μ_{cov} determines their relative weighting.

$$\mathbf{C}^{(g+1)} = (1 - c_{\text{cov}})\mathbf{C}^{(g)} + \frac{c_{\text{cov}}}{\mu_{\text{cov}}} \underbrace{\mathbf{p}_c^{(g+1)} \mathbf{p}_c^{(g+1)\,\mathrm{T}}}_{\text{rank-one update}} + c_{\text{cov}} \left(1 - \frac{1}{\mu_{\text{cov}}} \right)$$

$$\times \underbrace{\sum_{i=1}^{\mu} w_i \left(\frac{\mathbf{x}_{i:\lambda}^{(g+1)} - \mathbf{m}^{(g)}}{\sigma^{(g)}} \right) \left(\frac{\mathbf{x}_{i:\lambda}^{(g+1)} - \mathbf{m}^{(g)}}{\sigma^{(g)}} \right)^{\mathrm{T}}}_{\text{rank-}\mu \text{ update}} \tag{22}$$

where

$\mu_{\text{cov}} \geq 1$. Choosing $\mu_{\text{cov}} = \mu_{\text{eff}}$ is most appropriate.

$c_{\text{cov}} \approx \min(\mu_{\text{cov}}, \mu_{\text{eff}}, n^2)/n^2$.

Equation (22) reduces to (11) for $\mu_{\text{cov}} \to \infty$ and to (21) for $\mu_{\text{cov}} = 1$. The equation combines the advantages of (11) and (21). On the one hand, the information within the population of one generation is used efficiently by the rank-μ update. On the other hand, information of correlations between generations is exploited by using the evolution path for the rank-one update. The former is important in large populations, the latter is particularly important in small populations.

4 Step Size Control

We know two reasons to introduce a step size control in addition to the adaptation rule of (22) for $\mathbf{C}^{(g)}$.

1. The optimal overall step length cannot be well approximated by (22), in particular if μ_{eff} is chosen larger than one. For example, on $f_{\text{sphere}}(\mathbf{x}) = \sum_{i=1}^{n} x_i^2$, the optimal step size σ equals approximately $\mu \sqrt{f_{\text{sphere}}(\mathbf{x})}/n$, given $\mathbf{C}^{(g)} \approx \mathbf{I}$ and $\mu_{\text{eff}} = \mu \ll n$ [2, 17]. This dependency on μ cannot be realized by (11), and is also not well approximated by (22).

2. The largest reliable learning rate for the covariance matrix update in (22) is too slow to achieve competitive change rates for the overall step length. To achieve optimal performance on f_{sphere} with an evolution strategy, the overall step length must decrease by a factor of approximately $\exp(0.202) \approx 1.22$ within n function evaluations, as can be derived from progress formulas [2, p. 229]. That is, the time horizon for the step length change must be proportional to n or shorter. From the learning rate c_{cov} in (22) it follows that the adaptation is too slow to perform competitively on f_{sphere} whenever $\mu_{\text{eff}} \ll n$. This can be validated by simulations even for moderate dimensions, say, $n \geq 10$ and small μ_{eff}, say, $\mu_{\text{eff}} \leq 1 + \ln n$.

To control the step size $\sigma^{(g)}$ we utilize an evolution path, i.e. a sum of successive steps (see page 84). The method is denoted *cumulative path length control*, cumulative step size control, or *cumulative step size adaptation*. The length of an evolution path is exploited, based on the following reasoning.

- If the evolution path is long, the single steps are pointing to similar directions. Loosely speaking, they are correlated. Because the steps are similar, the same distance can be covered by fewer but longer steps in the same directions – consequently the step size should be increased.
- If the evolution path is short, single steps cancel each other out. Loosely speaking, they are anti-correlated. If steps annihilate each other, the step size should be decreased.
- In the desired situation, the steps are approximately perpendicular in expectation and therefore uncorrelated.

To define "long" and "short", we compare the length of the evolution path with its *expected length under random selection*.[8] Under random selection consecutive steps are independent and therefore uncorrelated. If selection biases the evolution path to be longer then expected, σ will be increased, and, vice versa. If selection biases the evolution path to be shorter than expected, σ will be decreased. In the ideal situation, selection does not bias the length of the evolution path at all.

Because, in general, the expected length of the evolution path $\mathbf{p}_c^{(g+1)}$ from (17) depends on its direction (compare (18)), a conjugate evolution path is constructed:

$$\mathbf{p}_\sigma^{(g+1)} = (1 - c_\sigma)\mathbf{p}_\sigma^{(g)} + \sqrt{c_\sigma(2 - c_\sigma)\mu_{\text{eff}}}\ \mathbf{C}^{(g)-\frac{1}{2}}\ \frac{\mathbf{m}^{(g+1)} - \mathbf{m}^{(g)}}{\sigma^{(g)}} \tag{23}$$

where

$\mathbf{p}_\sigma^{(g)} \in \mathbb{R}^n$ is the conjugate evolution path at generation g.

$c_\sigma < 1$. Again, $1/c_\sigma$ is the backward time horizon of the evolution path (compare (15)). For small μ_{eff}, a time horizon between $\sqrt{n}$ and n is reasonable.

$\sqrt{c_\sigma(2 - c_\sigma)\mu_{\text{eff}}}$ is a normalization constant, see (17).

$\mathbf{C}^{(g)-\frac{1}{2}} \stackrel{\text{def}}{=} \mathbf{B}^{(g)}\mathbf{D}^{(g)-1}\mathbf{B}^{(g)\mathrm{T}}$, where $\mathbf{C}^{(g)} = \mathbf{B}^{(g)}\left(\mathbf{D}^{(g)}\right)^2\mathbf{B}^{(g)\mathrm{T}}$ is an eigende-composition of $\mathbf{C}^{(g)}$, where $\mathbf{B}^{(g)}$ is an orthonormal basis of eigenvectors, and the diagonal elements of the diagonal matrix $\mathbf{D}^{(g)}$ are square roots of the corresponding positive eigenvalues.

For $\mathbf{C}^{(g)} = \mathbf{I}$, (23) replicates (17), because $\mathbf{C}^{(g)-\frac{1}{2}} = \mathbf{I}$ then. The transformation $\mathbf{C}^{(g)-\frac{1}{2}}$ re-scales the step $\mathbf{m}^{(g+1)} - \mathbf{m}^{(g)}$ within the coordinate system given by $\mathbf{B}^{(g)}$. The single factors of the transformation $\mathbf{C}^{(g)-\frac{1}{2}} = \mathbf{B}^{(g)}\mathbf{D}^{(g)-1}\mathbf{B}^{(g)\mathrm{T}}$ can be read as follows (from right to left):

$\mathbf{B}^{(g)\mathrm{T}}$ rotates the space such that the columns of $\mathbf{B}^{(g)}$, i.e. the principle axes of the distribution $\mathcal{N}(\mathbf{0}, \mathbf{C}^{(g)})$, rotate into the coordinate axes. Elements of the resulting vector relate to projections onto the corresponding eigenvectors.

$\mathbf{D}^{(g)-1}$ applies a (re-)scaling such that all axes become equally sized.

$\mathbf{B}^{(g)}$ rotates the result back into the original coordinate system. This last transformation ensures that directions of consecutive steps are comparable.

Consequently, the transformation $\mathbf{C}^{(g)-\frac{1}{2}}$ makes the expected length of $\mathbf{p}_\sigma^{(g+1)}$ independent of its direction, and for any sequence of realized covariance matrices $\mathbf{C}^{(g)}_{g=0,1,2,\dots}$ we have under random selection $\mathbf{p}_\sigma^{(g+1)} \sim \mathcal{N}(\mathbf{0}, \mathbf{I})$, given $\mathbf{p}_\sigma^{(0)} \sim \mathcal{N}(\mathbf{0}, \mathbf{I})$ [6].

To update $\sigma^{(g)}$, we "compare" $\|\mathbf{p}_\sigma^{(g+1)}\|$ with its expected length $\mathsf{E}\|\mathcal{N}(\mathbf{0}, \mathbf{I})\|$, that is

---

[8] Random selection means that the index $i : \lambda$ (compare (3)) is independent of the value of $\mathbf{x}_{i:\lambda}^{(g+1)}$ for all $i = 1, \dots, \lambda$, e.g. $i : \lambda = i$.

$$\ln \sigma^{(g+1)} = \ln \sigma^{(g)} + \frac{c_\sigma}{d_\sigma} \left(\frac{\|\mathbf{p}_\sigma^{(g+1)}\|}{\mathsf{E}\|\mathcal{N}(\mathbf{0}, \mathbf{I})\|} - 1 \right) , \tag{24}$$

where

$d_\sigma \approx 1$, damping parameter, scales the change magnitude of $\ln \sigma^{(g)}$. The factor c_σ/d_σ is based on in-depth investigations of the algorithm [6].

$\mathsf{E}\|\mathcal{N}(\mathbf{0}, \mathbf{I})\| = \sqrt{2}\, \Gamma(\frac{n+1}{2})/\Gamma(\frac{n}{2}) \approx \sqrt{n} + \mathcal{O}(1/n)$, expectation of the Euclidean norm of a $\mathcal{N}(\mathbf{0}, \mathbf{I})$ distributed random vector.

For $\|\mathbf{p}_\sigma^{(g+1)}\| = \mathsf{E}\|\mathcal{N}(\mathbf{0}, \mathbf{I})\|$ the second summand in (24) is zero, and $\sigma^{(g)}$ is unchanged, while $\sigma^{(g)}$ is increased for $\|\mathbf{p}_\sigma^{(g+1)}\| > \mathsf{E}\|\mathcal{N}(\mathbf{0}, \mathbf{I})\|$, and $\sigma^{(g)}$ is decreased for $\|\mathbf{p}_\sigma^{(g+1)}\| < \mathsf{E}\|\mathcal{N}(\mathbf{0}, \mathbf{I})\|$. The step size change is unbiased on the log scale, because $\mathsf{E}\left[\ln \sigma^{(g+1)} \mid \sigma^{(g)}\right] = \ln \sigma^{(g)}$ for $\mathbf{p}_\sigma^{(g+1)} \sim \mathcal{N}(\mathbf{0}, \mathbf{I})$. The role of unbiasedness is discussed in Sect. 6.

We show that successive steps taken by $\mathbf{m}^{(g)}$ are approximately $\mathbf{C}^{(g)^{-1}}$-conjugate. Equations (23) and (24) adapt σ such that the length of $\mathbf{p}_\sigma^{(g+1)}$ equals approximately $\mathsf{E}\|\mathcal{N}(\mathbf{0}, \mathbf{I})\|$. Starting from $(\mathsf{E}\|\mathcal{N}(\mathbf{0}, \mathbf{I})\|)^2 \approx \|\mathbf{p}_\sigma^{(g+1)}\|^2 = \mathbf{p}_\sigma^{(g+1)^{\mathsf{T}}} \mathbf{p}_\sigma^{(g+1)} = \mathrm{RHS}^{\mathsf{T}}\mathrm{RHS}$ of (23) and assuming that the expected squared length of $\mathbf{C}^{(g)^{-\frac{1}{2}}}(\mathbf{m}^{(g+1)} - \mathbf{m}^{(g)})$ is unchanged by selection we get

$$\mathbf{p}_\sigma^{(g)^{\mathsf{T}}} \mathbf{C}^{(g)^{-\frac{1}{2}}} (\mathbf{m}^{(g+1)} - \mathbf{m}^{(g)}) \approx 0 , \tag{25}$$

and

$$\left(\mathbf{C}^{(g)^{\frac{1}{2}}} \mathbf{p}_\sigma^{(g)} \right)^{\mathsf{T}} \mathbf{C}^{(g)^{-1}} \left(\mathbf{m}^{(g+1)} - \mathbf{m}^{(g)} \right) \approx 0 . \tag{26}$$

Given $1/c_{\mathrm{cov}} \gg 1$ and (25) we assume that $\mathbf{p}_\sigma^{(g-1)^{\mathsf{T}}} \mathbf{C}^{(g)^{-\frac{1}{2}}}(\mathbf{m}^{(g+1)} - \mathbf{m}^{(g)}) \approx 0$ and derive

$$\left(\mathbf{m}^{(g)} - \mathbf{m}^{(g-1)} \right)^{\mathsf{T}} \mathbf{C}^{(g)^{-1}} \left(\mathbf{m}^{(g+1)} - \mathbf{m}^{(g)} \right) \approx 0 . \tag{27}$$

That is, consecutive steps taken by the distribution mean become approximately $\mathbf{C}^{(g)^{-1}}$-conjugate.

Because $\sigma^{(g)} > 0$, (24) is equivalent to

$$\sigma^{(g+1)} = \sigma^{(g)} \exp \left(\frac{c_\sigma}{d_\sigma} \left(\frac{\|\mathbf{p}_\sigma^{(g+1)}\|}{\mathsf{E}\|\mathcal{N}(\mathbf{0}, \mathbf{I})\|} - 1 \right) \right) \tag{28}$$

The length of the evolution path is an intuitive and empirically well validated goodness measure for the overall step length. For $\mu_{\mathrm{eff}} > 1$ it is the best measure to our knowledge. Nevertheless, it fails to adapt nearly optimal step sizes on very noisy objective functions [3].

5 Simulations

The complete algorithm of the CMA evolution strategy is summarized in Appendix A, where all (default) parameter settings are given. We show single simulation runs of the CMA-ES on the test functions from Table 1, where $n = 8$.[9] All func-

Table 1. Convex-quadratic test functions. $\mathbf{y} = \mathbf{Ox}$, where $\mathbf{O}$ is an orthogonal matrix

Function	$\mathrm{cond}(\mathbf{H})$	f_{stop}	Initial interval
$f_{\mathrm{sphere}}(\mathbf{x}) = \frac{1}{2}\sum_{i=1}^{n} x_i^2$	1	10^{-9}	$[0.1, 0.3]^n$
$f_{\mathrm{elli}}(\mathbf{x}) = \frac{1}{2}\sum_{i=1}^{n} 10^{6\frac{i-1}{n-1}} y_i^2$	10^6	10^{-9}	$[0.1, 0.3]^n$
$f_{\mathrm{cigtab}}(\mathbf{x}) = \frac{1}{2}\left(y_1^2 + 10^4 \sum_{i=2}^{n-1} y_i^2 + 10^8 y_n^2\right)$	10^8	10^{-9}	$[5, 25]^n$
$f_{\mathrm{twoax}}(\mathbf{x}) = \frac{1}{2}\left(\sum_{i=1}^{\lfloor n/2 \rfloor} y_i^2 + 10^6 \sum_{i=\lfloor n/2 \rfloor + 1}^{n} y_i^2\right)$	10^6	10^{-9}	$[5, 25]^n$

tions are convex-quadratic and can be written in the form $f(\mathbf{x}) = \frac{1}{2}\mathbf{x}^{\mathrm{T}}\mathbf{Hx}$, where $\mathbf{H}$ is the positive definite Hessian matrix. For each function we run an *axis parallel* version and a *randomly oriented* version. In the axis parallel version the Hessian is diagonal, because we choose $\mathbf{O} = \mathbf{I}$ (see Table 1). For the randomly oriented version each column of $\mathbf{O}$ is uniformly distributed on the unit hypersphere [12], fixed for each run. The matrix $\mathbf{O}$ defines the coordinate system where the Hessian is diagonal. On f_{sphere}, instead of $\mathbf{O}$, we set $\mathbf{B}^{(0)}$ to an arbitrary orthogonal matrix in the "randomly oriented" version. Furthermore, the diagonal elements of $\mathbf{D}^{(0)}$ are set to $d_{ii} = 10^{-3+3\frac{i-1}{n-1}}$ and $\mathbf{C}^{(0)} = \mathbf{B}^{(0)}\mathbf{D}^{(0)}\mathbf{D}^{(0)}\mathbf{B}^{(0)}{}^{\mathrm{T}}$. That is, the condition number of $\mathbf{C}^{(0)}$ equals to 10^6 and $\mathbf{C}$ has to become spherical (condition number one) during the adaptation (see Fig. 3). Further settings and initial values for the CMA-ES are according to Fig. 7 and Table 2 in Appendix A.

By tracking eigenvalues and variances of the covariance matrix we can pursue, whether the objective of the covariance matrix adaptation is achieved, to approximate the inverse Hessian matrix of the objective function up to a constant factor. Eigenvalues of the Hessian correspond to the coefficients in Table 1 ($\{1, \ldots, 1\}$ for f_{sphere}, $\{10^{6\frac{i-1}{n-1}} \mid i = 1, \ldots, n\}$ for f_{elli}, $\{1, 10^4, 10^8\}$ for f_{cigtab}, and $\{1, 10^6\}$ for f_{twoax}).

The runs are shown in Fig. 3–6. The bottom figures show the square root of the eigenvalues of the covariance matrix, that is the lengths of the principal axes of the distribution ellipsoid, corresponding to diagonal elements, d_{ii}, of $\mathbf{D}$. After about 3500, 3500, 4000, and 5000 function evaluations, respectively, the adaptation has taken place and the axes lengths d_{ii} reflect the square root of the inverse eigenvalues of the Hessian, properly. Notice the striking connection between the matching of the lengths of the axes and the slope of the function value graph. Apart from effects of

---

[9] For exhaustive investigations of the CMA-ES on larger test function sets see [6,8,9,11,12] and for scale-up investigation up to $n = 320$ see [12].

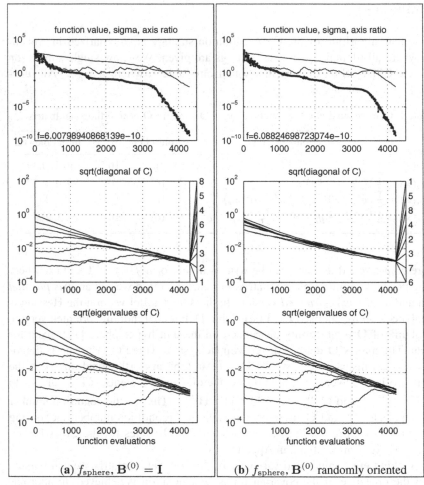

(a) f_{sphere}, $\mathbf{B}^{(0)} = \mathbf{I}$ **(b)** f_{sphere}, $\mathbf{B}^{(0)}$ randomly oriented

Fig. 3. Two runs on f_{sphere}, *where the initial covariance matrix,* $\mathbf{C}^{(0)}$, *is not spherical Above*: function value (*thick line*), σ (lower graph), $\sqrt{\text{cond}(\mathbf{C})}$ (upper graph). *Middle*: $\sqrt{\text{diag}(\mathbf{C})}$, index annotated. *Below*: square root of the eigenvalues of $\mathbf{C}$, i.e. $\text{diag}(\mathbf{D}) = [d_{11}, \ldots, d_{nn}]$, versus number of function evaluations

different $\mathbf{x}^{(0)}$ and different random seeds, the upper and lower figures are equivalent for the axis parallel (**a**) and the randomly oriented version (**b**).

On axis parallel functions, the principal axes of the search distribution should become axis parallel after the adaptation has taken place. The middle figures show the square root of the diagonal elements of the covariance matrix, $\sqrt{c_{ii}}$. The elements $\sqrt{c_{ii}}$ align to the principal axes lengths d_{ii} in the left figures. That means, the search ellipsoid becomes axis parallel oriented (apart from subspaces of equal eigenvalues, where the final orientation is irrelevant). The final ordering of the $\sqrt{c_{ii}}$ reflects the ordering of the coefficients in the objective function. In contrast, the ordering of the

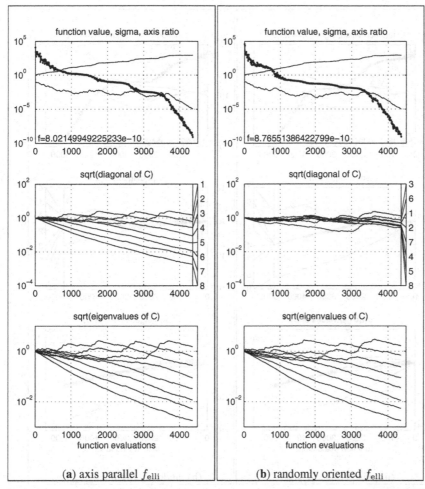

(a) axis parallel f_{elli} **(b)** randomly oriented f_{elli}

Fig. 4. Two runs on f_{elli} *Above*: function value (*thick line*), σ (lower graph), $\sqrt{\text{cond}(\mathbf{C})}$ (upper graph). *Middle*: $\sqrt{\text{diag}(\mathbf{C})}$, index annotated. *Below*: square root of the eigenvalues of $\mathbf{C}$, i.e. $\text{diag}(\mathbf{D}) = [d_{11}, \ldots, d_{nn}]$, versus number of function evaluations

$\sqrt{c_{ii}}$ on the randomly oriented functions is arbitrary. The course of $\sqrt{c_{ii}}$ depends on the given coordinate system and therefore is remarkably different between **(a)** and **(b)**. After the adaptation has taken place, in all cases the optimum is approached as fast as with an isotropic search distribution on f_{sphere}.

All the data give clear evidence that the inverse Hessian is well approximated. A measure for "well" can be derived from the runs on f_{sphere} (Fig. 3): the final condition number of $\mathbf{C}$ is smaller than five.

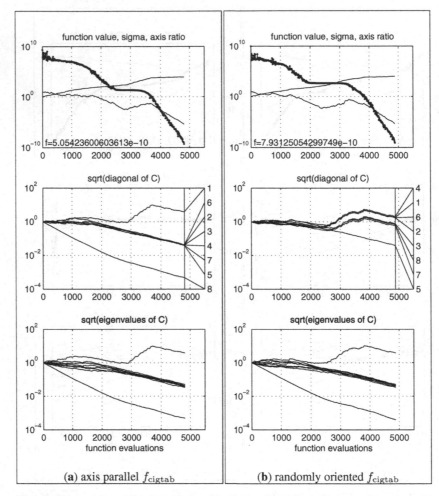

Fig. 5. Two runs on f_{cigtab} *Above*: function value (*thick line*), σ (lower graph), $\sqrt{\text{cond}(\mathbf{C})}$ (upper graph). *Middle*: $\sqrt{\text{diag}(\mathbf{C})}$, index annotated. *Below*: square root of the eigenvalues of $\mathbf{C}$, i.e. $\text{diag}(\mathbf{D}) = [d_{11}, \ldots, d_{nn}]$, versus number of function evaluations

6 Discussion

In effect, the CMA-ES transforms any ellipsoid function into a spherical function. It is highly competitive on a considerable number of test functions [6, 8, 9, 11, 12] and was successfully applied to real world problems.[10] We discuss a few basic design principles.

---

[10] To our knowledge a few dozen successful applications have been published up to now, see http://www.icos.ethz.ch/software/evolutionary_computation/cmaapplications.pdf

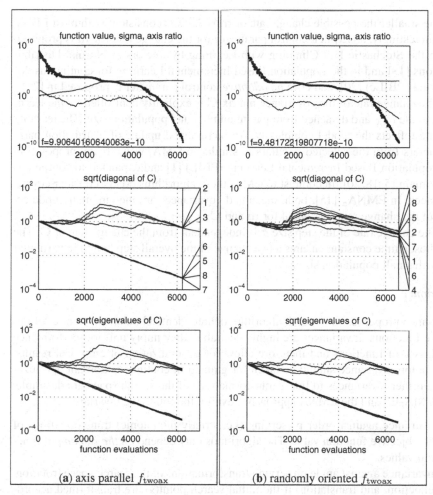

Fig. 6. Two runs on f_twoax *Above*: function value (*thick line*), σ (lower graph), $\sqrt{\text{cond}(\mathbf{C})}$ (upper graph). *Middle*: $\sqrt{\text{diag}(\mathbf{C})}$, index annotated. *Below*: square root of the eigenvalues of $\mathbf{C}$, i.e. $\text{diag}(\mathbf{D}) = [d_{11}, \ldots, d_{nn}]$, versus number of function evaluations

Change Rates

A great deal of differences between continuous domain EDAs with multiple dependencies and the CMA-ES can be found in the change rates of distribution parameters. We refer to a change rate as the expected parameter change *per sampled search point*, given a certain selection situation. The CMA-ES separately controls change rates for the mean value of the distribution, $\mathbf{m}$, the covariance matrix, $\mathbf{C}$, and the step size, σ.

- The change rate for the mean value $\mathbf{m}$, given a fixed sample distribution, is determined by the parent number and the recombination weights. The larger μ_eff, the smaller the possible change rate of **m** is. This is consistent with most EDAs. Interestingly, an explicit control parameter for the change rate for **m** is proposed in the Stochastic Hill Climbing with Learning by Vectors of Normal Distributions [18] and in the Population Based Incremental Learning for continuous domain (PBILc) [20], and even an *adaptive* control parameter is proposed in [21].

- The change rate of the covariance matrix **C** is explicitly controlled by the learning rate c_{cov} and detached from parent number and population size. The learning rate reflects the model complexity. An incremental update of distribution parameters from the selected population, similar to CMA, was already proposed in Population Based Incremental Learning (PBIL) [1] and expanded to continuous domain [5, 18, 20]. In contrast to CMA, these algorithms do not consider covariances. In EMNA$_a$ [15], both, mean and covariances are incrementally updated, but the change rates are equal for **m** and **C**.

- The change rate of the step size σ is independent from the change rate of **C**. The chosen time constant ensures a fast change of the overall step length in particular with small population sizes.

Invariance

Invariance properties of a search algorithm denote identical behavior on a set of objective functions. Invariances are highly desirable: they imply uniform performance on classes of functions and therefore allow for generalization of empirical results. Translation invariance should be taken for granted in continuous domain optimization. Further invariances to linear transformations of the search space are desirable. The CMA-ES and the EMNA approaches exhibit the following invariances.

- Invariance against order preserving (i.e. strictly monotonic) transformations of the objective function value. The algorithms only depend on *the ranking* of function values.

- Invariance against angle preserving transformations of the search space (rotation, reflection, and translation) if the initial search point(s) are transformed accordingly.

- Scale invariance if the initial scaling, e.g. $\sigma^{(0)}$, and the initial search point(s) are chosen accordingly.

- Invariance against any invertible linear transformation of the search space, **A**, if the initial covariance matrix $\mathbf{C}^{(0)} = \mathbf{A}^{-1} \left(\mathbf{A}^{-1} \right)^{\text{T}}$, and the initial search point(s) are transformed accordingly.

In our opinion, invariance should be a fundamental design criterion for any search algorithm.

Stationarity

An important design criterion for a stochastic search procedure is *unbiasedness* of variations of object and strategy parameters [12, 13]. Consider random selection, e.g. the objective function $f(\mathbf{x}) = \texttt{rand}$ to be independent of **x**. The population mean is unbiased if its expected value remains unchanged in the next generation, that is $\mathsf{E}\big[\mathbf{m}^{(g+1)}\,\big|\,\mathbf{m}^{(g)}\,\big] = \mathbf{m}^{(g)}$. For the population mean stationarity under random selection is a rather intuitive concept. In the CMA-ES, stationarity is respected for all parameters in (2). The distribution mean $\mathbf{m}$, the covariance matrix $\mathbf{C}$, and $\ln \sigma$ are unbiased. Unbiasedness of $\ln \sigma$ does not imply that σ is unbiased. Actually, under random selection, $\mathsf{E}\big[\sigma^{(g+1)}\,\big|\,\sigma^{(g)}\,\big] > \sigma^{(g)}$, compare (24).[11]

For variances (or step sizes) a bias toward increase or decrease will entail the danger of divergence or premature convergence, respectively, whenever the selection pressure is low. Nevertheless, on noisy problems a properly controlled bias toward increase, even on the log scale, can be beneficial.

7 Summary and Conclusion

We have compared the CMA evolution strategy with EDAs that estimate the complete covariance matrix of a multi-variate normal search distribution. We summarize identified key points.

- Estimation principle: Most EDAs estimate the distribution parameters from a set of *selected points*. The CMA estimates them from a set of *selected steps*. Using steps is much less prone to premature convergence and supports explorative search behavior.
- Step size control: Methods to estimate or adapt the covariance matrix do not achieve good overall step lengths. In EDAs, step size control is usually absent, making a potential increase of the overall step length almost impossible. In the CMA-ES, the adaptation of the covariance matrix is complemented with step size control. The adjustment of the step size is based on a *different adaptation principle*. Cumulative path length control often adapts nearly optimal step sizes usually leading to considerably larger step lengths. This improves convergence speed *and* global search capabilities at the same time.
- Population size, adaptation, and change rates: Choosing the population size λ is always a compromise. Small λ lead to faster convergence, and large λ help to avoid local optima. To achieve a fast learning scheme for a covariance matrix
  1. the population size λ must be comparatively small (see end of Sect. 3.2) and
  2. an adaptation procedure must be established, where parameters are updated rather than estimated from scratch in every generation.

  Appropriate time constants for change rates of the population mean, of the covariance matrix, and of the overall step length are essential for competitive performance. In the CMA-ES, learning rates can be adjusted independently and only the change rate of the population mean is (indirectly) associated with the population size λ (via μ_{eff}). Determining different change rates for different parameters by adjusting learning rates is an open issue in EDAs.

---

[11] Alternatively, if (28) would have been designed to be unbiased for $\sigma^{(g+1)}$, this would presumably imply $\mathsf{E}\big[\ln \sigma^{(g+1)}\,\big|\,\sigma^{(g)}\,\big] < \ln \sigma^{(g)}$, to our opinion a less desirable possibility.

- Invariances: To generalize empirical performance results, invariance properties are invaluable. Many EDAs use the given coordinate system to estimate the distribution, and are consequently not invariant to rotations of the search space. The CMA-ES is invariant under search space rotation and exhibits further invariances. Admittedly, a rotation invariant method cannot exploit separability of the objective function efficiently.[12]

Based on these key points the CMA can improve the performance on ill-conditioned and/or non-separable problems by orders of magnitude, leaving the performance on simple problems unchanged. In conclusion, the CMA evolution strategy is a state-of-the-art continuous domain evolutionary algorithm which is widely applicable and quasi parameter free.

Acknowledgments

The author wishes to gratefully thank Anne Auger, Christian Igel, Stefan Kern, and Fabrice Marchal for the many valuable comments on the manuscript.

References

1. S. Baluja and R. Caruana. Removing the genetics from standard genetic algorithm. In A. Prieditis and S. Russell, editors, *Proceedings of the International Conference on Machine Learning*, pp. 38–46. Morgan Kaufmann, 1995.
2. H.G. Beyer. *The Theory of Evolution Strategies*. Springer, 2001.
3. H.G. Beyer and D. Arnold. Qualms regarding the optimality of cumulative path length control in CSA/CMA-evolution strategies. *Evolutionary Computation*, 11(1):19–28, 2003.
4. P.A.N. Bosman and D. Thierens. Expanding from discrete to continuous estimation of distribution algorithms: The IDEA. In M. Schoenauer, K. Deb, G. Rudolph, X. Yao, E. Lutton, J. J. Merelo, and H.-P. Schwefel, editors, *Parallel Problem Solving from Nature – PPSN VI. Lecture Notes in Computer Science 1917*, pp. 767–776, 2000.
5. M. Gallagher and M. Frean. Population-based continuous optimization and probabilistic modeling. Technical Report MG-1-2001, echnical report, School of Information Technology and Electrical Engineering, University of Queensland, 2001.
6. N. Hansen. *Verallgemeinerte individuelle Schrittweitenregelung in der Evolutionsstrategie*. Mensch und Buch Verlag, 1998.
7. N. Hansen. Invariance, self-adaptation and correlated mutations in evolution strategies. In M. Schoenauer, K. Deb, G. Rudolph, X. Yao, E. Lutton, J.J. Merelo, and H.-P. Schwefel, editors, *Parallel Problem Solving from Nature - PPSN VI*, pp. 355–364. Springer, 2000.
8. N. Hansen and S. Kern. Evaluating the CMA evolution strategy on multimodal test functions. In Xin Yao et al., editor, *Parallel Problem Solving from Nature - PPSN VIII*, pp. 282–291. Springer, 2004.

---

[12] An n-dimensional *separable* problem can be solved by solving n 1-dimensional problems separately.

9. N. Hansen, S.D. Müller, and P. Koumoutsakos. Reducing the time complexity of the derandomized evolution strategy with covariance matrix adaptation (CMA-ES). *Evolutionary Computation*, 11(1):1–18, 2003.

10. N. Hansen and A. Ostermeier. Adapting arbitrary normal mutation distributions in evolution strategies: The covariance matrix adaptation. In *Proceedings of the 1996 IEEE Conference on Evolutionary Computation (ICEC '96)*, pp. 312–317, 1996.

11. N. Hansen and A. Ostermeier. Convergence properties of evolution strategies with the derandomized covariance matrix adaptation: The $(\mu/\mu_I, \lambda)$-CMA-ES. In *Proceedings of the 5th European Congresson Intelligent Techniques and Soft Computing*, pp. 650–654, 1997.

12. N. Hansen and A. Ostermeier. Completely derandomized self-adaptation in evolution strategies. *Evolutionary Computation*, 9(2):159–195, 2001.

13. K. Deb and H.G. Beyer. On self-adaptive features in real-parameter evolutionary algorithms. *IEEE Transactions on Evolutionary Computation*, 5(3):250–270, 2001.

14. S. Kern, S.D. Müller, N. Hansen, D. Büche, J. Ocenasek, and P. Koumoutsakos. Learning probability distributions in continuous evolutionary algorithms – a comparative review. *Natural Computing*, 3:77–112, 2004.

15. P. Larrañaga. A review on estimation of distribution algorithms. In P. Larrañaga and J. A. Lozano, editors, *Estimation of Distribution Algorithms. A New Tool for Evolutionary Computation*, pp. 80–90. Kluwer Academic Publishers, 2002.

16. P. Larrañaga, J. A. Lozano, and E. Bengoetxea. Estimation of distribution algorithms based on multivariate normal and Gaussian networks. Technical Report KZAA-IK-1-01, Dept. of Computer Science and Artificial Intelligence, University of the Basque Country, 2001.

17. I. Rechenberg. *Evolutionsstrategie '94*. Frommann-Holzboog, Stuttgart, Germany, 1994.

18. S. Rudlof and M. Köppen. Stochastic hill climbing by vectors of normal distributions. In *Proceedings of the First Online Workshop on Soft Computing (WSC1)*, 1996. Nagoya, Japan.

19. H.-P. Schwefel. *Evolution and Optimum Seeking*. John Wiley & Sons, Inc., 1995.

20. M. Sebag and A. Ducoulombier. Extending population-based incremental learning to continuos search spaces. In *Parallel Problem Solving from Nature - PPSN V*, pp. 418–427. Springer-Verlag, 1998. Berlin.

21. B. Yuan and M. Gallagher. Playing in continuous spaces: Some analysis and extension of population-based incremental learning. In Sarkar et al., editor, *Proc. Congress on Evolutionary Computation (CEC)*, pp. 443–450, 2003.

A Algorithm Summary: The (μ_W, λ)-CMA-ES

Figure 7 outlines the complete algorithm, summarizing (2), (3), (17), (22), (23), and (28). Symbols used are:

$\mathbf{x}_k^{(g+1)} \in \mathbb{R}^n$, for $k = 1, \ldots, \lambda$. Sample of λ search points of generation $g + 1$.

$\mathcal{N}(\mathbf{m}, \mathbf{C})$, multi-variate normal distribution with mean $\mathbf{m}$ and covariance matrix $\mathbf{C}$.

$\mathbf{x}_{i:\lambda}^{(g+1)}$, i-th best point out of $\mathbf{x}_1^{(g+1)}, \ldots, \mathbf{x}_\lambda^{(g+1)}$ from (29). The index $i : \lambda$ denotes the index of the i-th ranked point and $f(\mathbf{x}_{1:\lambda}^{(g+1)}) \leq f(\mathbf{x}_{2:\lambda}^{(g+1)}) \leq \cdots \leq f(\mathbf{x}_{\lambda:\lambda}^{(g+1)})$.

Set parameters

Set parameters λ, μ, $w_{i=1\ldots\mu}$, c_σ, d_σ, $c_{\rm c}$, $\mu_{\rm cov}$, and $c_{\rm cov}$ to their default values according to Table 2.

Initialization

Set evolution paths $\mathbf{p}_\sigma^{(0)} = \mathbf{0}$, $\mathbf{p}_{\rm c}^{(0)} = \mathbf{0}$, and covariance matrix $\mathbf{C}^{(0)} = \mathbf{I}$.
Choose step size $\sigma^{(0)} \in \mathbb{R}_+$ and distribution mean $\mathbf{m}^{(0)} \in \mathbb{R}^n$ problem dependent.[1]

For generation $g = 0, 1, 2, \ldots$, until stopping criterion met:

Sample new population of search points

$$\mathbf{x}_k^{(g+1)} \sim \mathcal{N}\left(\mathbf{m}^{(g)}, \left(\sigma^{(g)}\right)^2 \mathbf{C}^{(g)}\right) \qquad \text{for } k = 1, \ldots, \lambda \tag{29}$$

Selection and recombination

$$\mathbf{m}^{(g+1)} = \sum_{i=1}^{\mu} w_i\, \mathbf{x}_{i:\lambda}^{(g+1)}, \quad \sum_{i=1}^{\mu} w_i = 1,\; w_i > 0 \tag{30}$$

Step size control

$$\mathbf{p}_\sigma^{(g+1)} = (1 - c_\sigma)\mathbf{p}_\sigma^{(g)} + \sqrt{c_\sigma(2 - c_\sigma)\mu_{\rm eff}}\; \mathbf{C}^{(g)-\frac{1}{2}}\, \frac{\mathbf{m}^{(g+1)} - \mathbf{m}^{(g)}}{\sigma^{(g)}} \tag{31}$$

$$\sigma^{(g+1)} = \sigma^{(g)} \exp\left(\frac{c_\sigma}{d_\sigma}\left(\frac{\|\mathbf{p}_\sigma^{(g+1)}\|}{\mathsf{E}\|\mathcal{N}(\mathbf{0}, \mathbf{I})\|} - 1\right)\right) \tag{32}$$

Covariance matrix adaptation

$$\mathbf{p}_{\rm c}^{(g+1)} = (1 - c_{\rm c})\mathbf{p}_{\rm c}^{(g)} + h_\sigma^{(g+1)}\sqrt{c_{\rm c}(2 - c_{\rm c})\mu_{\rm eff}}\; \frac{\mathbf{m}^{(g+1)} - \mathbf{m}^{(g)}}{\sigma^{(g)}} \tag{33}$$

$$\mathbf{C}^{(g+1)} = (1 - c_{\rm cov})\mathbf{C}^{(g)} + \frac{c_{\rm cov}}{\mu_{\rm cov}}\left(\mathbf{p}_{\rm c}^{(g+1)}\mathbf{p}_{\rm c}^{(g+1)\,{\rm T}} + \delta(h_\sigma^{(g+1)})\mathbf{C}^{(g)}\right)$$

$$\qquad\qquad + c_{\rm cov}\left(1 - \frac{1}{\mu_{\rm cov}}\right)\sum_{i=1}^{\mu} w_i\, {\rm OP}\left(\frac{\mathbf{x}_{i:\lambda}^{(g+1)} - \mathbf{m}^{(g)}}{\sigma^{(g)}}\right) \tag{34}$$

---

[1] The optimum should presumably be within the cube $\mathbf{m}^{(0)} \pm 2\sigma^{(0)}(1, \ldots, 1)^{\rm T}$. If the optimum is expected to be in $[0, 1]^n$ (initial interval) we may choose the initial search point, $\mathbf{m}^{(0)}$, uniformly randomly in $[0, 1]^n$, and $\sigma^{(0)} = 0.5$. Different search intervals Δs_i for different variables can be reflected by a different initialization of $\mathbf{C}$, in that the diagonal elements of $\mathbf{C}^{(0)}$ obey $c_{ii}^{(0)} = (\Delta s_i)^2$.

Fig. 7. The $(\mu_{\rm W}, \lambda)$-CMA evolution strategy. Symbols: see text

$\mu_{\text{eff}} = \left(\sum_{i=1}^{\mu} w_i^2\right)^{-1}$ is the variance effective selection mass. It holds $1 \leq \mu_{\text{eff}} \leq \mu$.

$\mathbf{C}^{(g)-\frac{1}{2}} \stackrel{\text{def}}{=} \mathbf{B}^{(g)} \mathbf{D}^{(g)-1} \mathbf{B}^{(g)\mathrm{T}}$, where $\mathbf{C}^{(g)} = \mathbf{B}^{(g)} \mathbf{D}^{(g)} \mathbf{D}^{(g)} \mathbf{B}^{(g)\mathrm{T}}$ is an eigende-composition of the symmetric, positive definite covariance matrix $\mathbf{C}^{(g)}$. Columns of $\mathbf{B}^{(g)}$ are an orthonormal basis of eigenvectors, $\mathbf{B}^{(g)\mathrm{T}} \mathbf{B}^{(g)} = \mathbf{B}^{(g)} \mathbf{B}^{(g)\mathrm{T}} = \mathbf{I}$. Diagonal elements of the diagonal matrix $\mathbf{D}^{(g)}$ are square roots of the corresponding positive eigenvalues. The matrix $\mathbf{D}^{(g)}$ can be inverted by inverting its diagonal elements.

$\mathsf{E}\|\mathcal{N}(\mathbf{0}, \mathbf{I})\| = \sqrt{2}\,\Gamma(\frac{n+1}{2})/\Gamma(\frac{n}{2}) \approx \sqrt{n}\left(1 - \frac{1}{4n} + \frac{1}{21n^2}\right)$.

$$h_\sigma^{(g+1)} = \begin{cases} 1 \text{ if } \dfrac{\|\mathbf{p}_\sigma^{(g+1)}\|}{\sqrt{1-(1-c_\sigma)^{2(g+1)}}} < (1.5 + \frac{1}{n-0.5})\mathsf{E}\|\mathcal{N}(\mathbf{0}, \mathbf{I})\| \\ 0 \text{ otherwise} \end{cases}$$

the Heaviside function $h_\sigma^{(g+1)}$ stalls the update of $\mathbf{p}_c^{(g)}$ in (17) if $\|\mathbf{p}_\sigma^{(g+1)}\|$ is large. This prevents a too fast increase of axes of $\mathbf{C}$ in a linear surrounding, i.e. when the step size is far too small. This is useful when the initial step size chosen is far too small or when the objective function changes in time.

$\delta(h_\sigma^{(g+1)}) = (1 - h_\sigma^{(g+1)})c_c(2 - c_c) \leq 1$ is of minor relevance and can be set to 0. In the (unusual) case of $h_\sigma^{(g+1)} = 0$, it substitutes for the second term from (33) in (34).

$\mathrm{OP} : \mathbb{R}^n \to \mathbb{R}^{n \times n}, \mathbf{x} \mapsto \mathbf{x}\mathbf{x}^\mathrm{T}$, denotes the outer product of a vector with itself.

Default Parameters

The (external) strategy parameters are λ, μ, $w_{i=1\ldots\mu}$, c_σ, d_σ, c_c, μ_{cov}, and c_{cov}. Default strategy parameters values are given in Table 2. An in-depth discussion of most parameters is given in [12]. The default parameters of (37)–(39) are in particular chosen to be a robust setting and therefore, to our experience, applicable to a wide range of functions to be optimized. We do not recommend changing this setting. In contrast, the population size λ in (35) can be increased by the user.[13] If the λ-dependent default values for μ and w_i are used, the population size λ has a significant influence on the global search performance [8]. Increasing λ usually improves the global search capabilities and the robustness of the CMA-ES, at the price of a reduced convergence speed. The convergence speed decreases at most linearly with λ.

Implementation

We discuss a few implementational issues.

---

[13] Decreasing λ is not recommended. Too small values regularly have strong adverse effects on the performance.

Table 2. Default Strategy Parameters, where $\mu_{\text{eff}} = \frac{1}{\sum_{i=1}^{\mu} w_i^2} \geq 1$ and $\sum_{i=1}^{\mu} w_i = 1$

Selection and Recombination:

$$\lambda = 4 + \lfloor 3 \ln n \rfloor, \quad \mu = \lfloor \lambda/2 \rfloor, \tag{35}$$

$$w_i = \frac{\ln(\mu+1) - \ln i}{\sum_{j=1}^{\mu}(\ln(\mu+1) - \ln j)} \quad \text{for } i = 1, \ldots, \mu, \tag{36}$$

Step size control:

$$c_\sigma = \frac{\mu_{\text{eff}} + 2}{n + \mu_{\text{eff}} + 3}, \quad d_\sigma = 1 + 2 \max\left(0, \sqrt{\frac{\mu_{\text{eff}} - 1}{n+1}} - 1\right) + c_\sigma \tag{37}$$

Covariance matrix adaptation:

$$c_{\text{c}} = \frac{4}{n+4}, \quad \mu_{\text{cov}} = \mu_{\text{eff}} \tag{38}$$

$$c_{\text{cov}} = \frac{1}{\mu_{\text{cov}}} \frac{2}{(n+\sqrt{2})^2} + \left(1 - \frac{1}{\mu_{\text{cov}}}\right) \min\left(1, \frac{2\mu_{\text{eff}} - 1}{(n+2)^2 + \mu_{\text{eff}}}\right) \tag{39}$$

Multi-variate normal distribution: The distribution $\mathcal{N}(\mathbf{m}^{(g)}, \sigma^{(g)^2}\mathbf{C}^{(g)})$ in (29) is distributed as $\mathbf{m}^{(g)} + \sigma^{(g)}\mathbf{B}^{(g)}\mathbf{D}^{(g)}\mathcal{N}(\mathbf{0}, \mathbf{I})$ (see $\mathbf{C}^{(g)^{-\frac{1}{2}}}$ above for the definitions). This can be used to generate the random vector on the computer, because $\mathcal{N}(\mathbf{0}, \mathbf{I})$ is a vector with independent, $(0,1)$-normally distributed components that can be easily sampled on a computer.

Strategy internal numerical effort: In practice, the re-calculation of $\mathbf{B}^{(g)}$, $\mathbf{D}^{(g)}$, and $\mathbf{C}^{(g)^{-\frac{1}{2}}}$ does not need to be done until $\max(1, \lfloor 1/(10nc_{\text{cov}}) \rfloor)$ generations. For reasonable c_{cov} values, this reduces the numerical effort due to the eigendecomposition from $\mathcal{O}(n^3)$ to $\mathcal{O}(n^2)$ per generated search point. On a Pentium 4, 2.5 GHz processor the overall strategy internal time consumption is roughly $4(n+2)^2 \times 10^{-8}$ seconds per function evaluation [14].

Flat fitness: In the case of equal function values for several individuals in the population, it is feasible to increase the step size (see lines 92–96 in the source code below).

Constraints: A simple, and occasionally sufficient, way to handle any type of boundaries and constraints is re-sampling unfeasible $\mathbf{x}_k^{(g+1)}$ until they become feasible.

B MATLAB Code

```
001 function xmin=purecmaes
002 % CMA-ES: Evolution Strategy with Covariance Matrix Adaptation for
003 % nonlinear function minimization.
004 %
005 % This code is an excerpt from cmaes.m and implements the key parts
```

```
006   % of the algorithm. It is intendend to be used for READING and
007   % UNDERSTANDING the basic flow and all details of the CMA *algorithm*.
008   % Computational efficiency is sometimes disregarded.
009
010   % -------------------- Initialization --------------------------------
011
012   % User defined input parameters (need to be edited)
013   strfitnessfct = 'felli'; % name of objective/fitness function
014   N = 10;                  % number of objective variables/problem dimension
015   xmean = rand(N,1);       % objective variables initial point
016   sigma = 0.5;             % coordinate wise standard deviation (step size)
017   stopfitness = 1e-10;  % stop if fitness < stopfitness (minimization)
018   stopeval = 1e3*N^2;   % stop after stopeval number of function evaluations
019
020   % Strategy parameter setting: Selection
021   lambda = 4+floor(3*log(N)); % population size, offspring number
022   mu = floor(lambda/2);       % number of parents/points for recombination
023   weights = log(mu+1)-log(1:mu)'; % muXone array for weighted recombination
024   % lambda=12; mu=3; weights = ones(mu,1); % uncomment for (3_I,12)-ES
025   weights = weights/sum(weights);      % normalize recombination weights array
026   mueff=sum(weights)^2/sum(weights.^2); % variance-effective size of mu
027
028   % Strategy parameter setting: Adaptation
029   cc = 4/(N+4);      % time constant for cumulation for covariance matrix
030   cs = (mueff+2)/(N+mueff+3); % t-const for cumulation for sigma control
031   mucov = mueff;   % size of mu used for calculating learning rate ccov
032   ccov = (1/mucov) * 2/(N+1.4)^2 + (1-1/mucov) * ...  % learning rate for
033          ((2*mueff-1)/((N+2)^2+2*mueff));             % covariance matrix
034   damps = 1 + 2*max(0, sqrt((mueff-1)/(N+1))-1) + cs; % damping for sigma
035
036   % Initialize dynamic (internal) strategy parameters and constants
037   pc = zeros(N,1); ps = zeros(N,1);   % evolution paths for C and sigma
038   B = eye(N);                         % B defines the coordinate system
039   D = eye(N);                         % diagonal matrix D defines the scaling
040   C = B*D*(B*D)';                     % covariance matrix
041   eigeneval = 0;                      % B and D updated at counteval == 0
042   chiN=N^0.5*(1-1/(4*N)+1/(21*N^2));  % expectation of
043                                       %   ||N(0,I)|| == norm(randn(N,1))
044
045   % -------------------- Generation Loop --------------------------------
046
047   counteval = 0;  % the next 40 lines contain the 20 lines of interesting code
048   while counteval < stopeval
049
050     % Generate and evaluate lambda offspring
051     for k=1:lambda,
052       arz(:,k) = randn(N,1);  % standard normally distributed vector
053       arx(:,k) = xmean + sigma * (B*D * arz(:,k));   % add mutation    % Eq. 29
054       arfitness(k) = feval(strfitnessfct, arx(:,k)); % objective function call
055       counteval = counteval+1;
056     end
057
058     % Sort by fitness and compute weighted mean into xmean
059     [arfitness, arindex] = sort(arfitness); % minimization
060     xmean = arx(:,arindex(1:mu))*weights;    % recombination              % Eq. 30
061     zmean = arz(:,arindex(1:mu))*weights;    % == sigma^-1*D^-1*B'*(xmean-xold)
062
063     % Cumulation: Update evolution paths
064     ps = (1-cs)*ps + (sqrt(cs*(2-cs)*mueff)) * (B * zmean);             % Eq. 31
```

```
065    hsig = norm(ps)/sqrt(1-(1-cs)^(2*counteval/lambda))/chiN < 1.5+1/(N-0.5);
066    pc = (1-cc)*pc + hsig * sqrt(cc*(2-cc)*mueff) * (B*D*zmean);        % Eq. 33
067
068    % Adapt covariance matrix C
069    C = (1-ccov) * C ...                         % regard old matrix        % Eq. 34
070        + ccov * (1/mucov) * (pc*pc' ...  % plus rank one update
071                            + (1-hsig) * cc*(2-cc) * C) ...
072        + ccov * (1-1/mucov) ...             % plus rank mu update
073        * (B*D*arz(:,arindex(1:mu))) ...
074        *  diag(weights) * (B*D*arz(:,arindex(1:mu)))';
075
076    % Adapt step size sigma

077    sigma = sigma * exp((cs/damps)*(norm(ps)/chiN - 1));               % Eq. 32
078
079    % Update B and D from C
080    if counteval - eigeneval > lambda/ccov/N/10  % to achieve O(N^2)
081      eigeneval = counteval;
082      C=triu(C)+triu(C,1)'; % enforce symmetry
083      [B,D] = eig(C);        % eigen decomposition, B==normalized eigenvectors
084      D = diag(sqrt(diag(D))); % D contains standard deviations now
085    end
086
087    % Break, if fitness is good enough
088    if arfitness(1) <= stopfitness
089      break;
090    end
091
092    % Escape flat fitness
093    if arfitness(1) == arfitness(min(1+floor(lambda/2), 2+ceil(lambda/4)))
094      sigma = sigma * exp(0.2+cs/damps);
095      disp('escape flat fitness');
096    end
097
098    disp([num2str(counteval) ': ' num2str(arfitness(1))]);
099
100  end % while, end generation loop
101
102  % -------------------- Ending Message ---------------------------------
103
104  disp([num2str(counteval) ': ' num2str(arfitness(1))]);
105  xmin = arx(:, arindex(1)); % Return best point of last generation.
106                                 % Notice that xmean is expected to be even
107                                 % better.
108
109 % ---------------------------------------------------------------
110 function f=felli(x)
111   N = size(x,1); if N < 2 error('dimension must be greater one'); end
112   f=1e6.^((0:N-1)/(N-1)) * x.^2;  % condition number 1e6
```

Estimation of Distribution Programming:
EDA-based Approach to Program Generation

Kohsuke Yanai and Hitoshi Iba

Dept. of Frontier Informatics, Graduate School of Frontier Sciences,
The University of Tokyo,
5-1-5 Kashiwanoha, Kashiwa-shi, Chiba 277-8561, Japan
{yanai,iba}@iba.k.u-tokyo.ac.jp

Summary. We describe a framework for program evolution with an EDA-based approach. In this framework, the probability distribution of programs is estimated using a Bayesian network, and individuals are generated based on the estimated distribution. Considering that a dependency relationship of nodes in a program tree is explicit, i.e. the dependency relationship is strong between a parent node and its child node in a program expressed as a tree structure, we have chosen a Bayesian network as the distribution model of programs.

In order to demonstrate the effectiveness of our approach, this chapter shows results of comparative experiments with Genetic Programming. Thereafter, we discuss how Estimation of Distribution Programming works and the transitions of the evolved programs that are the forte of our methods. We also analyze the performance of a hybrid system which combines Estimation of Distribution Programming and Genetic Programming.

1 Introduction

In this chapter, we describe a program evolution method based on a probabilistic model and investigate the behavior of the proposed system.

A well-known technique for a program search is Genetic Programming (GP) [9]. Although various types of crossover and mutation operators were proposed for GP[1] there have been very few basic algorithms comparable to GP. We use a program evolution method which has different mechanisms from GP, and show that some of the GP difficulties can be solved effectively[2].

This chapter proposes Estimation of Distribution Programming (EDP) based on a probability distribution expression using a Bayesian network. EDP is a search

---

[1] For example, uniform crossover and one-point crossover [16], homologous crossover and size fair crossover [10], depth-dependent crossover [8] [7], macromutation [2], self-adaptive crossover [1], recombinative guidance crossover [6], and so on.

[2] It is well known that GP search space is significantly constrained [5], and that the bloat control is difficult [11]. Other GP difficulties have been reported in solving a royal tree problem [18] and a max problem [17].

K. Yanai and H. Iba: *Estimation of Distribution Programming: EDA-based Approach to Program Generation*, StudFuzz
192, 103–122 (2006)
www.springerlink.com

method that uses an EDA-like approach to solve GP-applicable problems. In EDA, it is important to assume a gene locus dependency relationship. In a program tree this relationship is strong between the parent node and its child node, so that it is expected that the EDA approach will be effective for solving tree structure search problems [21]. We compare the performance of EDP and GP on several benchmark tests, and discuss the trends of problems that are the forte of EDP.

We also discuss the performance of a hybrid system which consists of EDP and GP. Applying the hybrid system of EDP and GP to a function regression problem, we discover some important tendencies in the behavior of this hybrid system. The hybrid system is not only superior to pure GP in a search performance but also have interesting features in program evolution. More tests reveal how and when EDP and GP compensate for each other.

2 Estimation of Distribution Programming

2.1 Algorithm of EDP

We give an outline of the proposed algorithm. EDP starts with a randomly generated population. Secondly, each individual in the current population is evaluated by a fitness function and assigned its fitness value. Next, superior individuals with high fitness values are selected, and a new distribution is estimated based on those selected individuals (see Sect. 2.3). We use the elitist strategy and then individuals are generated by using a newly acquired distribution (see Sect. 2.4). The estimation of distribution and the program generation are repeated until a termination criterion is met. Figure 1 indicates a pseudo code of EDP.

Initial Population

According to function node generation probability P_F and terminal node generation probability P_T $(1 - P_F)$, initial M individuals are generated randomly, where M is

---

Let P be a population, S a set of selected individuals, D a distribution, E_S an elite size, and M a population size.

1. $P := $ Generate_Programs_Randomly
2. While (True)
3.   Evaluate_Individuals(P)
4.   If (termination_criterion) Return(P)
5. $S := $ Selection(P)
6. $D := $ Estimate_Distribution(S)
7. $P := $ Elite_Selection(P, E_S)
8. $P := $ P + Generate_Individuals(D, $M - E_S$)

Fig. 1. Pseudo code of EDP

the population size. However, if tree size limitation is reached, terminal nodes are generated. Let F be the function node set and let T be the terminal node set. For example, the probabilities of function node "+" and terminal node "x" are given:

If tree size limitation is not reached,

$$\begin{cases} P(X = \text{"}+\text{"}) & = P_F \times \frac{1}{|F|} \\ P(X = \text{"}x\text{"}) & = P_T \times \frac{1}{|T|} \end{cases} \tag{1}$$

If tree size limitation is reached,

$$\begin{cases} P(X = \text{"}+\text{"}) & = 0 \\ P(X = \text{"}x\text{"}) & = \frac{1}{|T|} \end{cases} \tag{2}$$

EDP Operator

Superior individuals with high fitness values are selected within sampling size S_S, and a new distribution is estimated based on those selected individuals. We use the elitist strategy, i.e. elite E_S individuals are selected from the population in the order of fitness superiority and copied to the new population, where E_S is the elite size.

Then the remaining population, that is $M - E_S$ individuals, is generated by using a newly acquired distribution. This new distribution is considered better than the previous one because it samples superior individuals in the population.

2.2 Distribution Model

We use a Bayesian network as the distribution model of programs. Values of probabilistic variables are symbols for each node in the program tree. Assign the index numbers to each node of evolving programs as in Fig. 2, the range of probabilistic variable X_i is the symbols of node i, that is, $X_i \in T \cup F$.

For instance, assume $F = \{+, -, *, /\}$ and $T = \{x_1, x_2\}$,

$$P(X_5 = \text{"}+\text{"}|X_2 = \text{"}/\text{"}) = \frac{2}{7} \tag{3}$$

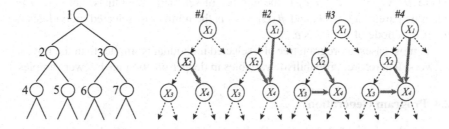

Fig. 2. Program tree   **Fig. 3.** Efficient network topology

means that the conditional probability that node 5 becomes " $+$ " is $\frac{2}{7}$ if node 2 is "$/$". C_i is the set of probabilistic variables which X_i is dependent on. In the former example, $C_5 = \{X_2\}$.

The topology of a Bayesian network is fixed during evolution, and only conditional probability tables are learned by sampling superior individuals in a population. Let d_{max} be the depth of a Bayesian network. We assume that the max arity of node symbols is 2 in this chapter. Although EDP cannot generate a larger program than a complete binary tree with a depth d_{max}, it can generate a smaller one.

There are several efficient topologies of a Bayesian network as indicated in Fig. 3. The simplest one, that is, #1 in Fig. 3, is used for our experiments. The topology of a Bayesian network is tree-like and it is the same as program's topology. In this model, the probability of each node in a program tree is dependent on only its parent node symbol. This is based on the assumption that a dependency relationship is strong between the parent node and its child nodes.

2.3 Estimation of Distribution

The probability distribution is updated incrementally [3] as follows:

$$P_{t+1}(X_i = x | C_i = c) = (1 - \eta)\hat{P}(X_i = x | C_i = c) + \eta P_t(X_i = x | C_i = c) \quad (4)$$

where $P_t(X_i = x | C_i = c)$ is the distribution of the tth generation and $\hat{P}(X_i = x | C_i = c)$ is the distribution estimated based on superior individuals in the $(t + 1)$th population, η is the learning rate which means dependence degree on the previous generation. The closer η is to 1, the less a change of distribution is. Especially in case of $\eta = 0$, the distribution is updated based on the population at only the $(t + 1)$th generation without referring to the past distribution.

$\hat{P}(X_i = x | C_i = c)$ is estimated as follows. At first, S_S individuals are sampled by tournament selection with tournament size T_{edp}, and maximum likelihood estimation is performed based on these selected individuals. Therefore,

$$\hat{P}(X_i = x | C_i = c) = \frac{\#(X_i = x, C_i = c)}{\#(C_i = c)} \quad (5)$$

where $\#(X_i = x, C_i = c)$ is the number of selected individuals that node i is x when its parent node is c, and $\#(C_i = c)$ is the number of selected individuals that the parent node of node i is c.

In most cases, a program tree of a selected individual is smaller than the Bayesian network. Therefore, probabilistic variables in deeper position have fewer samples.

2.4 Program Generation

At first, the acquired distribution $P_t(X_i = x | C_i = c)$ is modified applying Laplace correction [4] by

$$P'_t(X_i = x | C_i = c) = (1 - \alpha)P_t(X_i = x | C_i = c) + \alpha P_{bias}(X_i = x | C_i = c) \quad (6)$$

where α is a constant that expresses the Laplace correction rate, $P_{bias}(X_i = x|C_i = c)$ is the probability to bias distribution. For instance, if it is already known that $X_2 = $ " $+$ " is desirable, adjusting $P_{bias}(X_i = x|C_i = c)$ as the probability of $X_2 = $ "$+$" is high would lead to more effective evolution. In this way, the system can incorporate preknowledge by Laplace correction. For our experiments, the Laplace correction rate α is decided as

$$\alpha = 0.01(|F| + |T|) \tag{7}$$

This modification also makes all occurrence probabilities of node symbols positive. Next, according to $P'_t(X_i = x|C_i = c)$, node symbols are decided in sequence from root to terminals. If the size of generated tree reaches d_{max}, only terminal node symbols are selected. Therefore, a larger program tree than the Bayesian network is not generated.

3 Performance of EDP

3.1 Comparative Experiments with GP

The performance was compared for EDP and GP in standard benchmark problems, i.e. a max problem [17], a boolean 6-multiplexer problem [9], and a function regression problem [9]. Let $prog_i$ be a program tree of the ith individual in a population. If the program tree has some variables, $prog_i(X)$ represents the value obtained by substituting X. If the program tree has no variable, $prog_i$ represents the value returned by the program tree. Let fit_i be the fitness value of the ith individual.

Max Problem

In a max problem, the purpose is to create the maximum value, based on the assumption that $T = \{0.5\}$ and $F = \{+, *\}$, and the maximum tree depth is 7. For a tree produces the largest value, the $+$ nodes must be used with 0.5 to assemble subtrees A with the value 2.0. These can then be connected via $*$, as shown in Fig. 4. Hence, 65536 is the optimum solution [3]. The fitness value for ith individual is the value of tree, that is,

$$fit_i = prog_i \tag{8}$$

The parameters of EDP and GP are indicated in Table 1.

Figure 5 and Table 2 show the results of a comparative test using EDP, GP and a random search. The vertical axis represents the max fitness value in a population at each generation: fit_{max}, i.e.

$$fit_{max} = \max_{i \in M} fit_i \tag{9}$$

---

[3] The maximum node size for the depth of 7 is 127 in a complete binary tree. Within the node size of 127, it is proved that the maximum value is not 65536, but 123596.1914.

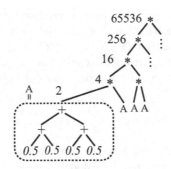

Fig. 4. The maximum value by a tree of limited depth

Table 1. Parameters for a max problem

Common parameters for EDP and GP	
M: population size	200
E_S: elite size	5
F: function node sets	$\{+, *\}$
T: terminal node sets	$\{0.5\}$
P_F: generation probability of function node	$\frac{2}{3}$
P_T: generation probability of terminal node	$\frac{1}{3}$
Tree size limitation in initializing population	max depth = 7

EDP parameters	
α: Laplace correction rate	0.03
P_{bias}: the probability to bias distribution	$\frac{1}{|F|+|T|} = \frac{1}{3}$
η: learning rate	0.2
S_S: sampling size	200
T_{edp}: tournament size for sampling	20
Tree size limitation	max depth = 7

GP parameters	
P_M: mutation probability	0.1
P_C: crossover probability	0.9
T_{gp}: tournament size for GP operator	5
Tree size limitation	max depth = 7

The mean and the standard deviation for 100 runs are indicated in Fig. 5. Note that they are not a mean fitness value and a standard deviation of a population. The solution in an evolutionary computing is given by an individual who has the maximum fitness value in a population. Therefore, system performances should be compared in maximum fitness values. It can be seen that EDP method produces a higher mean fitness value at each generation and also higher performance on the average. In ad-

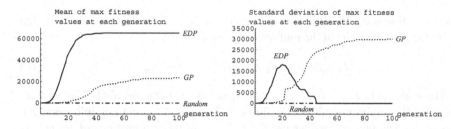

Fig. 5. Comparative results with a max problem

dition, the standard deviation of EDP, i.e. the deviation due to the search runs, is so small that the likelihood of the search being successful is higher.

As presented in Table 2, EDP was able to find the optimal solution in all runs, whereas only 34 runs (out of 100 runs) resulted in evolving the optimal solution with GP. These results suggest intrinsic difference between EDP and GP.

Next, the experiment was carried out with the addition of "0" to the terminal node set. In this problem "0" is completely useless and harmful as a node, and produces non-functional code segments, i.e. introns. As shown in Fig. 6, although the performance of GP was low, with EDP algorithm the most suitable solution was found successfully.

Boolean 6-Multiplexer Problem

Consider the problem of learning the Boolean 6-multiplexer function F_{6mp} : $\{0,1\}^6 \rightarrow \{0,1\}$. The input to the Boolean 6-multiplexer function consists of 2 address bits and 2^2 data bits, where $6 = 2 + 2^2$. The value of the Boolean multi-

Table 2. Percentage of runs finding the optimal solution

Method	Max problem	Multiplexer problem	Max problem adding "0" terminal node
EDP	100	23	86
GP	34	82	0
Random	0	0	0

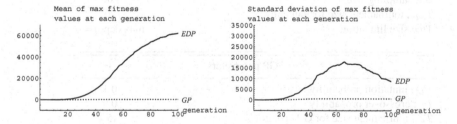

Fig. 6. Comparative results when "0" terminal node was added with a max problem

plexer function is the Boolean value (0 or 1) of the particular data bit that is singled out by the 2 address bits of the multiplexer. Formally,

$$F_{6mp}(a_0, a_1, d_0, d_1, d_2, d_3) = d_{2^1 a_1 + a_0} \qquad (10)$$

The node set is $T = \{x_0, x_1, x_2, x_3, x_4, x_5\}$, $F = \{and, or, not\}$. The parameters of EDP and GP are indicated in Table 3. There are $2^6 = 64$ possible combinations of the 6 arguments, and we use the entire set of 64 combinations of arguments as the fitness cases for evaluating fitness. That is, we do not use sampling. The fitness values are simply the number of fitness cases for which the individual tree returns a correct Boolean value. Let X_i be an input data set, i.e. $X_j = \{x_{j1}, \ldots, x_{j6}\}$, where x_{jk} is the kth digit of the number j. Then, the fitness value is given with the following formula:

$$fit_i = \sum_{j=0}^{63} \text{match}(prog_i(X_j), F_{6mp}(X_j)) \qquad (11)$$

where

Table 3. Parameter for a boolean 6-multiplexer problem

Common parameters for EDP and GP	
M: population size	500
E_S: elite size	5
F: function node sets	$\{and, or, not\}$
T: terminal node sets	$\{x_0, x_1, x_2, x_3, x_4, x_5\}$
P_F: generation probability of function node	$\frac{3}{9}$
P_T: generation probability of terminal node	$\frac{6}{9}$
Tree size limitation in initializing population	max depth = 6

EDP parameters	
α: Laplace correction rate	0.09
P_{bias}: the probability to bias distribution	$\frac{1}{\lvert F\rvert + \lvert T\rvert} = \frac{1}{9}$
η: learning rate	0.2
S_S: sampling size	200
T_{edp}: tournament size for sampling	20
Tree size limitation	max depth = 6

GP parameters	
P_M: mutation probability	0.1
P_C: crossover probability	0.9
T_{gp}: tournament size for GP operator	5
Tree size limitation	max depth = 6

$$\text{match}(a, b) = \begin{cases} 1 & \text{if } a = b \\ 0 & \text{else} \end{cases} \tag{12}$$

Figure 7 shows the results of a comparative test using EDP, GP and a random search. We cannot confirm the superiority of EDP with this experiment. In 6-multiplexer problem, EDP could not search more efficiently than GP. However, EDP was superior to a random search.

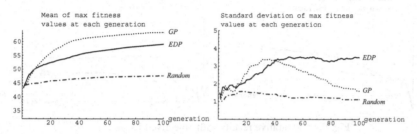

Fig. 7. Comparative results with a boolean 6-multiplexer problem

Function Regression Problem

Consider a function regression problem. f_{obj} is the function to be approximated. The fitness value is given with the following formula:

$$fitness = 1000 - 50 \sum_{j=1}^{30} |prog(X_j) - f_{obj}(X_j)| \tag{13}$$

where

$$X_j = 0.2(j - 1) \tag{14}$$

i.e. training examples are the real values at intervals of 0.2 from 0 to 5.8. Objective functions are

$$\text{A} : f_{obj}(x) = (2 - 0.3x)\sin(2x)\cos(3x) + 0.01x^2 \tag{15}$$

$$\text{B} : f_{obj}(x) = x\cos(x)\sin(x)(\sin^2(x)\cos(x) - 1) \tag{16}$$

$$\text{C} : f_{obj}(x) = x^3\cos(x)\sin(x)e^{-x}(\sin^2(x)\cos(x) - 1) \tag{17}$$

which are plotted in Fig. 8. Objective function C is cited from [19]. Although B is obtained from simplification of C, B is more difficult to search. A is our original function and the most difficult of the three objective functions.

As indicated in Figs. 9, 10 and 11, EDP's performance was worse than GP's in a function regression problem. This result seems to suggest that EDP is not always superior. However, as we can see later, the EDP operator plays an inevitable role in combination with GP. The effectiveness of the hybrid system of EDP and GP is described in Sect. 4.

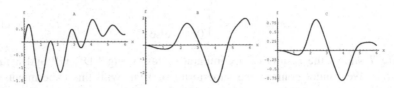

Fig. 8. Objective functions

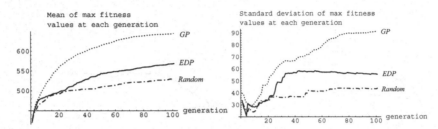

Fig. 9. Comparative results with objective function A

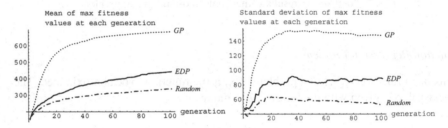

Fig. 10. Comparative results with objective function B

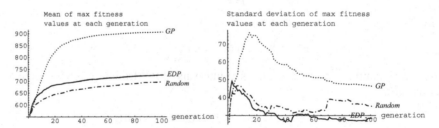

Fig. 11. Comparative results with objective function C

3.2 Summaries of EDP Performance

EDP was able to search for a solution effectively in a GP-hard problem, i.e. a max problem. On the other hand, in both a boolean 6-multiplexer problem and a function regression problem, it has been shown that EDP's performance was worse than GP's. In order to conclude that the differences of these values are statistically significant and reliable, not only mean but also standard deviation and sample size (100) should be taken into consideration. We used Welch's test for the obtained experimental re-

Table 4. Parameter for a function regression problem

Common parameters for EDP and GP	
M: population size	1000
E_S: elite size	5
F: function node sets	$\{+, -, *, /, \cos, \sin\}$
T: terminal node sets	$\{x, 0.05, 0.10, 0.15, \ldots, 1.00\}$
P_F: generation probability of function node	0.8
P_T: generation probability of terminal node	0.2
Tree size limitation in initializing population	max depth = 6

EDP parameters	
α: Laplace correction rate	0.27
P_{bias}: the probability to bias distribution	$\frac{1}{|F|+|T|} = \frac{1}{27}$
η: learning rate	0.2
S_S: sampling size	200
T_{edp}: tournament size for sampling	20
Tree size limitation	max depth = 6

GP parameters	
P_M: mutation probability	0.1
P_C: crossover probability	0.9
T_{gp}: tournament size for GP operator	5
Tree size limitation	max depth = 6

Table 5. P-values on Welch's test

Problem	EDP and GP	EDP and Random
Max problem	3.49×10^{-25}	1.10×10^{-340}
Multiplexer problem	4.53×10^{-21}	2.52×10^{-59}
Regression A	2.53×10^{-11}	1.80×10^{-7}
Regression B	8.52×10^{-30}	5.44×10^{-19}
Regression C	6.96×10^{-75}	7.03×10^{-10}

sults. By means of Welch's test, it can be judged whether 2 data sets are samples from the same statistical population or not. As a result of Welch's test with 5% significance level, the differences between EDP and GP at the 100th generation were significant in all cases. Statistically speaking, the null hypothesis that data in EDP and in GP were sampled from the same statistical population was rejected (the probability that the null hypothesis is correct is less than 5%). Welch's test concluded that the differences were significant. Table 5 indicates the p-values obtained in the test. This seems to indicate that EDP works intrinsically differently from the traditional GP.

In a max problem, in order to produce better solutions, it is necessary for EDP to increase the generation probability of "$*$" from the depth 1 to 4, and the probability of "$+$" at the depth 6. In the early stage of the evolution, the generation probability of "$+$" is expected to become high in a shallow part. Then, more frequently subtrees identical to $(+ (+ 0.5\ 0.5) (+ 0.5\ 0.5))$ (subtree A shown in Fig. 4) are produced in a deep part, the higher the generation probability of "$*$" becomes.

In a boolean 6-multiplexer problem, a positional restriction of EDP operator seems to have caused the worse performance. Using the 3-multiplexer function F_{3mp}, it is easy to compose the 6-multiplexer function in the following way:

$$F_{6mp}(a_0, a_1, d_0, d_1, d_2, d_3) =$$
$$(or\ (and\ F_{3mp}(a_1, d_0, d_1)\ (not\ a_0))$$
$$(and\ F_{3mp}(a_1, d_2, d_3)\ a_0)) \tag{18}$$

Furthermore, it has been reported that the 11-multiplexer function and the 6-multiplexer function were easily acquired by GP with the 6-multiplexer and the 3-multiplexer structures respectively [9]. An individual equivalent to the 3-multiplexer function would be assigned a high fitness value, i.e. $32 + 16 = 48$. Therefore, the composition of the 3-multiplexer functions is so important for the effective evolution of the 6-multiplexer function that they are expected to prosper in a population. Note that useful subtrees, i.e. so-called building blocks, cannot shift their position with EDP because the probability distribution is dependent on the position within a tree, while GP crossover can move them to an arbitrary position. In other words, EDP imposes a positional restriction. Consequently, EDP could not always use the generated structure of the 3-multiplexer function efficiently in order to compose the 6-multiplexer function. This is the reason why EDP operator failed to generate better individuals in some cases.

4 Hybrid System of EDP and GP

4.1 Algorithm of Hybrid System

We research the hybrid system which consists of EDP and GP. Figure 12 indicates a pseudo code of our hybrid system.

The most important parameter in this algorithm is "r", it decides the system behavior and the ratio of GP to EDP in an individual generation, called the hybrid ratio. Through the combination of EDP and GP, the difficulty indicated in Sect. 3.2 might be overcome. However, it is not obvious whether GP gains anything from hybridization. In this section, we test the system performance in a function regression problem changing the hybrid ratio r from 0 to 1.

4.2 Performance Difference Due to the Hybrid Ratio

Figures 13, 14, and 15 show the mean of max fitness values for 100 runs. Note that it is not a mean fitness value of a population, but a mean value of the maximum fitness value.

Let P be a population, S a set of selected individuals, D a distribution, r a hybrid ratio, E_S an elite size, and M a population size.

1. $P :=$ Generate_Programs_Randomly
2. While (True)
3.   Evaluate_Individuals(P)
4.   If (termination_criterion) Return(P)
5. $S :=$ Selection(P)
6. $D :=$ Estimate_Distribution(S)
7. $P :=$ Elite_Selection(P, E_S)
8. $P :=$ P + Crossover&Mutation($P, rM - E_S$)
9. $P :=$ P + Generate_Individuals($D, (1 - r)M$)

Fig. 12. Pseudo code of the hybrid system

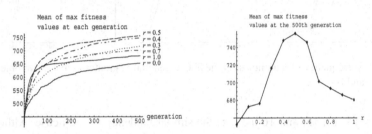

Fig. 13. Results for objective function A

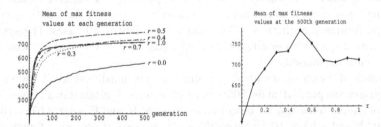

Fig. 14. Results for objective function B

Figure 16 shows the frequency of runs in which the maximum fitness value at the 500th generation is over x, that is,

$$F(x) = \sum_{k=1}^{100} \delta(x \le f_{max\,k,500}) \qquad (19)$$

where $f_{max\,k,500}$ is the maximum fitness value in a population of the 500th generation at the kth run, and

$$\delta(x \le a) = \begin{cases} 1 & : x \le a \\ 0 & : x > a \end{cases} \qquad (20)$$

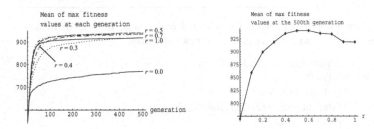

Fig. 15. Results for objective function C

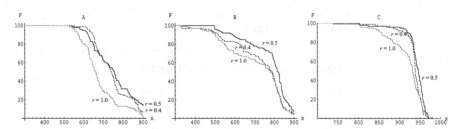

Fig. 16. $F(x)$: frequency of max fitness at the 500th generation greater than x, with objective functions A and B

Figures 13, 14, 15, and 16 indicate the similar tendency in each case. Although the $r = 1.0$ system which is pure GP, demonstrated the best performance in younger generations, gradually hybrid systems overtook pure GP one after another. The "overtaking" was conspicuous when $r = 0.3$ or $r = 0.4$. At the 500th generation, the performance of the $r = 0.5$ system was the best in all cases. The system performances at the 500th generation reached a peak at $r = 0.5$, and got worse as the hybrid ratio was biased.

As a result of Welch's test with 5% significance level, the differences between the $r = 0.5$ system and pure GP at the 500th generation were significant in all cases. The p-values obtained in the test for objective function A, B, and C were 2.57×10^{-7}, 1.23×10^{-4}, and 1.52×10^{-27} respectively. In the case of objective function C, although the difference in values was slight, standard deviation was negligible (see Fig. 16); Welch's test concluded that the differences were significant.

Mean cannot give adequate information for system performances, hence we showed Fig. 16. Figure 16 demonstrates that the hybrid system is also superior to pure GP in the success rate of a search. For instance, in the case of A, the probabilities that the maximum fitness value at the 500th generation is over 700 are $\frac{63}{100}$ with $r = 0.5$ and $\frac{30}{100}$ with pure GP respectively.

4.3 Analysis of the Behavior of EDP

This section investigates the hybrid system's performance, changing the hybrid ratio r at each generation. In Fig. 13, until the 50th generation, the higher the GP ratio of the system is, the better its performance. Therefore, the system that has a high GP

Table 6. Systems with changing r, where i is the generation number

System	r
A: classical hybrid	$r = 0.3$
B: classical hybrid	$r = 0.5$
C: pure GP	$r = 1.0$
D: linear increasing	$r = \dfrac{i}{500}$
E: linear decreasing	$r = 1 - \dfrac{i}{500}$
F: random	r is a random value at each generation
G: switching	$r = \begin{cases} 1.0 & : i < 205 \\ 0.3 & : i \geq 205 \end{cases}$
H: switching	$r = \begin{cases} 1.0 & : i < 40 \\ 0.5 & : i \geq 40 \end{cases}$

ratio in younger generations and decreases the ratio later is expected to have higher performance.

Comparative experiments were carried out with 8 variations of systems, as shown in Table 6. The objective function is the first one used in Sect. 3.1, i.e. (15). In the system D, the GP ratio is linearly increased from 0, at the initial generation, to 1.0, at the 500th generation, whereas it is linearly decreased in the system E. In the system G, the ratio is changed from 1.0 to 0.3 at the 205th generation. Note that the $r = 0.3$ system overtook the pure GP at the 205th generation (see Fig. 13). In the system H, the ratio is tuned in the same manner as G. Therefore, H and G are supposed to be the top favorites among these systems.

Figures 17 and 18 show the results of comparative experiments. Surprisingly, system A overtook G. As a result of Welch's test with 5% significance level, the differences were significant. The p-value obtained in the test was 0.026. This result means that population states of A and G are far different in spite of close performance at the 205th generation. In other words, EDP's behavior before the 205th generation likely has a good influence later. Although B also overtook H, the result was not significant statistically. The p-value obtained in the test for system B and H was 0.364.

Another interesting result is that system D was superior to all other systems, especially E. As a result of Welch's test with 5% significance level, the differences were significant. The p-value was 0.0473. Although it was expected that D would be worse than E, judging from Fig. 13, the result was quite the opposite. This point is evidence that EDP functions well in early generations.

In order to test the hypothesis that the probability distribution memorizes the past EDP's work, the system of $\eta = 0$ was simulated. This system estimates distribution without referring to the past distribution (see Sect. 2.3). Objective function A was used.

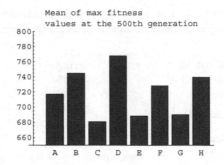

Fig. 17. Mean of max fitness values at the 500th generation

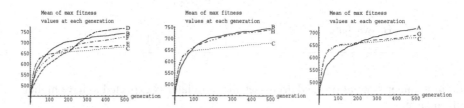

Fig. 18. Mean of max fitness values at each generation

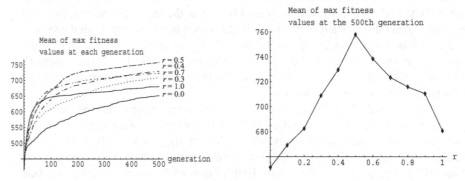

Fig. 19. System of $\eta = 0$

As indicated in Fig. 19, the characteristic of the hybrid system was kept. The "overtaking" still took place and the $r = 0.5$ system was the best. Therefore, the past information accumulated in the probability distribution does not cause the high performance of the hybrid system.

5 Discussion

The previous experimental results revealed the following aspects of EDP:

- EDP's search was intrinsically different from GP's.

- EDP's search was successful in a max problem with the addition of "0" to the terminal node set.
- Subtrees were not easily shifted in EDP.
- The hybrid system outperformed the pure GP in later generations.
- The hybrid system with linearly increasing hybrid ratio gave the best performance.

EDP does not refer to the previous generation directly, but abandon all individuals in previous generation and generate new individuals based on the distribution at an every generation. Thus, a random search is regarded as EDP with an uniform distribution. In 6-multiplexer problem and a regression problem, although EDP could not search more efficiently than GP, EDP was superior to a random search. Therefore, the probability distribution could be estimated effectively. The estimation of a distribution was done to some extent for the program search.

In the $r = 0.5$ hybrid system, the updating times of the maximum fitness values at each generation of the EDP operator and the GP operator are counted respectively. Surprisingly, the EDP operator hardly contributes to construction of the best individual directly, and only the GP operator does. In addition, as shown in Fig. 17, system D, which has linearly increasing hybrid ratio, gave the best performance of all. System D cannot benefit from EDP in later generations. These results suggest individuals constructed by EDP have more multifarious sub-structures in an early stage, and these various structures are put together in later generations. It is GP that can build better individuals, but not EDP.

The hybrid algorithm was tested in a function regression problem where the behavior of the EDP algorithm was bad. We also research how the hybrid system degrades in a max problem where previously EDP behaved properly. Figure 20 shows the performance of the hybrid system in a max problem. Although the performance of the hybrid system was a little worse than pure EDP's, the search by the hybrid system was successful.

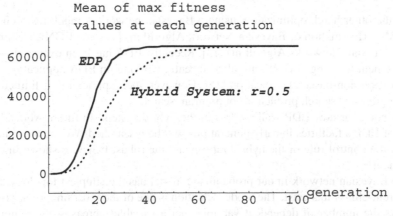

Fig. 20. Performance of the hybrid system in a max problem

6 Conclusion

This paper presented a new EDA-based approach, i.e. EDP, to program evolution and have shown the experimental results with EDP and GP.

When "0" was added to a set of terminal nodes, EDP performed much better than GP. We cannot always know what are effective nodes for problems before. This result suggests that EDP can perform evolution skillfully even if harmful nodes are included in a node set. Thus, it is expected that the occurrence probability of this harmful node is kept lower by the EDP method due to the obtained distribution. This indicates that EDP can control introns effectively, while GP may suffer from increasing introns and allow them to cause a bloat [10].

The experimental results clearly indicated that EDP worked effectively in early generations and contributed to later high performance. It turned out that pure GP could not generate enough kinds of subtrees in early generations to build better solutions. On the other hand, useful subtrees are not easily shifted by EDP to another position in the tree. We conclude that hybridization helps EDP and GP compensate for their defects and build a better evolutionary system.

Future and Related Works

Probabilistic Incremental Program Evolution (PIPE) [19] was used to perform a program search based on a probabilistic model. However, PIPE assumes the independence of program nodes and differs from our approach using a Bayesian network in this assumption. The merits of having probabilistic dependency relationship are as follows:

1. Because an occurrence probability of a node symbol is dependent on its parent node, estimation and generation are serial from a parent node to a child. Therefore, it can derive and generate building blocks.
2. The past dominant structure can survive after switching the probability distribution based on a parent node symbol.

On the other hand, optimization using a Bayesian network is much researched, e.g., EBNA (Estimation of Bayesian Network Algorithm) [12] and EGNA (Estimation of Gaussian Networks Algorithm) [13]. Recently, EDA has been extended with reinforcement learning [14]. We are also currently working on EDA application for a gene expression-based classification [15]. However, their application is limited to fixed length array search problems, not program search.

It is not clear how EDP really works in the hybrid system. In future works, the details of EDP's facilities in early generations will be researched. We are also interested in the control rule of the hybrid ratio r and the robust behavior shown in our experiments.

The Bayesian network in our probabilistic model has the simplest topology, i.e. only parent-child links exist. The model selection is one of the most important problems. As the number of dependent variables per a variable increases, the required memory size is exponentially increasing. The adequate sampling size for updating a distribution is also proportional to the exponential of the number of dependency links per a node. Therefore, the trade-off exists between the performance and calculation costs. Our future research will be on the study of the system performance with other topologies. We also plan to improve EDP in order to shift subtrees within a program tree, independently from the hybridization with GP.

This chapter discussed the program evolution on the premise that program representation consists of a single parse tree. However, the validity of the representation depends on the problem class. Without recursion and memory, the expressiveness of a parse tree is not Turing-complete. It is suggested that the different choice of representation will result in the different program evolution [20]. The extension of the program representation should be considered for the sake of establishing a probabilistic model-based evolution.

References

1. P. J. Angeline. Two selfadaptive crossover operations for genetic programming. In *Advances in Genetic Programming II*. MIT Press, 1995.
2. P. J. Angeline. Subtree crossover causes bloat. In *Genetic Programming 1998: Proceedings of the Third Annual Conference*, pp. 745–752. Morgan Kaufmann, 22–25 July 1998.
3. S. Baluja. Population-based incremental learning: A method for integrating genetic search based function optimization and competitive learning. Technical Report CMU-CS-94-163, Carnegie Mellon University, Pittsburgh, PA, 1994.
4. B. Cestnik. Estimating probabilities: A crucial task in machine learning. In *Proceedings of the 9th European Conference on Artificial Intelligence*, pp. 147–149, 1990.
5. J. M. Daida, H. Li, R. Tang, and A. M. Hilss. What makes a problem GP-hard? validating a hypothesis of structural causes. In *Proceedings of the Genetic and Evolutionary Computation Conference GECCO-2004*, pp. 1665–1677. Springer-Verlag, 2003.
6. H. Iba and H. de Garis. Extending genetic programming with recombinative guidance. In *Advances in Genetic Programming 2*, pp. 69–88. MIT Press, 1995.
7. C. Igel and K. Chellapilla. Investigating the influence of depth and degree of genotypic change on fitness in genetic programming. In *Proceedings of the Genetic and Evolutionary Computation Conference GECCO-1999*, volume 2, pp. 1061–1068. Morgan Kaufmann, 13-17 July 1999.
8. T. Ito, H. Iba, and S. Sato. Depth-dependent crossover for genetic programming. In *Proceedings of the 1998 IEEE World Congress on Computational Intelligence*, pp. 775–780. IEEE Press, 5-9 May 1998.
9. J. R. Koza. *Genetic Programming: On the Programming of Computers by Means of Natural Selection*. MIT Press, 1992.
10. W. B. Langdon. Size fair and homologous tree genetic programming crossovers. In *Proceedings of the Genetic and Evolutionary Computation Conference GECCO-1999*, volume 2, pp. 1092–1097. Morgan Kaufmann, 13-17 1999.
11. W. B. Langdon and R. Poli. Genetic programming bloat with dynamic fitness. In *Proceedings of the First European Workshop on Genetic Programming*. Springer-Verlag, 1998.
12. P. Larrañaga, R. Etxeberria, J. A. Lozano, and J. M. Peña. Combinatorial optimization by learning and simulation of Bayesian networks. In *Proceedings of the Conference in Uncertainty in Artificial Intelligence: UAI-2000*, pp. 343–352, 2000.

13. P. Larrañaga, R. Etxeberria, J. A. Lozano, and J. M. Peña. Optimization in continuous domains by learning and simulation of Gaussian networks. In *Proceedings of the Workshop in Optimization by Building and Using Probabilistic Models*, pp. 201–204, 2000.

14. T. K. Paul and H. Iba. Reinforcement learning estimation of distribution algorithm. In *Proceedings of Genetic and Evolutionary Computation Conference GECCO-2003*. Springer-Verlag, 2003.

15. T. K. Paul and H. Iba. Selection of the most useful subset of genes for gene expression-based classification. In *Proceedings of Congress on Evolutionary Computation: CEC-2004*, 2004.

16. R. Poli and W. B. Langdon. On the search properties of different crossover operators in genetic programming. In *Genetic Programming 1998: Proceedings of the Third Annual Conference*, pp. 293–301. Morgan Kaufmann, 22–25 1998.

17. R. Poli and W. B. Langdon. *Foundations of Genetic Programming*. Springer-Verlag, 2002.

18. W. F. Punch, D. Zongker, and E. D. Goodman. The royal tree problem, a benchmark for single and multiple population genetic programming. In *Advances in Genetic Programming II*, pp. 299–316. MIT Press, 1995.

19. R. P. Salustowicz and J. Schmidhuber. Probabilistic incremental program evolution: Stochastic search through program space. In M. van Someren and G. Widmer, editors, *Machine Learning: ECML-97*, volume 1224, pp. 213–220. Springer-Verlag, 1997.

20. T. Yabuki and H. Iba. Genetic programming using a Turing complete representation: recurrent network consisting of trees. In L. Nunes de Castro and F. J. Von Zuben, editors, *Recent Developments in Biologically Inspired Computing*. Idea Group Inc., 2004. (to be published).

21. K. Yanai and H. Iba. Estimation of distribution programming based on Bayesian networks. In *Proceedings of Congress on Evolutionary Computation: CEC-2003*, pp. 1618–1625, 2003.

Multi–objective Optimization
with the Naive MIDEA

Peter A.N. Bosman[1] and Dirk Thierens[2]

[1] National Research Institute for Mathematics and Computer Science P.O. Box 94079 1090
GB Amsterdam, The Netherlands
Peter.Bosman@cwi.nl

[2] Institute of Information and Computing Sciences, Utrecht University, P.O. Box 80.089,
3508 TB Utrecht, The Netherlands
Dirk.Thierens@cs.uu.nl

Summary. EDAs have been shown to perform well on a wide variety of single-objective
optimization problems, for binary and real-valued variables. In this chapter we look into the
extension of the EDA paradigm to multi-objective optimization. To this end, we focus the
chapter around the introduction of a simple, but effective, EDA for multi-objective optimization: the naive MIDEA (mixture-based multi-objective iterated density-estimation evolutionary algorithm). The probabilistic model in this specific algorithm is a mixture distribution.
Each component in the mixture is a univariate factorization. As will be shown in this chapter,
mixture distributions allow for wide-spread exploration of a multi-objective front, whereas
most operators focus on a specific part of the multi-objective front. This wide-spread exploration aids the important preservation of diversity in multi-objective optimization. To further
improve and maintain the diversity that is obtained by the mixture distribution, a specialized
diversity preserving selection operator is used in the naive MIDEA. We verify the effectiveness of the naive MIDEA in two different problem domains and compare it with two other
well-known efficient multi-objective evolutionary algorithms (MOEAs).

1 Introduction

In this chapter, we apply the EDA paradigm to multi-objective optimization. We
put the focus on a specific EDA, which we call the naive mixture-based multi-
objective iterated density-estimation evolutionary algorithm (naive MIDEA). The
naive MIDEA is an instance of the MIDEA framework for multi-objective optimization using EDAs. We will show how the naive MIDEA can be implemented for
both binary as well as real problem variables.

The remainder of this chapter is organized as follows. In Sect. 2, we first discuss multi-objective optimization. In Sect. 3 we develop the MIDEA framework
and specifically focus on the naive MIDEA instance. In Sect. 4 we validate the
performance of MIDEAs on eight test problems and compare the results with two

P.A.N. Bosman and D. Thierens: *Multi–objective Optimization with the Naive MIDEA*, StudFuzz **192**, 123–157 (2006)
www.springerlink.com   © Springer-Verlag Berlin Heidelberg 2006

other state-of-the-art MOEAs and discuss our findings. We present our conclusions in Sect. 5.

2 Multi-objective Optimization

Multi-objective optimization differs from single-objective optimization in that we have a multiple of objectives that we wish to optimize simultaneously without an expression of weight or preference for any of the objectives. Often, these multiple objectives are conflicting. Such problems naturally arise in many real world situations. An example of conflicting objectives that often arises in industry, is when we want to minimize the costs of some production process while at the same time we also want to minimize the pollution caused by the same production process. Such conflicting objectives give rise to a key characteristic of multi-objective optimization problems, which is the existence of sets of solutions that cannot be ordered in terms of preference when only considering the objective function values simultaneously. To formalize this notion, four relevant concepts exist. Assuming that we have m objectives $f_i(\boldsymbol{x})$, $i \in \mathcal{M} = \{0, 1, \ldots, m-1\}$, that, without loss of generality, we seek to minimize, these four concepts can be defined as follows:

1. **Pareto dominance**
   A solution $\boldsymbol{x}$ is said to (Pareto) dominate a solution $\boldsymbol{y}$ (denoted $\boldsymbol{x} \succ \boldsymbol{y}$)
   iff $(\forall i \in \mathcal{M} : f_i(\boldsymbol{x}) \leq f_i(\boldsymbol{y})) \wedge (\exists i \in \mathcal{M} : f_i(\boldsymbol{x}) < f_i(\boldsymbol{y}))$
2. **Pareto optimal**
   A solution $\boldsymbol{x}$ is said to be Pareto optimal **iff** $\neg \exists \boldsymbol{y} : \boldsymbol{y} \succ \boldsymbol{x}$
3. **Pareto optimal set**
   The set $\mathcal{P}_S$ of all Pareto optimal solutions: $\mathcal{P}_S = \{\boldsymbol{x} | \neg \exists \boldsymbol{y} : \boldsymbol{y} \succ \boldsymbol{x}\}$
4. **Pareto optimal front**
   The set $\mathcal{P}_F$ of all objective function values corresponding to the solutions in $\mathcal{P}_S$: $\mathcal{P}_F = \{(f_0(\boldsymbol{x}), f_1(\boldsymbol{x}), \ldots, f_{m-1}(\boldsymbol{x})) | \boldsymbol{x} \in \mathcal{P}_S\}$

The Pareto optimal set $\mathcal{P}_S$ is a definition of all trade-off optimal solutions in the parameter space. The Pareto optimal front $\mathcal{P}_F$ is the same set of solutions, only regarded in the objective space. The size of either set can be infinite, in which case it is impossible to find the optimal set or front with a finite number of solutions. Regardless of the size of $\mathcal{P}_S$ or $\mathcal{P}_F$, it is commonly accepted that we are interested in finding a good representation of these sets with a finite number of solutions. The definition of a good representation, is difficult however. The reason for this is that it is desirable to obtain a diverse set of solutions as well as it is desirable to obtain a front or set that is close to the optimal one. Furthermore, it depends on the mapping between the parameter space and the objective space whether a good spread of the solutions in the parameter space is also a good spread of the solutions in the objective space. However, it is common practice [9] to search for a good diversity of the solutions along the Pareto *front*. The reason for this is that a decision-maker will ultimately have to pick a single solution. Therefore, it is often best to present a wide variety of trade-off solutions for the specified goals.

The notion of searching a space by maintaining a population of solutions is characteristic of evolutionary algorithms (EAs), which makes them natural candidates for multi-objective optimization aiming to cover a good approximation of the Pareto optimal front. A strongly increasing amount of research has indeed been done in the field of evolutionary multi-objective optimization in recent years [9] with very promising results.

3 The Naive MIDEA

To obtain EDAs that are well-suited for multi-objective optimization, we propose to instantiate two steps in the framework. Firstly, to stimulate the preservation of diversity along the Pareto front, we instantiate the selection mechanism by using a diversity preserving truncation selection operator. Secondly, we partially instantiate the search for a probability distribution to use by enforcing the use of mixture distributions.

3.1 Diversity-preserving Truncation Selection

Background and Motivation

Selection in evolutionary algorithms is meant to select the better solutions of the population to perform variation with. In multi-objective optimization however, the notion of "a better solution" has two sides to it. On the one hand we want the solutions to be as close to the Pareto optimal front as possible. On the other hand, we want a good diverse representation of the Pareto optimal front. A good selection operator in a MOEA must thus exert selection pressure with respect to both of these aspects.

Selection Pressure towards the Pareto Optimal Front

In a practical application, we have no indication of how close we are to the Pareto optimal front. To ensure selection pressure towards the Pareto optimal front in the absence of such information, the best we can do is to find solutions that are dominated as little as possible by any other solution.

A straightforward way to obtain selection pressure towards non-dominated solutions is therefore to count for each solution in the population the number of times it is dominated by another solution in the population, which is called the domination count of a solution [3, 16]. The rationale behind the domination count approach is that ultimately we would like no solution to be dominated by any other solution, so the less times a solution is dominated, the better. A lower domination count is preferable. Using this value we can apply truncation selection or tournament selection to obtain solid pressure towards non-dominated solutions.

Another approach to ensuring a preference for solutions that are dominated as little as possible, is to assign a preference to different domination *ranks* [12, 17].

The solutions that are in the jth rank are those solutions that are non-dominated if the solutions of all ranks $i < j$ are disregarded. Note that the best domination rank contains all solutions that are non-dominated in the complete population. A lower rank is preferable. Using this value we can again apply for instance either truncation selection or tournament selection. Similar to the domination count approach, this approach effectively prefers solutions that are closer to the set of non-dominated solutions. It has been observed that in practice the difference between domination-counting and the domination-ranking schemes in practice is only very small [5].

Selection Pressure towards Diversity

In most multi-objective selection schemes, diversity is used as a second comparison key in selection. This prohibits tuning the amount of selection pressure towards diversity to the amount of selection pressure towards getting close to the Pareto optimal front. An example is the approach taken in the NSGA-II in which solutions are selected based on their non-domination rank using tournament selection [12]. If the ranks of two solutions are equal, the solution that has the largest total distance between its two neighbors summed over each objective, is preferred. This gives a preference to non-crowded solutions.

The explicit selection pressure towards diversity may serve more than just the purpose of ensuring that a diverse subset is selected from a certain set of non-dominated solutions. If we only apply selection pressure to finding the non-dominated solutions and enable diversity preservation only to find a good spread of solutions in current Pareto front, we increase the probability that we only find a subset of a discontinuous Pareto optimal front. Selection pressure towards diversity will most likely be too late in helping out to find the other parts of the discontinuous Pareto optimal front as well. Therefore, we may need to spend more attention on diversity preservation during optimization and perhaps even increase the amount of diversity preservation. Another reason why we may need to increase the selection pressure towards diversity is that a variation operator is used that can find many more non-dominated solutions, which could cause a MOEA to converge prematurely onto subregions of a Pareto optimal front or onto locally optimal sets of non-dominated solutions, unless the population size is increased. However, given a fixed number of evaluations, this can be a significant drawback in approaching the Pareto optimal front. This problem can be alleviated by placing more emphasis on selection pressure towards diversity and by consequently reducing the effort in the selection pressure towards getting close to the Pareto optimal front. By doing so, the variation operator is presented with a more diverse set of solutions from which a more diverse set of offspring will result. Furthermore, solutions that are close to each other will now have a smaller joint chance that they will both be selected, which improves the ability to approach the Pareto optimal front since premature convergence is less likely.

Combining Selection Pressures

Concluding, to ensure pressure towards the Pareto optimal front and towards diversity at the same time, the selection procedure must be provided with a component that prefers a diverse selection of solutions. However, since the goal is to preserve diversity *along* the Pareto front, rather than to preserve diversity in general, the selection on the basis diversity should not precede selection on the basis of getting close to the Pareto optimal front.

Selection Operator

In the selection operator that we propose, the ratio of the amount of selection pressure towards the Pareto optimal front and the amount of selection pressure towards diversity can be tuned using a single parameter δ to better fit the specific needs of the problem solver. In most selection operators this ratio is fixed beforehand. Ultimately, the selection operator selects $\lfloor \tau n \rfloor$ solutions, where n is the population size and $\tau \in [\frac{1}{n}; 1]$ is the selection percentile. Just as there are two forms of selection pressure to be exerted by the selection operator as discussed above, there are two phases in our selection operator.

1. In the first phase, the domination count [16] of all solutions is first computed as mentioned above. Subsequently, a pre-selection $\boldsymbol{S}^P$ is made of $\lfloor \delta \tau n \rfloor$ solutions ($\delta \in [1; \frac{1}{\tau}]$) using truncation selection on the domination count (select the best $\lfloor \delta \tau n \rfloor$ solutions). However, if the solution with the largest domination count to end up in $\boldsymbol{S}^P$ by truncation selection has a domination count of 0, *all* solutions with a domination count of 0 are selected instead, resulting in $|\boldsymbol{S}^P| \geq \lfloor \delta \tau n \rfloor$. This ensures that once the search starts to converge onto a certain Pareto front, we enforce diversity over all of the available solutions on the front.

2. In the second phase, the final selection $\boldsymbol{S}$ is obtained from $\boldsymbol{S}^P$. To do so, a nearest neighbor heuristic is used to promote diversity. First, a solution with an optimal value for a randomly chosen objective is deleted from $\boldsymbol{S}^P$ and added to $\boldsymbol{S}$. Note that the choice of objective is arbitrary as the key is to find a diverse selection of solutions. To stimulate this, we can select a solution that is optimal along any objective. For all solutions in $\boldsymbol{S}^P$, the nearest neighbor distance is computed to the single solution in $\boldsymbol{S}$. The distance that we use is the Euclidean distance scaled to the sample range in each objective. The solution in $\boldsymbol{S}^P$ with the *largest* distance is then deleted from $\boldsymbol{S}^P$ and added to $\boldsymbol{S}$. The distances in $\boldsymbol{S}^P$ are updated by investigating whether the distance to the newly added point in $\boldsymbol{S}$ is smaller than the currently stored distance. These last two steps are repeated until $\lfloor \tau n \rfloor$ solutions are in the final selection.

An example application of this operator is presented in Fig. 1. This selection operator has a running time complexity of $\mathcal{O}(n^2)$. This is no worse than the minimum of $\mathcal{O}(n^2)$ for computing the domination counts which is required in all MOEAs.

3.2 Mixture Distributions

A mixture probability distribution is a weighted sum of $k > 1$ probability distributions. Each probability distribution in the mixture probability distribution is called

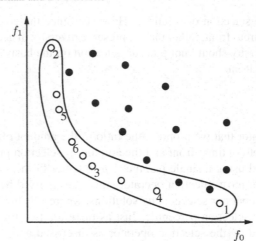

Fig. 1. An example of the application of the diversity preserving selection operator with $n = 22$, $\delta = \frac{5}{3}$, $\tau = \frac{3}{10}$, which gives $\lfloor \delta\tau n \rfloor = 11$ and $\lfloor \tau n \rfloor = 6$. Objectives f_0 and f_1 should both be minimized. The dominated solutions are black whereas the non-dominated solutions are white. The solutions that belong to the preselection are outlined. The solutions that are finally selected are numbered in the order in which they are chosen from the preselection. Here objective f_0 has been chosen to initiate the selection process

a mixture component. Let $\boldsymbol{\mathcal{Z}} = (Z_0, Z_1, \ldots, Z_{l-1})$ be a vector for all random variables involved in the EDA (i.e. Z_i is a random variable associated with the ith problem variable). A mixture probability distribution for random variables $\boldsymbol{\mathcal{Z}}$ is then defined as follows:

$$P^{mixture}(\boldsymbol{\mathcal{Z}}) = \sum_{i=0}^{k-1} \beta_i P^i(\boldsymbol{\mathcal{Z}}) \tag{1}$$

where $\beta_i \geq 0$, $i \in \{0, 1, \ldots, k-1\}$, and $\sum_{i=0}^{k-1} \beta_i = 1$. The β_i with which the mixture components are weighted in the sum are called mixing coefficients.

The Benefit of Mixture Distributions

The general advantage of mixture probability distributions is that a larger class of independence relations between the random variables can be expressed than when using non-mixture probability distributions since a mixture probability distribution makes a combination of multiple probability distributions. In many cases, simple probability distributions can be estimated to get accurate descriptions of the data in different parts of the sample space. By adding the k "simple" probability distributions into the mixture probability distribution, an accurate description of the data in the complete sample space can be obtained. This allows for the modelling of quite complex dependencies between the problem variables. By using mixture probability distributions, a powerful, yet computationally tractable type of probability distribution can be used within EDAs, that provides for processing complicated interactions between a problem's variables.

For multi-objective optimization, mixture distributions can have a specific advantage that renders them particularly useful. The specific advantage is geometrical in nature. If we for instance cluster the solutions as observed in the objective space and then estimate a simpler probability distribution in each cluster, the probability distributions in these clusters can portray specific information about the different regions along the Pareto optimal front that we are ultimate interested in multi-objective optimization. Each simpler probability distribution to be used in each cluster can for instance be a factorized probability distribution as is used in most EDAs. Drawing new solutions from the resulting mixture probability distribution gives solutions that are more likely to be well spread along the front as each mixture component delivers a subset of new solutions. The use of such a mixture distribution thus results in a parallel exploration along the current Pareto front. This parallel exploration may very well provide a better spread of new solutions along the Pareto front than when a single non-mixture distribution is used to capture information about the complete Pareto front. In Fig. 2 an example is given of what the result of clustering the selected solutions in the objective space typically looks like. The effect of splitting up the solutions along the Pareto front, thereby facilitating parallel exploration along the front, can clearly be seen.

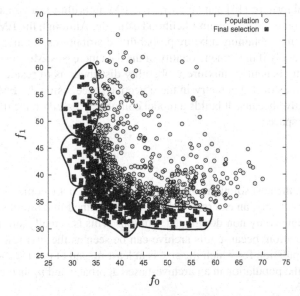

Fig. 2. An example of the breaking up the front of selected solutions using clustering. Objectives f_0 and f_1 should both be minimized. The four individual clusters that are defined in this example are outlined

Estimating Mixture Distributions

From the previous subsection describing the specific advantages of mixture probability distributions for multi-objective, we already have a straightforward manner to estimate mixture probability distributions from data using clustering. To actually build the mixture distribution from the simpler distributions, the mixing coefficients β_i must still be chosen. This can be done in various ways. A common approach is to set β_i to the proportion of the size of the ith cluster with respect to the sum of the sizes of all clusters. For the specific application of multi-objective optimization however, we propose to assign each cluster an equally large mixing coefficient, i.e. $\beta_i = 1/k$. The reason for this is that we want to distribute the solutions as good as possible along the Pareto front. Giving each cluster an equal probability of producing new solutions maximizes parallel exploration along the Pareto front. The only thing left to choose then is which clustering algorithm to use. Exact algorithms for partitioning (i.e. clustering into mutually disjoint subsets) exist [20], but the running times for these algorithms are of no practical use for building EDAs. What we require, is a fast approximate assessment of clusters such that we can estimate a relatively simple probability distribution in each cluster in a good way. Computationally efficient clustering algorithms exist that provide useful results [20]. Examples are the leader algorithm and the K-means algorithm.

A different approach to estimating a mixture probability distribution from data is to compute a maximum likelihood estimation. To this end, the Expectation Maximization (EM) algorithm [14] can be used. The EM algorithm is a general iterative approach to computing a maximum likelihood estimate. Although the EM algorithm is a valid approach to obtaining mixture probability distributions, it tends to be time-consuming, especially if the dimensionality of the data increases. Moreover, since we expect the specific benefit of mixture probability distributions to reside in dividing the data on the basis of its geometry in the objective space, using the EM algorithm seems less attractive because it builds a model completely based on the data as given in the parameter space.

3.3 Elitism

If elitism is used, the best solutions of the current generation are copied into the next generation. Alternatively, an external archive of a predefined maximum size n_a may be used that contains only non-dominated solutions. This is actually similar to using elitism in a population, because this archive can be seen as the first few population members in a population for which the size is at least n_p and at most $n_p + n_a$, where n_p is the size of the population in an archive-based approach and n_a is the size of the external archive.

Elitism plays an important role in multi-objective optimization since many solutions exist that are all equally preferable. It is important to have access to many of them during optimization to advance the complete set of non-dominated solutions further. An ideal variation operator is capable of generating solutions that are closer to the Pareto optimal front, but also spread out across the entire current set of non-dominated solutions as well as possibly outside it to extend the diversity of the set of non-dominated solutions even further. However, obtaining new and diverse non-dominated solutions is hard, especially as the set of non-dominated solutions approaches the Pareto optimal front. If a non-dominated solution gets lost in a certain generation, it may take quite some effort before a new non-dominated solution in its vicinity is generated again. For this reason, elitism is commonly accepted [24, 36] to be a very important tool for improving the results obtained by any MOEA.

Elitism can be used within the MIDEA framework in a straightforward manner because truncation selection is already used (Sect. 3.1). An elitist MIDEA selects the best $\lfloor \tau n \rfloor$ solutions using the diversity-preserving truncation selection operator. Subsequently, only the worst $n - \lfloor \tau n \rfloor$ solutions are replaced with new offspring that result from sampling the estimated probability distribution. The best $\lfloor \tau n \rfloor$ solutions that were selected, are thus kept in the population.

3.4 The MIDEA Framework

The MIDEA variant that we use in our experiments is described in pseudo-code in Fig. 3.

MIDEA
1 Initialize a population of n random solutions and evaluate their objectives
2 Iterate until termination
2.1 Compute the domination counts
2.2 Select $\lfloor \tau n \rfloor$ solutions with the diversity preserving selection operator
2.3 Estimate a mixture probability distribution $P^{mixture}(\mathcal{Z})$
2.4 Replace the non-selected solutions with new solutions drawn from $P^{mixture}(\mathcal{Z})$
2.5 Evaluate the objectives of the new solutions

Fig. 3. Pseudo-code for the MIDEA framework

The Naive MIDEA Instance

Probability Distributions in Each Cluster

In Sect. 3.2 we have argued that mixture distributions can play an important role in multi-objective optimization. Moreover, we have argued that a simple, but effective approach to estimating mixture distributions is to cluster the selected solutions on the bases of the geometry of their objective values. We therefore suggest keeping the probability distributions to be estimated in each cluster as simple as possible. This suggestion leads to the choice of using univariate factorized probability distributions in each cluster in the naive MIDEA. In a factorized probability distribution, each random variable is regarded separately, meaning that a probability distribution is estimated for each random variable separately. For discrete random variables, this amounts to repeatedly counting frequencies and computing proportions for a single random variable. For real-valued random variables this implies estimating for instance the mean and variance of a one-dimensional normal distribution repeatedly. The mathematical formulation of the univariate factorization is:

$$P^{univariate}(\boldsymbol{\mathcal{Z}}) = \prod_{i=0}^{l-1} P(Z_i) \qquad (2)$$

Since in each cluster we thus disregard all dependencies between random variables, we call this specific MIDEA instance naive in analogy with the well-known naive Bayes classifier. However, the clusters are expected to already provide a large benefit for multi-objective optimization. Moreover, algorithms such as UMDA [27] and the compact GA [19] that use same probability distribution as in (2) (without clustering) have provided good results on many interesting single-objective optimization problems. Hence, we already expect good optimization behavior for the naive MIDEA.

Clearly, non-naive instances of MIDEA can be made directly by estimating more involved probability distributions in each cluster, such as Bayesian factorized probability distributions. Although we will present the results of some experiments with such more involved probability distributions for comparison reasons, we refer the interested reader for more details to the literature on either these probability distributions (e.g. [7, 10, 25]) or to the relevant literature on single-objective EDAs (e.g. [1, 2, 4, 18, 23, 28–33]).

Clustering Algorithm

Since we are interested in obtaining useful results in as little time as possible, we suggest the use of a fast clustering algorithm. Possibly this adds to the naiveness of our naive MIDEA instance, but other clustering algorithms are easily implemented if required.

The algorithm that we propose to use is the leader algorithm. The leader algorithm is one of the fastest partitioning algorithms [20]. The use of it can thus be beneficial if the amount of overhead that is introduced by factorization mixture selection methods is desired to remain small. There is no need to specify in advance how many partitions there should be. The first solution to make a new partition is appointed to be its leader. The leader algorithm goes over the solutions exactly once. For each solution it encounters, it finds the first partition that has a leader being closer to the solution than a given threshold $\mathfrak{T}_d$. If no such partition can be found, a new partition is created containing only this single solution. To prevent the first partitions from becoming quite a lot larger than the later ones, we randomize the order in which the partitions are inspected. The asymptotic running time for finding the first partition with a leader closer than $\mathfrak{T}_d$ is the same as going over all partitions and finding the closest partition. Therefore, we prefer to find the closest partition.

One of the drawbacks of the (randomized) leader algorithm is that it is not invariant given the sequence of the input solutions. Most partitioning algorithms do not have this property, but not as strongly as the leader algorithm. Therefore, to be sure that the ordering of the solutions is not subject to large repeating sequences of solutions, we randomize the ordering of the solutions each time the leader algorithm is applied.

Pseudo-Code

The naive MIDEA is an instance of the general MIDEA framework. Figure 4 shows how the naive MIDEA can be obtained from the general MIDEA framework by using a specific instantiation of lines 2.3 and 2.4.

naive MIDEA
(instantiation of steps 2.3 and 2.4 of the general MIDEA framework)

1 $(c^0, c^1, \ldots, c^{k-1}) \leftarrow$ LeaderAlgorithm$(\mathfrak{T}_d)$
2 **for** $i \leftarrow 0$ **to** $k-1$ **do**
  2.1 $\beta_i \leftarrow 1/k$
  2.2 **for** $j \leftarrow 0$ **to** $l-1$ **do**
     2.2.1 Estimate a one-dimensional probability distribution $P^{i,j}(Z_j)$ for
        random variable Z_j from the solutions in the ith cluster (i.e. c^i)
3 **for** $i \leftarrow \lfloor \tau n \rfloor$ **to** $n-1$ **do**
  3.1 Initialize a new solution z
  3.2 Choose an index $q \in \{0, 1, \ldots, k-1\}$ with probability β_q
  3.3 **for** $j \leftarrow 0$ **to** $l-1$ **do**
     3.3.1 Draw a value for z_j from the one-dimensional probability distribution
        $P^{q,j}(Z_j)$ associated with the qth-cluster
  3.4 Add z to the set of new offspring.

Fig. 4. Pseudo-code for the naive MIDEA

4 Experiments

In this section we compare MIDEA instances to two well-known state-of-the-art MOEAs that aim at obtaining a diverse set of solutions along the Pareto front. The SPEA algorithm by Zitzler and Thiele [38] and the NSGA-II algorithm by Deb et al. [12] showed superior performance compared to most other MOEAs [12, 36]. The test suite we used consists of eight multi-objective optimization problems. We varied the dimensionality of these problems to get a total of sixteen problem instances to test the MOEAs on. The multi-objective optimization problems are described in Sect. 4.1. The performance measures we use to score the results of the algorithms with are described in Sect. 4.2. In Sect. 4.3 we present our experiment setup. In Sect. 4.4 we discuss the obtained results. Finally, in Sect. 4.5 we give a short summary for the EA practitioner.

4.1 Multi-objective Optimization Problems

Our test suite consists of problems with real-valued variables as well as with binary variables. To make a clear distinction between these two cases, we write real-valued variables as y_i and binary variables as x_i. In both cases we have used four different optimization problems and two different dimensionalities for these problems to obtain a total test suite size of 16 problems. In the following we give a brief description of the problems in our test suite.

Real-valued Multi-objective Optimization Problems

A variety of test problems for real-valued variables has been proposed that may cause different types of problems for multi-objective optimization algorithms [11, 13, 36]. From this set of problems, we have selected three problems that are commonly used to benchmark multi-objective optimization algorithms. The fourth real-valued test problem is a new test problem we have designed to test the performance of MOEAs if there are strong interactions between the problem variables. These problems represent a spectrum of multi-objective problem difficulty as they make it difficult for a multi-objective optimization algorithm to progress towards the global optimal front and to maintain a diverse spread of solutions due to properties such as discontinuous fronts and multi-modality. The problems with real-valued variables that we use in our experiments are all defined for two objectives. An overview of our test problems is given in Fig. 5.

BT_1

Function BT_1 differs from the other three functions in that it has multivariate (linear) interactions between the problem variables. Therefore, more complex factorizations are required to exploit these interactions, whereas the other functions are well-suited to be optimized using the univariate factorization. The Pareto optimal front is given by $f_1(\boldsymbol{y}) = 1 - y_0$.

ZDT_4

Function ZDT_4 was introduced by Zitzler et al. [36]. It is very hard to obtain the optimal front $f_1(\boldsymbol{y}) = 1 - \sqrt{y_0}$ in ZDT_4 since there are many local fronts. Moreover, the number of local fronts increases as we get closer to the Pareto optimal front. The main problem that a MOEA should be able to overcome to optimize this problem is thus strong multi-modality.

ZDT_6

Function ZDT_6 was also introduced by Zitzler et al. [36]. The density of solutions in ZDT_6 increases as we move away from the Pareto optimal front. Furthermore, ZDT_6 has a non-uniform density of solutions *along* the Pareto optimal front as there are more solutions as $f_0(\boldsymbol{y})$ goes up to 1. Therefore, a good diverse spread of solutions along the Pareto front is hard to obtain. The Pareto front for ZDT_6 is given by $f_1(\boldsymbol{y}) = 1 - f_0(\boldsymbol{y})^2$ with $f_0(\boldsymbol{y}) \in [1 - e^{-1/3}; 1]$.

Name	Definition	Range

BT_1	**Minimize** $(f_0(\boldsymbol{y}), f_1(\boldsymbol{y}))$
	
	Where • $f_0(\boldsymbol{y}) = y_0$
	
	    • $f_1(\boldsymbol{y}) = 1 - f_0(\boldsymbol{y}) +$
	
	       $10^7 - \frac{100}{\left(10^{-5} + \sum_{i=1}^{l-1} \left\|\sum_{j=1}^{i} y_i\right\|\right)}$
		• $y_0 \in [0; 1]$
		
		• $y_i \in [-3; 3]$
		
		  $(1 \le i < l)$

ZDT_4	**Minimize** $(f_0(\boldsymbol{y}), f_1(\boldsymbol{y}))$
	
	Where • $f_0(\boldsymbol{y}) = y_0$
	
	    • $f_1(\boldsymbol{y}) = \gamma \left(1 - \sqrt{\frac{f_0(\boldsymbol{y})}{\gamma}}\right)$
	
	    • $\gamma = 1 + 10(l-1) + \sum_{i=1}^{l-1} \left(y_i^2 - 10\cos(4\pi y_i)\right)$
		• $y_0 \in [0; 1]$
		
		• $y_i \in [-5; 5]$
		
		  $(1 \le i < l)$

ZDT_6	**Minimize** $(f_0(\boldsymbol{y}), f_1(\boldsymbol{y}))$
	
	Where • $f_0(\boldsymbol{y}) = 1 - e^{-4y_0}\sin^6(6\pi y_0)$
	
	    • $f_1(\boldsymbol{y}) = \gamma \left(1 - \left(\frac{f_0(\boldsymbol{y})}{\gamma}\right)^2\right)$
	
	    • $\gamma = 1 + 9 \left(\sum_{i=1}^{l-1} \frac{y_i}{9}\right)^{0.25}$
		• $y_i \in [0; 1]$
		
		  $(0 \le i < l)$

CTP_7	**Minimize** $(f_0(\boldsymbol{y}), f_1(\boldsymbol{y}))$
	
	Where • $f_0(\boldsymbol{y}) = y_0$
	
	    • $f_1(\boldsymbol{y}) = \gamma \left(1 - \frac{f_0(\boldsymbol{y})}{\gamma}\right)$
	
	    • $\gamma = 1 + 10(l-1) + \sum_{i=1}^{l-1} \left(y_i^2 - 10\cos(4\pi y_i)\right)$
	
	Such that • $\cos(-\frac{5\pi}{100})f_1(\boldsymbol{y}) - \sin(-\frac{5\pi}{100})f_0(\boldsymbol{y}) \ge$
	
	    $40\|\sin(5\pi \left[\sin(-\frac{5\pi}{100})f_1(\boldsymbol{y}) + \right.$
	
	      $\left.\cos(-\frac{5\pi}{100})f_0(\boldsymbol{y})\right])\|^6$
		• $y_0 \in [0; 1]$
		
		• $y_i \in [-5; 5]$
		
		  $(1 \le i < l)$

Fig. 5. Real-valued multi-objective optimization test problems

CTP_7

Function CTP_7 was introduced by Deb et al. [13]. Its Pareto optimal front differs slightly from that of ZDT_4, but otherwise shares the multi-modal front problem. In addition, this problem has constraints in the objective space, which makes finding a diverse representation of the Pareto front more difficult since the Pareto front is discontinuous and it is hard to obtain an approximation that has a few solutions in each feasible part of that front.

Binary Multi-objective Optimization Problems

In Fig. 6, we have specified four binary multi-objective optimization problems. Next to being binary, these problems are also multi-objective variants of well-known com-

Name	Definition
MS $\left(\begin{array}{c}\text{Maximum}\\ \text{Satisfiability}\end{array}\right)$	**Maximize** $(f_0(\boldsymbol{x}), f_1(\boldsymbol{x}), \ldots, f_{m-1}(\boldsymbol{x}))$ **Where** $\bullet \ \forall_{i\in\mathcal{M}} : f_i(\boldsymbol{x}) = \sum_{j=0}^{c_i-1} \mathrm{sgn}\left(\left[\sum_{k=0}^{l-1}(C_i)_{jk} \otimes x_k\right]\right)$ $\bullet \ \mathrm{sgn}(x) = \left\{ \begin{array}{rl} 1 & \text{if } x > 0 \\ 0 & \text{if } x = 0 \\ -1 & \text{if } x < 0 \end{array}\right.$ $\bullet$ $\begin{array}{c\|cc}\otimes & 0 & 1 \\ \hline -1 & 1 & 0\end{array}$ $\begin{array}{c\|cc}\otimes & 0 & 1 \\ \hline 0 & 0 & 0\end{array}$ $\begin{array}{c\|cc}\otimes & 0 & 1 \\ \hline 1 & 0 & 1\end{array}$
KN (Knapsack)	**Maximize** $(f_0(\boldsymbol{x}), f_1(\boldsymbol{x}), \ldots, f_{m-1}(\boldsymbol{x}))$ **Where** $\bullet \ \forall_{i\in\mathcal{M}} : f_i(\boldsymbol{x}) = \sum_{j=0}^{l-1} P_{ij}x_j$ **Such that** $\bullet \ \forall_{i\in\mathcal{M}} : \sum_{j=0}^{l-1} W_{ij}x_j \leq c_i$
SC (Set Covering)	**Minimize** $(f_0(\boldsymbol{x}), f_1(\boldsymbol{x}), \ldots, f_{m-1}(\boldsymbol{x}))$ **Where** $\bullet \ \forall_{i\in\mathcal{M}} : f_i(\boldsymbol{x}) = \sum_{j=0}^{l-1} C_{ij}x_j$ **Such that** $\bullet \ \forall_{i\in\mathcal{M}} : \forall_{0\leq j<r} : \sum_{k=0}^{l-1}(A_i)_{jk}x_k \geq 1$
MST $\left(\begin{array}{c}\text{Minimal}\\ \text{Spanning}\\ \text{Tree}\end{array}\right)$	**Minimize** $(f_0(\boldsymbol{x}), f_1(\boldsymbol{x}), \ldots, f_{m-1}(\boldsymbol{x}))$ **Where** $\bullet \ \forall_{i\in\mathcal{M}} : f_i(\boldsymbol{x}) = \sum_{j=0}^{l-1} W_{ij}x_j$ **Such that** $\bullet \ \forall_{S\subseteq V} : \sum_{x_j\in(S\times(V-S))} x_j \geq 1$ $\bullet \ \forall_{S\subseteq V} : \sum_{x_j\in(S\times S)} x_j \leq |S| - 1$

Fig. 6. Binary multi-objective combinatorial optimization test problems

binatorial optimization problems. The number of objectives for these problems is not restricted to two and is denoted by m.

It is important to note that we have used random instances for the combinatorial optimization problems. In the case of only a single objective, random instances may on average be easy for some combinatorial problems. However, in the case of multiple objectives, finding the Pareto front is usually much more difficult, even if efficient algorithms are available for the single-objective case [15]. Therefore, the instances used in our test suite are not expected to be over-easy. Furthermore, the problems also serve to indicate differences between the different multi-objective algorithmic approaches other than the fact that dependencies between problem variables can be exploited. This relative performance of the algorithms may be well observed using our proposed test-suite. On the other hand, the degree of interaction between the problem variables in randomly generated problem instances may not be too large, which may cause optimization algorithms that regard the problem variables independently of each other to be the most efficient.

Maximum Satisfiability

In the maximum satisfiability problem, we are given a propositional formula in conjunctive normal form. The goal is to satisfy as many clauses as possible. The solution string is a truth assignment to the involved literals. These formulas can be represented by a matrix in which row i specifies what literals appear either positive (1) or negative (-1) in clause i. In the multi-objective variant of this problem, we have m of such matrices and only a single solution to satisfy as many clauses as possible in each objective at the same time.

Knapsack

The multi-objective knapsack problem was first used to test MOEAs on by Zitzler and Thiele [38]. We are given m knapsacks with a specified capacity and n items. Each item can have a different weight and profit in every knapsack. Selecting item i in a solution implies placing it in every knapsack. A solution may not cause exceeding the capacity of any knapsack.

Set Covering

In the set covering problem, we are given l locations at which we can place some service at a specified cost. Furthermore, associated with each location is a set of regions $\subseteq \{0, 1, \ldots r - 1\}$ that can be serviced from that location. The goal is to select locations such that *all* regions are serviced against minimal costs. In the multi-objective variant of set covering, m services are placed at a location. Each service however covers its own set of regions when placed at a certain location and has its own cost associated with a certain location. A binary solution indicates at which locations the services are placed.

Minimal Spanning Tree

In the minimal spanning tree problem we are given an undirected graph (V, E) such that each edge has a certain weight. We are interested in selecting edges $E_T \subseteq E$ such that (V, E_T) is a spanning tree. The objective is to find a spanning tree such that the weight of all its edges is minimal. In the multi-objective variant of this problem, each edge can have a different weight in each objective.

4.2 Performance Indicators

To measure the performance of a MOEA we only consider the subset of all non-dominated solutions that is contained in the final population that results from running the MOEA. We call such a subset an approximation set and denote it by $\mathcal{S}$. The size of the approximation set depends on the settings used to run the MOEA with.

To actually measure performance, performance indicators are used. A performance indicator is a function that, given an approximation set $\mathcal{S}$, returns a real value that indicates how good $\mathcal{S}$ is with respect to a certain feature that is measured by the performance indicator. Performance indicators are commonly used to determine the performance of a MOEA and to compare this performance with other MOEAs if the number of evaluations is fixed beforehand. More detailed information regarding the importance of using good performance indicators to evaluate MOEAs may be found in dedicated literature [5, 22, 37].

Since we are interested in performance as measured in the objective space, we define the distance between two multi-objective solutions z^0 and z^1 to be the Euclidean distance between their objective values $f(z^0)$ and $f(z^1)$:

$$d(z^0, z^1) = \sqrt{\sum_{i=0}^{m-1} (f_i(z^1) - f_i(z^0))^2} \tag{3}$$

If we only want to measure diversity, we can use the **FS** (*Front Spread*) indicator. This performance indicator was first used by Zitzler [35]. The **FS** indicator indicates the size of the objective space covered by an approximation set. A larger **FS** indicator value is preferable. The **FS** indicator for an approximation set S is defined to be the maximum Euclidean distance inside the smallest m-dimensional bounding-box that contains S. This distance can be computed using the maximum distance among the solutions in S in each dimension separately:

$$\mathbf{FS}(S) = \sqrt{\sum_{i=0}^{m-1} \max_{(z^0, z^1) \in S \times S}\{(f_i(z^0) - f_i(z^1))^2\}} \tag{4}$$

In combination with the **FS** indicator, it is also important to know how many points are available in the set of non-dominated solutions, because a larger set of trade-off points is more desirable. This quantity is called the **FO** (*Front Occupation*) indicator and was first used by Van Veldhuizen [34]. A larger **FO** indicator value is preferable.

$$\mathbf{FO}(S) = |S| \tag{5}$$

The ultimate goal is to cover the Pareto optimal front. An intuitive way to define the distance between an approximation set S and the Pareto optimal front is to average the minimum distance between a solution and the Pareto optimal front over each solution in S. We refer to this distance as the distance from a set of non-dominated solutions to the Pareto optimal front and it serves as an indicator of how close an approximation set has come to the Pareto optimal front. We denote it by $D_{S \to \mathcal{P}_F}$. This performance indicator was first used by Van Veldhuizen [34]. A smaller value for this performance indicator is preferable.

$$D_{S \to \mathcal{P}_F}(S) = \frac{1}{|S|} \sum_{z^0 \in S} \min_{z^1 \in \mathcal{P}_S}\{d(z^0, z^1)\} \tag{6}$$

An approximation set with a good $D_{S \to \mathcal{P}_F}$ indicator value does not imply that a good diverse representation of the Pareto optimal set has been obtained, since the indicator only reflects how far away the obtained points are from the Pareto optimal front on average. An approximation set consisting of only a single solution can already have a low value for this indicator. To include the goal of diversity, the reverse of the $D_{\mathcal{S} \to \mathcal{P}_F}$ indicator is a better guideline for evaluating MOEAs. In the reverse distance indicator, we compute for each solution in the Pareto optimal set the distance to the closest solution in an approximation set $\mathcal{S}$ and take the average as the indicator value. We denote this indicator by $D_{\mathcal{P}_F \to \mathcal{S}}$ and refer to it as the distance from the Pareto optimal front to an approximation set. A smaller value for this performance indicator is preferable. In the definition of this indicator, we must realize that the Pareto optimal front may be continuous. For an exact definition, we therefore have to use a line integration over the entire Pareto front. For a 2-dimensional multi-objective problem we obtain the following expression:

$$D_{\mathcal{P}_F \to \mathcal{S}}(\mathcal{S}) = \int_{\mathcal{P}_F} \min_{z^0 \in \mathcal{S}}\{d(z^0, z^1)\} df(z^1) \tag{7}$$

In most practical experiments, it is easier to compute a uniformly sampled set of many solutions along the Pareto optimal front and to use this discretized representation of $\mathcal{P}_F$ instead. A discretized version of the Pareto optimal front is also available if a discrete multi-objective optimization problem is being solved. In the discrete case, the $D_{\mathcal{S} \to \mathcal{P}_F}$ indicator is defined by:

$$D_{\mathcal{P}_F \to \mathcal{S}}(\mathcal{S}) = \frac{1}{|\mathcal{P}_S|} \sum_{z^1 \in \mathcal{P}_S} \min_{z^0 \in \mathcal{S}}\{d(z^0, z^1)\} \tag{8}$$

An illustration of the $D_{\mathcal{P}_F \to \mathcal{S}}$ indicator is presented in Fig. 7. The $D_{\mathcal{P}_F \to \mathcal{S}}$ indicator represents both the goal of getting close to the Pareto optimal front as well as

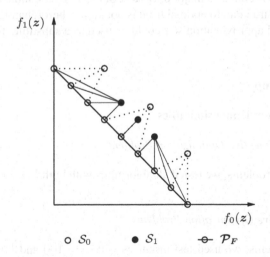

Fig. 7. The approximation set $\mathcal{S}_1$ is closer to the (discretized) Pareto optimal front but has less diversity, while approximation set $\mathcal{S}_0$ is further away from the front but has greater diversity: both sets have approximately the same $D_{\mathcal{P}_F \to \mathcal{S}}$ indicator value though

the goal of getting a diverse, wide-spread front of solutions. The $D_{\mathcal{P}_F \to \mathcal{S}}$ indicator for an approximation set $\mathcal{S}$ is zero if and only if all points in $\mathcal{P}_F$ are contained in $\mathcal{S}$ as well. Furthermore, a single solution from the Pareto optimal set will lead to the same $D_{\mathcal{P}_F \to \mathcal{S}}$ indicator as a more diverse set of solutions that has objective values that are slightly further away from the Pareto optimal front. Moreover, a similarly diverse approximation set of solutions that is closer to the Pareto optimal front, will have a lower $D_{\mathcal{P}_F \to \mathcal{S}}$ indicator value. However, an approximation set of solutions that is extremely diverse but far away from the Pareto optimal front, such as the non-dominated solutions of a randomly generated set of solutions, has a bad $D_{\mathcal{P}_F \to \mathcal{S}}$ indicator value. This underlines the important point that diversity is *not* equally important as is getting close to the Pareto optimal front because a larger diversity is often not hard to come by. What is important is the diversity *along* the objectives of a set of non-dominated solutions that is as close as possible to the Pareto optimal front.

A performance indicator that is closely related to the $D_{\mathcal{P}_F \to \mathcal{S}}$ indicator, is the hypervolume indicator by Knowles and Corne [22]. In the hypervolume indicator, a point in the objective space is picked such that it is dominated by all points in the approximation sets that need to be evaluated. The indicator value is then equal to the hypervolume of the multi-dimensional region enclosed by the approximation set and the picked reference point. This value is an indicator of the region in the objective space that is dominated by the approximation set. The main difference between the hypervolume indicator and the $D_{\mathcal{P}_F \to \mathcal{S}}$ indicator is that for the hypervolume indicator a reference point has to be chosen. Different reference points lead to different indicator values. Moreover, different reference points can lead to indicator values that indicate a preference for different approximation sets. Since in the $D_{\mathcal{P}_F \to \mathcal{S}}$ indicator the true Pareto optimal front is used, the $D_{\mathcal{P}_F \to \mathcal{S}}$ indicator does not suffer from this drawback. Of course, a major drawback of the $D_{\mathcal{P}_F \to \mathcal{S}}$ indicator is that in a real application the true Pareto optimal front is not known beforehand. In that case, the Pareto front of all approximation sets could be used as a substitute for the actual Pareto optimal front.

4.3 Experiment Setup

Optimization Problem Dimensionalities

Real-Valued Multi-Objective Optimization Problems

For the real-valued problems, we tested all algorithms with both $l = 10$ and $l = 100$ problem variables.

Binary Multi-Objective Optimization Problems

For the binary problems, we used test instances with $l = 100$ and $l = 1000$. For the maximum satisfiability problem, we generated the test instances by generating 2500 clauses for $l = 100$ and 12500 clauses for $l = 1000$ with a random number of literals between 1 and 5. For the knapsack problem, we generated instances by generating random weights in $[1; 10]$ and random profits in $[1; 10]$. The capacity of a knapsack was set at half of the total weight of all the items, weighted according to that knapsack objective. For set covering, the costs were generated at random in $[1; 10]$. We used 250 regions and 2500 regions to be serviced for $l = 100$ and $l = 1000$ respectively. We varied the problem difficulty through the region-location adjacency relation. This relation was generated by making each location adjacent to 70 and 50 randomly selected regions for $l = 100$ and $l = 1000$ respectively. Finally, for the minimum spanning tree problem, we used full graphs with 105 edges (15 vertices) and 1035 edges (46 vertices). The dimensionality of these problems is therefore not precisely 100 and 1000. The weights of the edges were generated randomly in $[1; 10]$.

Optimization Problem Constraints

Problems CTP_7, set covering, knapsack and minimal spanning tree have constraints. To deal with them, we can use a repair mechanism to transform infeasible solutions into feasible solutions. Another approach is based on the notion of constraint-domination introduced by Deb et al. [13]. This notion allows to deal with constrained multi-objective problems in a general fashion. A solution z^0 is said to constraint-dominate solution z^1 if any of the following is true:

1. Solution z^0 is feasible and solution z^1 is infeasible
2. Solutions z^0 and z^1 are both infeasible, but z^0 has a smaller overall constraint violation
3. Solutions z^0 and z^1 are both feasible and $z^0 \succ z^1$

The overall constraint violation is the amount by which a constraint is violated, summed over all constraints. We have used this principle for problems CTP_l and set covering. For the knapsack problem, an elegant repair mechanism was proposed earlier by Zitzler and Thiele [38]. For the minimal spanning tree problem, the number of constraints grows exponentially with the problem size l. We therefore propose to use repair mechanisms for these latter two problems.

Knapsack Repair Mechanism

If a solution violates a constraint, the repair mechanism iteratively removes items until all constrains are satisfied. The order in which the items are investigated, is determined by the maximum profit/weight ratio. The items with the lowest profit/weight ratio are removed first.

Minimal Spanning Tree Repair Mechanism

First the edges are removed from the currently constructed graph and they are sorted according to their weight. Next, they are added to the graph so that no cycles are introduced. This is done by only allowing edges to be introduced between the connected components in the graph. If after this phase, the number of connected components has not been reduced to 1, all edges between the connected components are regarded in increasing weight and again the connected components are merged until a single component is left.

General Algorithmic Setup

We ran every algorithm 50 times on each problem. In any single run we chose to allow a maximum of $20 \cdot 10^3$ evaluations for the real-valued problems of dimensionality $l = 10$ and the binary problems of dimensionality $l = 100$ and a maximum of $100 \cdot 10^3$ evaluations for the real-valued problems of dimensionality $l = 100$ and the binary problems of dimensionality $l = 1000$. As a result of imposing the restriction of a maximum of evaluations, a value for the population size n exists for each MOEA such that the MOEA will perform best. For too large population sizes, the search will move towards a random search and for too small population sizes, there is not enough information to perform adequate model selection and induction. We therefore increased the population size in steps of 25 to find the best results. To actually select the best population size, we selected the result with the lowest value for the $D_{\mathcal{P}_F \to \mathcal{S}}$ indicator.

Algorithms

We tested a few variants of three MOEAs. In the following we will describe the details that are required in addition to the details given in earlier sections for constructing the actual MOEAs that we will use for testing.

SPEA

For SPEA, we used uniform crossover and one-point crossover with a probability of 0.8. Bit-flipping mutation was used in combination with either of these recombination operators with a probability of 0.01. These settings were used previously by the SPEA authors [36]. We allowed the size of the external storage in SPEA to become as large as the population size. For the real problems, we encoded every variable with 30 bits.

NSGA-II

For NSGA-II, we used the same crossover and mutation operators and the same encoding for the real variables.

MIDEA

For MIDEA, we used the leader clustering algorithm in the objective space such that four clusters were constructed on average. If the number of clusters becomes too large, the requirements for the population size increases in order to facilitate proper factorization selection in each cluster. We do not suggest that the number of clusters we use is optimal, but it will serve to indicate the effectiveness of parallel exploration along the Pareto front as well as diversity preservation. In each cluster, we either used the univariate factorization (i.e. naive MIDEA) or we estimated a Bayesian factorization based upon normal distributions in the case of real variables. For details on how the Bayesian factorization is learned, see [1]. However, in the case of 100-dimensional real-valued problems, we allowed only at most a single parent for any variable. In the case of binary variables, we used the optimal dependency tree algorithm by Chow and Liu [8] to estimate a tree factorization in each cluster. To further investigate the influence of the different components in the MIDEA algorithm, we also performed tests in which only a single cluster is used. Furthermore, we also replaced the use of estimating probability distributions by the use of one-point crossover and uniform crossover with mutation as used in the SPEA and NSGA-II algorithms. In the case of clustering in combination with the use of crossover operators, restricted mating was employed in order to ensure clustered exploration along the front. In restricted mating crossover, an offspring is produced using two parent solutions that are picked from the same cluster. For the truncation percentile, we used the rule of thumb by Mühlenbein and Mahnig [26] and set τ to 0.3. Furthermore, for the comparison benchmarks, we set the diversity preservation parameter to $\delta = 1.5$, which was experimentally determined to give good results both with respect to diversity preservation as well as selective pressure. For an investigation of the influence of δ on the performance of MIDEA, we also varied δ and observed the results in some additional experiments, the results of which are reported below.

Overview of Abbreviations

In presenting the results, the different evolutionary algorithms that were tested are abbreviated to save space. For reference, a list of abbreviations that we have used is presented in Fig. 8.

Abbrev.	Meaning
UX	Uniform crossover (prob. 1) + bit-flipping mutation (prob. 0.01)
1X	One-point crossover (prob. 1) + bit-flipping mutation (prob. 0.01)
Univariate	The univariate factorization (2)
Learning	A more advanced Bayesian factorization is learned
1 Cluster	No clustering because everything is placed in a single cluster
Par. Clust.	Clustering in the parameter space
Obj. Clust.	Clustering in the objective space
M	An instance of the MIDEA framework

Fig. 8. List of abbreviations used in the presentation of the results

4.4 Results

To compare the MOEAs, we investigate their average performance with respect to performance indicators introduced in Sect. 4.2. The performance indicators that we use are the $D_{\mathcal{P}_F \to \mathcal{S}}$ indicator, the **FS** indicator and the **FO** indicator. For the $D_{\mathcal{P}_F \to \mathcal{S}}$ performance indicator, we used different sets to represent the Pareto optimal front for the real-valued optimization problems and the binary optimization problems. For the real-valued optimization problems we used a uniformly sampled set of 5000 solutions along the Pareto optimal front. Since we do not know the Pareto optimal front for the binary optimization problems, we used the Pareto front over all results obtained by all MOEAs.

For each of the performance indicators, we computed their average and standard deviation over the 50 runs to get an assessment of their performance. The averages are tabulated in Figs. 9 through 14. The best results are written in boldface. For each algorithm, the type of variation is indicated as a superscript. The MIDEA algorithms are indicated by a single $\mathbb{M}$ symbol. For all tested MIDEA algorithms, the subscript indicates whether only a single cluster was used or whether clustering was performed in either the parameter space or the objective space. The population sizes that led to the best performance, are tabulated in Figs. 15 and 16. For the standard deviations, we refer the interested reader to a technical report [6]. Although the average behavior is the most interesting, the standard deviations are vital to determine whether the differences in the average behavior of the different algorithms are significant. To investigate these significances, we have performed Aspin-Welch-Satterthwaite (AWS) statistical hypothesis T-tests at a significance level of $\alpha = 0.05$. The AWS T-test is a statistical hypothesis test for the equality of means in which the equality of variances is not assumed [21]. For each problem, we verified for each pair of algorithms whether the average obtained performance indicator values differ significantly. We assigned a value of 1 if an algorithm scored significantly better and a value of -1 if an algorithm scored significantly worse. We summed the so obtained matrices over all problems to get the statistically significant improvement matrices that are shown in Figs. 17 through 19. We also computed the sum for each algorithm of its significant improvement values over all other algorithms to indicate the summed relative statistically significant performance of the algorithms. A less detailed summary of the statistical significance tests is shown in Fig. 21. In this figure histograms are used to indicate the sum of the results of the statistical significance tests for each algorithm compared with all other algorithms. The histogram represents the sums for the real-valued problems and the combinatorial problems for the different tested dimensionalities and the average of these four sums.

Influence of Problem Dimensionality

Although the MIDEA variants already mostly outperform the other tested algorithms in the case in which the dimensionality of the problem is smaller ($l = 10$ for the real-valued problems, $l = 100$ for the binary problems), they perform even better in the case in which the dimensionality of the problem is larger. This is most likely due to

	$D_{\mathcal{P}_F \to S}$							
EA	BT_1^{10}	ZDT_4^{10}	ZDT_6^{10}	CTP_7^{10}	BT_1^{100}	ZDT_4^{100}	ZDT_6^{100}	CTP_7^{100}
SPEA^{UX}	$100 \cdot 10^5$	4.62	0.193	7.97	$100 \cdot 10^5$	470	7.64	499
SPEA^{1X}	$100 \cdot 10^5$	3.90	0.172	7.31	$100 \cdot 10^5$	447	7.06	476
NSGA-II^{UX}	$100 \cdot 10^5$	4.39	0.303	7.25	$100 \cdot 10^5$	360	5.99	348
NSGA-II^{1X}	$100 \cdot 10^5$	**1.40**	0.328	**3.32**	$100 \cdot 10^5$	297	6.59	303
M$_{1\ Cluster}^{UX}$	$100 \cdot 10^5$	4.43	0.358	6.63	$100 \cdot 10^5$	374	6.72	378
M$_{1\ Cluster}^{1X}$	$100 \cdot 10^5$	1.89	0.291	4.13	$100 \cdot 10^5$	336	6.81	345
M$_{Par.\ Clust.}^{UX}$	$100 \cdot 10^5$	4.01	0.368	6.42	$100 \cdot 10^5$	400	6.98	394
M$_{Par.\ Clust.}^{1X}$	$100 \cdot 10^5$	1.65	0.298	3.77	$100 \cdot 10^5$	332	7.01	340
M$_{Obj.\ Clust.}^{UX}$	$100 \cdot 10^5$	3.98	0.354	7.27	$100 \cdot 10^5$	311	5.96	326
M$_{Obj.\ Clust.}^{1X}$	$100 \cdot 10^5$	2.03	0.311	3.95	$100 \cdot 10^5$	328	6.74	335
M$_{1\ Cluster}^{Univariate}$	$100 \cdot 10^5$	14.0	1.08	16.5	$100 \cdot 10^5$	774	3.06	875
M$_{1\ Cluster}^{Learning}$	$100 \cdot 10^5$	11.2	**0.00239**	15.3	$100 \cdot 10^5$	597	**0.434**	600
M$_{Par.\ Clust.}^{Univariate}$	$999 \cdot 10^4$	5.36	0.798	7.93	$100 \cdot 10^5$	168	3.70	192
M$_{Par.\ Clust.}^{Learning}$	$999 \cdot 10^4$	14.0	0.159	17.1	$100 \cdot 10^5$	416	0.470	523
naive MIDEA	$100 \cdot 10^5$	5.00	0.306	8.64	$100 \cdot 10^5$	157	4.60	**161**
M$_{Obj.\ Clust.}^{Learning}$	**$998 \cdot 10^4$**	11.5	0.287	12.6	**$100 \cdot 10^5$**	144	1.30	165

Fig. 9. Average of the $D_{\mathcal{P}_F \to S}$ performance indicator on all real-valued problems. Note: naive MIDEA could also have been abbreviated as M$_{Obj.\ Clust.}^{Univariate}$.

	$D_{\mathcal{P}_F \to S}$							
EA	MS^{100}	KN^{100}	SC^{100}	MST^{105}	MS^{1000}	KN^{1000}	SC^{1000}	MST^{1035}
SPEA^{UX}	12.7	10.4	2.93	2.10	181	83.9	550	6.78
SPEA^{1X}	11.8	9.14	2.99	2.12	270	105	484	6.40
NSGA-II^{UX}	11.5	8.29	1.79	1.88	180	76.4	289	7.15
NSGA-II^{1X}	11.7	9.33	2.64	2.22	283	114	360	6.60
M$_{1\ Cluster}^{UX}$	9.65	6.20	**0.931**	2.76	80.4	52.3	**72.4**	5.14
M$_{1\ Cluster}^{1X}$	12.4	7.34	1.9	2.72	135	93.0	109	4.66
M$_{Par.\ Clust.}^{UX}$	10.6	6.96	1.23	2.69	104	58.8	75.4	5.42
M$_{Par.\ Clust.}^{1X}$	13.4	8.13	1.54	2.86	169	107	101	4.96
M$_{Obj.\ Clust.}^{UX}$	7.50	**3.71**	1.49	**1.30**	69.0	**18.8**	189	3.33
M$_{Obj.\ Clust.}^{1X}$	10.5	5.98	1.89	1.54	116	46.3	305	3.11
M$_{1\ Cluster}^{Univariate}$	18.8	16.4	1.48	3.18	141	117	76.5	9.60
M$_{1\ Cluster}^{Learning}$	11.4	7.25	1.50	2.70	262	77.6	94.2	5.89
M$_{Par.\ Clust.}^{Univariate}$	18.3	13.2	1.54	3.26	168	118	105	9.68
M$_{Par.\ Clust.}^{Learning}$	12.5	7.56	1.85	2.54	262	115	269	7.69
naive MIDEA	**7.20**	4.32	1.24	1.54	**36.9**	28.1	181	3.58
M$_{Obj.\ Clust.}^{Learning}$	9.37	5.91	2.52	1.72	52.4	37.4	650	**2.64**

Fig. 10. Average of the $D_{\mathcal{P}_F \to S}$ performance indicator on all combinatorial problems. Note: naive MIDEA could also have been abbreviated as M$_{Obj.\ Clust}^{Univariate}$

	Front Spread **FS**							
EA	BT_1^{10}	ZDT_4^{10}	ZDT_6^{10}	CTP_7^{10}	BT_1^{100}	ZDT_4^{100}	ZDT_6^{100}	CTP_7^{100}
SPEA^{UX}	225	51.4	5.22	44.9	2.06	692	1.85	733
SPEA^{1X}	369	55.8	5.26	46.3	2.31	736	3.02	773
NSGA-II^{UX}	179	3.60	1.09	1.76	0.413	35.2	0.756	29.3
NSGA-II^{1X}	23.4	8.93	1.03	1.31	1.02	33.4	0.665	13.9
$M_{1\ \text{Cluster}}^{UX}$	655	8.55	2.90	39.1	2.18	395	3.43	365
$M_{1\ \text{Cluster}}^{1X}$	78.6	2.46	1.92	1.41	2.27	94.0	1.40	88.6
$M_{\text{Par. Clust.}}^{UX}$	357	12.2	5.05	4.85	2.11	384	3.10	345
$M_{\text{Par. Clust.}}^{1X}$	199	2.45	5.33	1.66	2.31	129	1.53	93.1
$M_{\text{Obj. Clust.}}^{UX}$	685	40.8	4.11	41.8	2.15	740	4.75	737
$M_{\text{Obj. Clust.}}^{1X}$	262	3.38	3.94	58.9	2.29	359	2.30	371
$M_{1\ \text{Cluster}}^{\text{Univariate}}$	293	70.8	1.15	84.7	1.82	393	0.180	347
$M_{1\ \text{Cluster}}^{\text{Learning}}$	$129 \cdot 10^1$	84.9	3.00	87.4	2.12	635	2.20	342
$M_{\text{Par. Clust.}}^{\text{Univariate}}$	$508 \cdot 10^1$	24.0	2.47	28.8	2.19	231	0.05	306
$M_{\text{Par. Clust.}}^{\text{Learning}}$	$112 \cdot 10^2$	142	5.15	116	1.91	577	7.01	588
naive MIDEA	$209 \cdot 10^1$	90.4	**5.29**	114	2.45	636	**8.10**	619
$M_{\text{Obj. Clust.}}^{\text{Learning}}$	**$164 \cdot 10^2$**	**197**	3.68	**188**	**3.28**	**$175 \cdot 10^1$**	3.97	**$183 \cdot 10^1$**

Fig. 11. Average of the **FS** performance indicator on all real-valued problems. Note: naive MIDEA could also have been abbreviated as $M_{\text{Obj. Clust.}}^{\text{Univariate}}$

	Front Spread **FS**							
EA	MS^{100}	KN^{100}	SC^{100}	MST^{105}	MS^{1000}	KN^{1000}	SC^{1000}	MST^{1035}
SPEA^{UX}	116	69.5	**64.6**	30.6	288	254	631	52.1
SPEA^{1X}	126	82.6	50.1	**32.3**	399	308	**636**	50.8
NSGA-II^{UX}	120	78.3	17.3	26.3	370	288	144	33.7
NSGA-II^{1X}	129	79.0	12.8	23.9	364	291	107	36.1
$M_{1\ \text{Cluster}}^{UX}$	132	92.6	20.7	17.8	304	285	112	40.1
$M_{1\ \text{Cluster}}^{1X}$	141	91.9	18.3	19.3	329	247	105	47.9
$M_{\text{Par. Clust.}}^{UX}$	129	90.8	20.1	18.4	265	289	125	40.7
$M_{\text{Par. Clust.}}^{1X}$	132	91.4	17.3	20.1	277	261	112	46.8
$M_{\text{Obj. Clust.}}^{UX}$	187	119	21.9	30.1	600	483	199	58.7
$M_{\text{Obj. Clust.}}^{1X}$	183	103	21.1	26.0	579	430	155	58.0
$M_{1\ \text{Cluster}}^{\text{Univariate}}$	79.2	43.3	16.1	16.9	122	98.4	10.8	22.7
$M_{1\ \text{Cluster}}^{\text{Learning}}$	143	90.0	18.2	19.7	124	214	135	37.5
$M_{\text{Par. Clust.}}^{\text{Univariate}}$	90.8	57.4	16.7	16.7	72.9	85.2	10.7	23.1
$M_{\text{Par. Clust.}}^{\text{Learning}}$	143	106	18.4	20.5	124	109	19.2	32.1
naive MIDEA	**192**	116	27.6	32.1	665	503	313	**65.2**
$M_{\text{Obj. Clust.}}^{\text{Learning}}$	191	**125**	22.4	30.3	**784**	**512**	66.2	60.2

Fig. 12. Average of the **FS** performance indicator on all combinatorial problems. Note: naive MIDEA could also have been abbreviated as $M_{\text{Obj. Clust.}}^{\text{Univariate}}$

EA	BT_1^{10}	ZDT_4^{10}	ZDT_6^{10}	CTP_7^{10}	BT_1^{100}	ZDT_4^{100}	ZDT_6^{100}	CTP_7^{100}
			Front Occupation **FO**					
SPEA^{UX}	**60.9**	99.0	50.0	43.5	49.8	27.6	18.7	26.7
SPEA^{1X}	38.7	**187**	49.6	43.2	48.8	27.4	29.3	26.8
NSGA-II^{UX}	5.42	59.7	47.5	59.3	100	5.80	6.00	4.00
NSGA-II^{1X}	29.5	32.7	31.2	9.98	75.0	5.00	6.60	3.00
$M_{1\ Cluster}^{UX}$	9.92	41.7	8.06	9.00	14.4	12.8	14.4	12.6
$M_{1\ Cluster}^{1X}$	13.4	30.3	6.52	11.9	16.5	7.10	6.64	5.94
$M_{Par.\ Clust.}^{UX}$	7.46	25.4	8.02	18.2	15.4	12.9	15.2	12.4
$M_{Par.\ Clust.}^{1X}$	9.78	24.7	7.80	11.9	17.5	7.20	8.12	6.68
$M_{Obj.\ Clust.}^{UX}$	13.9	10.0	8.48	8.62	19.1	20.0	19.6	21.7
$M_{Obj.\ Clust.}^{1X}$	9.94	31.4	7.32	15.6	17.4	12.2	9.76	12.2
$M_{1\ Cluster}^{Univariate}$	5.74	6.88	4.90	4.14	36.7	6.9	2.55	3.20
$M_{1\ Cluster}^{Learning}$	6.06	8.36	**258**	4.96	13.1	5.25	**369**	3.75
$M_{Par.\ Clust.}^{Univariate}$	29.6	98.8	30.0	**82.0**	33.4	69.4	3.70	18.3
$M_{Par.\ Clust.}^{Learning}$	52.7	65.4	104	69.2	**149**	105	92.0	**112**
naive MIDEA	12.5	68.7	56.3	34.0	64.5	**106**	27.7	78.9
$M_{Obj.\ Clust.}^{Learning}$	30.1	26.4	197	32.1	111	50.8	163	43.0

Fig. 13. Average of the **FO** performance indicator on all real-valued problems. Note: naive MIDEA could also have been abbreviated as $M_{Obj.\ Clust.}^{Univariate}$.

EA	MS^{100}	KN^{100}	SC^{100}	MST^{105}	MS^{1000}	KN^{1000}	SC^{1000}	MST^{1035}
			Front Occupation **FO**					
SPEA^{UX}	46.8	46.5	**25.1**	42.8	49.4	49.5	26.2	48.8
SPEA^{1X}	46.1	**77.6**	24.3	**93.2**	49.9	49.7	**26.5**	**95.0**
NSGA-II^{UX}	33.5	35.5	12.0	32.3	35.4	33.1	7.50	64.7
NSGA-II^{1X}	41.1	35.4	6.80	24.5	42.0	36.4	7.20	64.8
$M_{1\ Cluster}^{UX}$	100	28.1	11.3	20.8	197	46.8	12.4	25.4
$M_{1\ Cluster}^{1X}$	130	43.8	14.9	20.3	212	43.1	16.1	38.5
$M_{Par.\ Clust.}^{UX}$	112	32.2	10.6	23.7	171	46.9	13.0	26.0
$M_{Par.\ Clust.}^{1X}$	136	50.2	13.2	24.5	179	44.1	17.8	37.1
$M_{Obj.\ Clust.}^{UX}$	**165**	48.4	11.1	29.3	269	78.1	15.0	44.2
$M_{Obj.\ Clust.}^{1X}$	160	61.1	16.2	33.5	325	52.3	13.2	48.5
$M_{1\ Cluster}^{Univariate}$	56.9	15.6	8.56	17.6	37.5	20.6	3.92	16.7
$M_{1\ Cluster}^{Learning}$	105	37.5	10.0	20.9	48.5	64.2	19.2	61.0
$M_{Par.\ Clust.}^{Univariate}$	59.8	21.9	8.87	16.6	85.4	15.9	4.90	16.3
$M_{Par.\ Clust.}^{Learning}$	104	40.9	9.60	20.9	48.5	47.5	8.67	58.7
naive MIDEA	147	36.1	11.9	25.9	129	65.1	16.1	41.9
$M_{Obj.\ Clust.}^{Learning}$	143	51.8	10.0	25.3	**411**	**101**	8.0	65.9

Fig. 14. Average of the **FO** performance indicator on all combinatorial problems. Note: naive MIDEA could also have been abbreviated as $M_{Obj.\ Clust.}^{Univariate}$.

EA	Population Size n							
	BT_1^{10}	ZDT_4^{10}	ZDT_6^{10}	CTP_7^{10}	BT_1^{100}	ZDT_4^{100}	ZDT_6^{100}	CTP_7^{100}
SPEA^{UX}	50	50	25	25	25	25	25	25
SPEA^{1X}	25	100	25	25	25	25	25	25
NSGA-II^{UX}	200	200	100	100	100	200	200	150
NSGA-II^{1X}	200	375	75	300	75	200	150	300
$\mathbb{M}_{1\,Cluster}^{UX}$	75	100	25	25	100	125	200	125
$\mathbb{M}_{1\,Cluster}^{1X}$	100	450	25	300	125	325	100	175
$\mathbb{M}_{Par.\,Clust.}^{UX}$	175	75	25	100	75	125	150	175
$\mathbb{M}_{Par.\,Clust.}^{1X}$	125	450	25	275	150	175	100	175
$\mathbb{M}_{Obj.\,Clust.}^{UX}$	225	25	25	25	125	200	200	300
$\mathbb{M}_{Obj.\,Clust.}^{1X}$	150	475	25	725	125	200	100	150
$\mathbb{M}_{1\,Cluster}^{Univariate}$	150	50	75	50	100	75	375	50
$\mathbb{M}_{1\,Cluster}^{Learning}$	150	75	425	75	175	100	700	100
$\mathbb{M}_{Par.\,Clust.}^{Univariate}$	175	125	175	125	225	150	450	150
$\mathbb{M}_{Par.\,Clust.}^{Learning}$	400	250	275	250	200	150	550	125
naive $\mathbb{M}$IDEA	275	125	200	125	250	200	800	200
$\mathbb{M}_{Obj.\,Clust.}^{Learning}$	450	200	250	150	225	300	400	250

Fig. 15. Population sizes used for the real-valued problems. Note: naive $\mathbb{M}$IDEA could also have been abbreviated as $\mathbb{M}_{Obj.\,Clust.}^{Univariate}$

EA	Population Size n							
	MS^{100}	KN^{100}	SC^{100}	MST^{105}	MS^{1000}	KN^{1000}	SC^{1000}	MST^{1035}
SPEA^{UX}	25	25	25	25	25	25	25	25
SPEA^{1X}	25	50	25	125	25	25	25	50
NSGA-II^{UX}	350	325	300	200	200	200	200	250
NSGA-II^{1X}	100	325	250	200	150	250	150	200
$\mathbb{M}_{1\,Cluster}^{UX}$	575	350	550	1250	775	775	325	1000
$\mathbb{M}_{1\,Cluster}^{1X}$	550	400	300	1200	800	625	500	1050
$\mathbb{M}_{Par.\,Clust.}^{UX}$	500	525	500	2600	650	775	350	1100
$\mathbb{M}_{Par.\,Clust.}^{1X}$	525	575	425	2375	650	650	475	1200
$\mathbb{M}_{Obj.\,Clust.}^{UX}$	550	425	550	1975	750	775	775	1800
$\mathbb{M}_{Obj.\,Clust.}^{1X}$	475	425	825	1400	825	500	650	1750
$\mathbb{M}_{1\,Cluster}^{Univariate}$	700	200	450	5000	1375	800	225	800
$\mathbb{M}_{1\,Cluster}^{Learning}$	850	700	700	1850	1350	850	500	1600
$\mathbb{M}_{Par.\,Clust.}^{Univariate}$	750	600	525	7000	300	375	250	700
$\mathbb{M}_{Par.\,Clust.}^{Learning}$	1075	950	1050	1850	1350	700	700	2900
naive $\mathbb{M}$IDEA	500	300	900	2500	875	750	900	1850
$\mathbb{M}_{Obj.\,Clust.}^{Learning}$	1000	925	1050	4000	1400	1500	1100	2350

Fig. 16. Population sizes used for the combinatorial problems. Note: naive $\mathbb{M}$IDEA could also have been abbreviated as $\mathbb{M}_{Obj.\,Clust.}^{Univariate}$

Statistically Significant Improvement Matrix	SPEA^{UX}	SPEA^{1X}	$\text{NSGA-II}^{\text{UX}}$	$\text{NSGA-II}^{\text{1X}}$	$\text{M}^{\text{UX}}_{\text{1 Cluster}}$	$\text{M}^{\text{1X}}_{\text{1 Cluster}}$	$\text{M}^{\text{UX}}_{\text{Par. Clust.}}$	$\text{M}^{\text{1X}}_{\text{Par. Clust.}}$	$\text{M}^{\text{UX}}_{\text{Obj. Clust.}}$	$\text{M}^{\text{1X}}_{\text{Obj. Clust.}}$	$\text{M}^{\text{Univariate}}_{\text{1 Cluster}}$	$\text{M}^{\text{Learning}}_{\text{1 Cluster}}$	$\text{M}^{\text{Univariate}}_{\text{Par. Clust.}}$	$\text{M}^{\text{Learning}}_{\text{Par. Clust.}}$	naive MIDEA	$\text{M}^{\text{Learning}}_{\text{Obj. Clust.}}$	Sum
SPEA^{UX}	0	-8	-7	-4	-9	3	-10	3	-11	3	4	-4	-2	-1	-12	-7	-62
SPEA^{1X}	8	0	-7	-3	-8	5	-8	6	-9	5	6	-3	-1	0	-12	-8	-29
$\text{NSGA-II}^{\text{UX}}$	7	7	0	2	-3	5	-4	4	-10	4	6	-2	-2	3	-13	-7	-3
$\text{NSGA-II}^{\text{1X}}$	4	3	-2	0	-1	11	-1	9	-6	12	5	-5	0	-1	-11	-8	9
$\text{M}^{\text{UX}}_{\text{1 Cluster}}$	9	8	3	1	0	3	6	3	-8	2	11	7	6	6	-7	-6	44
$\text{M}^{\text{1X}}_{\text{1 Cluster}}$	-3	-5	-5	-11	-3	0	-1	-4	-7	-8	-4	-7	-10	-8	-11	-11	-98
$\text{M}^{\text{UX}}_{\text{Par. Clust.}}$	10	8	4	1	-6	1	0	2	-7	1	11	6	6	5	-9	-8	25
$\text{M}^{\text{1X}}_{\text{Par. Clust.}}$	-3	-6	-4	-9	-3	4	-2	0	-7	3	-4	-7	-10	-8	-10	-11	-77
$\text{M}^{\text{UX}}_{\text{Obj. Clust.}}$	11	9	10	6	8	7	7	7	0	6	8	6	2	8	-3	1	93
$\text{M}^{\text{1X}}_{\text{Obj. Clust.}}$	-3	-5	-4	-12	-2	8	-1	-3	-6	0	-4	-8	-10	-8	-11	-11	-80
$\text{M}^{\text{Univariate}}_{\text{1 Cluster}}$	-4	-6	-6	-5	-11	4	-11	4	-8	4	0	-10	-4	-7	-12	-11	-83
$\text{M}^{\text{Learning}}_{\text{1 Cluster}}$	4	3	2	5	-7	7	-6	7	-6	8	10	0	1	2	-8	-7	15
$\text{M}^{\text{Univariate}}_{\text{Par. Clust.}}$	2	1	2	0	-6	10	-6	10	-2	10	4	-1	0	-1	-6	-7	10
$\text{M}^{\text{Learning}}_{\text{Par. Clust.}}$	1	0	-3	1	-6	8	-5	8	-8	8	7	-2	1	0	-8	-6	-4
naive MIDEA	12	12	13	11	7	11	9	10	3	11	12	8	6	8	0	5	**138**
$\text{M}^{\text{Learning}}_{\text{Obj. Clust.}}$	7	8	7	8	6	11	8	11	-1	11	11	7	7	6	-5	0	102

The table is headed by $D_{\mathcal{P}_F \to \mathcal{S}}$.

Fig. 17. Number of times an improvement was found to be statistically significant in the $D_{\mathcal{P}_F \to \mathcal{S}}$ performance indicator, summed over all tested problems. The numbers in a single row indicate the summed number of significantly better or worse results compared to the algorithms in the different columns. Note: naive MIDEA could also have been abbreviated as $\text{M}^{\text{Univariate}}_{\text{Obj. Clust.}}$.

the more powerful diversity exploration and preservation in MIDEA. As the dimensionality of the problem goes up, the parameter space (i.e. the search space) becomes larger. In the case of the binary combinatorial problems, the number of solutions in the objective space becomes larger as well. If clustering in the objective space is used in MIDEA, better results are obtained on average as the dimensionality of the problem increases. In Fig. 20 the Pareto fronts over 50 runs for a selection of algorithms are plotted on one problem from each problem class and dimensionality. The better diversity preservation and proper distribution of the points along the front can be seen clearly for the problems of larger dimensionality. For the lower dimensionality problems, better diversity preservation can also be observed, which is most exemplified by the fact that MIDEA obtains non-dominated solutions at the outer ends of the front for the knapsack problem with $l = 100$.

Statistically Significant Improvement Matrix	SPEA^{UX}	SPEA^{IX}	NSGA-II^{UX}	NSGA-II^{IX}	$\mathbb{M}^{\text{UX}}_{\text{1 Cluster}}$	$\mathbb{M}^{\text{IX}}_{\text{1 Cluster}}$	$\mathbb{M}^{\text{UX}}_{\text{Par. Clust.}}$	$\mathbb{M}^{\text{IX}}_{\text{Par. Clust.}}$	$\mathbb{M}^{\text{UX}}_{\text{Obj. Clust.}}$	$\mathbb{M}^{\text{IX}}_{\text{Obj. Clust.}}$	$\mathbb{M}^{\text{Univariate}}_{\text{1 Cluster}}$	$\mathbb{M}^{\text{Learning}}_{\text{1 Cluster}}$	$\mathbb{M}^{\text{Univariate}}_{\text{Par. Clust.}}$	$\mathbb{M}^{\text{Learning}}_{\text{Par. Clust.}}$	naive $\mathbb{M}\text{IDEA}$	$\mathbb{M}^{\text{Learning}}_{\text{Obj. Clust.}}$	**Sum**
SPEA^{UX}	0	-7	8	8	2	13	5	11	-3	11	11	3	13	2	-7	-9	61
SPEA^{IX}	7	0	16	14	8	15	10	13	0	13	11	5	13	3	-6	-7	115
NSGA-II^{UX}	-8	-16	0	5	-8	3	-8	3	-15	1	3	-8	1	-6	-16	-14	-83
NSGA-II^{IX}	-8	-14	-5	0	-8	1	-8	1	-16	-1	1	-9	0	-6	-16	-14	-102
$\mathbb{M}^{\text{UX}}_{\text{1 Cluster}}$	-2	-8	8	8	0	12	3	11	-12	5	9	-2	10	-4	-16	-12	10
$\mathbb{M}^{\text{IX}}_{\text{1 Cluster}}$	-13	-15	-3	-1	-12	0	-12	-2	-13	-6	0	-11	-4	-7	-15	-14	-128
$\mathbb{M}^{\text{UX}}_{\text{Par. Clust.}}$	-5	-10	8	8	-3	12	0	12	-11	8	9	0	8	-3	-15	-11	7
$\mathbb{M}^{\text{IX}}_{\text{Par. Clust.}}$	-11	-13	-3	-1	-11	2	-12	0	-11	-6	0	-9	-5	-7	-15	-13	-115
$\mathbb{M}^{\text{UX}}_{\text{Obj. Clust.}}$	3	0	15	16	12	13	11	11	0	9	11	9	14	6	-9	-8	113
$\mathbb{M}^{\text{IX}}_{\text{Obj. Clust.}}$	-11	-13	-1	1	-5	6	-8	6	-9	0	2	-8	2	-8	-16	-12	-74
$\mathbb{M}^{\text{Univariate}}_{\text{1 Cluster}}$	-11	-11	-3	-1	-9	0	-9	0	-11	-2	0	-11	1	-13	-15	-16	-111
$\mathbb{M}^{\text{Learning}}_{\text{1 Cluster}}$	-3	-5	8	9	2	11	0	9	-9	8	11	0	12	-4	-13	-13	23
$\mathbb{M}^{\text{Univariate}}_{\text{Par. Clust.}}$	-13	-13	-1	0	-10	4	-8	5	-14	-2	-1	-12	0	-14	-15	-16	-110
$\mathbb{M}^{\text{Learning}}_{\text{Par. Clust.}}$	-2	-3	6	6	4	7	3	7	-6	8	13	4	14	0	-8	-11	42
naive $\mathbb{M}\text{IDEA}$	7	6	16	16	16	15	15	15	9	16	15	13	15	8	0	-2	**180**
$\mathbb{M}^{\text{Learning}}_{\text{Obj. Clust.}}$	9	7	14	14	12	14	11	13	8	12	16	13	16	11	2	0	172

Front Spread **FS**

Fig. 18. Number of times an improvement was found to be statistically significant in the FS performance indicator, summed over all tested problems. The numbers in a single row indicate the summed number of significantly better or worse results compared to the algorithms in the different columns. Note: naive $\mathbb{M}\text{IDEA}$ could also have been abbreviated as $\mathbb{M}^{\text{Univariate}}_{\text{Obj. Clust.}}$.

Influence of Mixtures by Clustering the Objective Space

The fact that the use of mixtures by clustering the objective space allows for enhanced diversity exploration and preservation, can also be observed by the difference between the spread obtained by $\mathbb{M}\text{IDEA}$ with crossover operators using only a single cluster versus the case in which on average four clusters are used. A wider spread of solutions is found when clustering in the objective space is enabled. Furthermore, although clustering in the parameter space is a powerful approach to enhance the learning of probabilistic models, it does not immediately lead to better results in multi-objective optimization.

Influence of the Problem Structure Exploitation Capabilities of EDAs

On the BT_1 problem, modelling interactions in $\mathbb{M}\text{IDEA}$ clearly leads to better results than those obtained by the other MOEAs. Thus, exploiting interactions can be beneficial in multi-objective optimization. For the BT_1 problem with $l = 10$, if we allow for $5 \cdot 10^5$ evaluations, the $\mathbb{M}\text{IDEA}$ variant that learns Bayesian factorizations is even

Statistically Significant Improvement Matrix	Front Occupation **FO**																
	$SPEA^{UX}$	$SPEA^{1X}$	$NSGA\text{-}II^{UX}$	$NSGA\text{-}II^{1X}$	$M_{1\,Cluster}^{UX}$	$M_{1\,Cluster}^{1X}$	$M_{Par.\,Clust.}^{UX}$	$M_{Par.\,Clust.}^{1X}$	$M_{Obj.\,Clust.}^{UX}$	$M_{Obj.\,Clust.}^{1X}$	$M_{1\,Cluster}^{Univariate}$	$M_{1\,Cluster}^{Learning}$	$M_{Par.\,Clust.}^{Univariate}$	$M_{Par.\,Clust.}^{Learning}$	naive MIDEA	$M_{Obj.\,Clust.}^{Learning}$	**Sum**
$SPEA^{UX}$	0	-3	10	12	11	16	11	16	7	16	14	5	7	-3	1	-4	116
$SPEA^{1X}$	3	0	11	13	12	16	12	16	10	16	14	7	8	-1	3	0	140
$NSGA\text{-}II^{UX}$	-10	-11	0	2	-1	8	1	6	-3	8	11	-3	2	-8	-4	-6	-8
$NSGA\text{-}II^{1X}$	-12	-13	-2	0	-3	9	-3	8	-6	7	10	-4	0	-11	-8	-11	-39
$M_{1\,Cluster}^{UX}$	-11	-12	1	3	0	10	0	10	-10	9	14	2	3	-7	-14	-10	-12
$M_{1\,Cluster}^{1X}$	-16	-16	-8	-9	-10	0	-11	-4	-11	-8	-2	-5	-14	-16	-15	-15	-160
$M_{Par.\,Clust.}^{UX}$	-11	-12	-1	3	0	11	0	10	-10	8	14	2	3	-5	-14	-12	-14
$M_{Par.\,Clust.}^{1X}$	-16	-16	-6	-8	-10	4	-10	0	-12	-4	-1	-5	-14	-16	-16	-15	-145
$M_{Obj.\,Clust.}^{UX}$	-7	-10	3	6	10	11	10	12	0	12	14	7	3	-2	-2	-8	59
$M_{Obj.\,Clust.}^{1X}$	-16	-16	-8	-7	-9	8	-8	4	-12	0	-2	-4	-13	-16	-16	-15	-130
$M_{1\,Cluster}^{Univariate}$	-14	-14	-11	-10	-14	2	-14	1	-14	2	0	-12	-7	-16	-16	-16	-153
$M_{1\,Cluster}^{Learning}$	-5	-7	3	4	-2	5	-2	5	-7	4	12	0	2	-3	-6	-8	-5
$M_{Par.\,Clust.}^{Univariate}$	-7	-8	-2	0	-3	14	-3	14	-3	13	7	-2	0	-11	-10	-9	-10
$M_{Par.\,Clust.}^{Learning}$	3	1	8	11	7	16	5	16	2	16	16	3	11	0	2	-2	115
naive MIDEA	-1	-3	4	8	14	15	14	16	2	16	16	6	10	-2	0	-3	112
$M_{Obj.\,Clust.}^{Learning}$	4	0	6	11	10	15	12	15	8	15	16	8	9	2	3	0	**134**

Fig. 19. Number of times an improvement was found to be statistically significant in the FO performance indicator, summed over all tested problems. The numbers in a single row indicate the summed number of significantly better or worse results compared to the algorithms in the different columns. Note: naive MIDEA could also have been abbreviated as $M_{Obj.\,Clust.}^{Univariate}$.

capable of finding near optimal solutions whereas the other MOEAs were observed not to be able to produce comparable results. Furthermore, if we compare the results of the MIDEA without clustering and with learning interactions with the MIDEA without clustering and also without learning interactions (i.e. $M_{1\,Cluster}^{Learning}$ vs. $M_{1\,Cluster}^{Univariate}$), exploiting interactions often leads to better results and thus enhances the quality of the multi-objective search process. However, the same can be said for clustering the objective space in general. Moreover, the much cheaper operation of clustering the objective space can lead to significant improvements, regardless of the type of recombination used inside each cluster. Concordantly, the naive MIDEA in which objective clustering is used obtains good results overall. In fact, summarized over all problems, the naive MIDEA is arguably the best algorithm that we have tested. Moreover, the naive MIDEA runs quickly, even for problems with many variables. Hence, learning dependencies between a problems' variables does not necessarily lead to advanced information about the trade-off in objective space that is the most important in multi-objective optimization problems. Clustering the objective space on the other hand does seem to help directly.

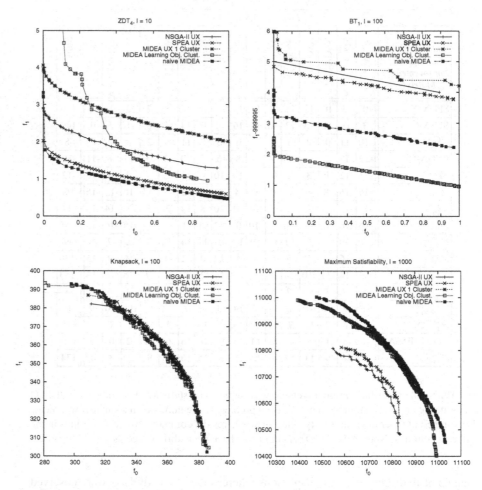

Fig. 20. Pareto fronts over 50 runs on a few of the tested problems. For clarity only a selection of all tested algorithms is shown. Note: naive MIDEA could also have been written as MIDEA Univariate Obj. Clust.

Using more advanced factorizations to further exploit a problem's structure in the form of dependencies between a problem's variables can lead to the generation of more solutions on a less preferred front. Although such an approximation set is a result that can be found more efficiently by estimating involved probability distributions instead of using classical recombination operators, such a result is intuitively less desirable. More research is required to investigate the issue of exploiting dependencies between a problem's variables in an EDA for multi-objective optimization further. On the one hand it would be interesting to attempt to overcome this problem and ensure that the added complexity of the inductive capabilities of estimating probability distributions results in a more effective exploration towards the Pareto optimal front. On the other hand it would be interesting to investigate what type of

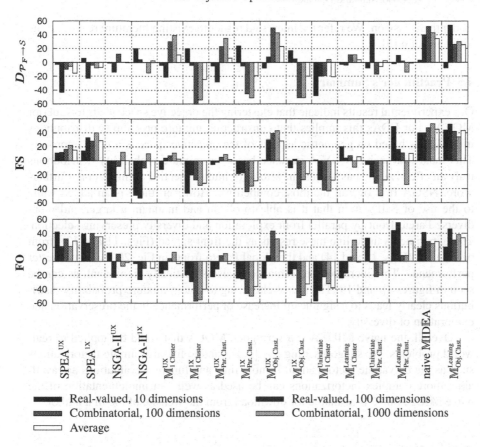

Fig. 21. A summary of the results of the statistical hypothesis tests performed for each pair of algorithms. For each algorithm, the sum of the outcome of the statistical hypothesis tests is shown for the real-valued problems and the combinatorial problems for each dimensionality separately. Furthermore, the average of these values is also shown, which serves as a global indicator of the performance of an algorithm relative to the other tested algorithms. Note: naive MIDEA could also have been abbreviated as $M_{Obj.\ Clust.}^{Univariate}$.

(real-world) multi-objective optimization problems can be solved more efficiently using MIDEA instances because of difficulties such as non-linear dependencies between the problem variables.

The Influence of δ

In our benchmarks, we have picked a specific value for $δ$. However, the $δ$ parameter is a unique parameter that determines the balance between non-domination selection pressure and diversity preservation selection pressure. Although we acknowledge the influence of this parameter, we find it outside the scope of this chapter for an in-depth discussion. We refer the interested reader to existing literature regarding the influence of δ [5].

4.5 Practitioner's Summary

Our experimental results indicate that clustering the objective space leads to superior MOEAs. For EDAs this implies that constructing mixture probability distributions in MIDEAs based on geometric aspects of the objective space is a good approach. This makes the naive MIDEA instance based on mixture probability distributions truly an effective and easy-to-use new tool for multi-objective optimization. Furthermore, NSGA-II is overall the most competitive. However, there is an added value to the use of MIDEA in that it is able to obtain and maintain a larger and more diverse Pareto front by parallel front exploration and diversity preserving selection. The experiments underline these results as the front spread (Figs. 11 and 12), front occupation (Figs. 13 and 14) and the global Pareto fronts in Fig. 20 indicate a better performance. This increased performance is also statistically significant, as can be seen in figures 18 and 19. The use of clustering to obtain mixture probability distributions clearly leads to a significant increase of performance in the preservation and exploration of diversity.

Overall, the naive MIDEA is a very good MOEA that could be applied to real-world problems. We suggest setting $\delta \in [1; 1\frac{1}{2}]$ and to first use simple factorizations such as the univariate factorization. If more time and function evaluations are available, more complex factorizations can be used as well. An implementation of the naive MIDEA in C is available for download from the website of the first author.

5 Conclusions

In this paper we have presented the naive MIDEA for multi-objective optimization. The naive MIDEA clusters the selected solutions in the objective space, after which it estimates a univariate factorization in each cluster separately. New solutions are then drawn from the so-obtained mixture probability distribution. The naive MIDEA is a specific instance of the algorithmic framework MIDEA which is a general form of an EDA for multi-objective optimization in which a probabilistic model is learned. For the specific task of multi-objective optimization, the use of mixture distributions obtained by clustering the objective space has been observed to stimulate the desirable parallel exploration along the Pareto front. The naive MIDEA has only little computational overhead since clustering in the objective space can be done very fast as can the estimation of a univariate factorization. Furthermore, although no further exploitation of dependencies between a problem's variables is used in the naive MIDEA, the results obtained for the naive MIDEA are already superior to results obtained with algorithms in which clustering the objective space is not used. Concluding, the naive MIDEA has been found to be a fast, easy-to-use and effective tool for multi-objective optimization.

References

1. P. A. N. Bosman and D. Thierens. Advancing continuous IDEAs with mixture distributions and factorization selection metrics. In M. Pelikan and K. Sastry, editors, *Proceedings of the Optimization by Building and Using Probabilistic Models OBUPM Workshop at the GECCO-2001 Genetic and Evolutionary Computation Conference*, pp. 208–212. Morgan Kaufmann Publishers, 2001.

2. P. A. N. Bosman and D. Thierens. Exploiting gradient information in continuous iterated density estimation evolutionary algorithms. In B. Kröse, M. de Rijke, G. Schreiber, and M. van Someren, editors, *Proceedings of the 13th Belgium-Netherlands Artificial Intelligence Conference BNAIC'01*, pp. 69–76, 2001.

3. P. A. N. Bosman and D. Thierens. Multi-objective optimization with diversity preserving mixture-based iterated density estimation evolutionary algorithms. *International Journal of Approximate Reasoning*, 31:259–289, 2002.

4. P. A. N. Bosman and D. Thierens. Permutation optimization by iterated estimation of random keys marginal product factorizations. In J. J. Merelo, P. Adamidis, H.-G. Beyer, J.-J. Fernández-Villicañas, and H.-P. Schwefel, editors, *Parallel Problem Solving from Nature - PPSN VII*, pp. 331–340, Berlin, 2002. Springer-Verlag.

5. P. A. N. Bosman and D. Thierens. The balance between proximity and diversity in multi-objective evolutionary algorithms. *IEEE Transactions on Evolutionary Computation*, 7:174–188, 2003.

6. P.A.N. Bosman and D. Thierens. A thorough documentation of obtained results on real-valued continuous and combinatorial multi-objective optimization problems using diversity preserving mixture-based iterated density estimation evolutionary algorithms. Technical report UU-CS-2002–52, Institute of Information and Computing Sciences, Utrecht University, Utrecht, 2002.

7. W. Buntine. Operations for learning with graphical models. *Journal of Artificial Intelligence Research*, 2:159–225, 1994.

8. C. K. Chow and C. N. Liu. Approximating discrete probability distributions with dependence trees. *IEEE Transactions on Information Theory*, 14:462–467, 1968.

9. C. A. Coello Coello. A comprehensive survey of evolutionary-based multiobjective optimization techniques. *Knowledge and Information Systems. An International Journal*, 1(3):269–308, 1999.

10. A. P. Dawid and S. L. Lauritzen. Hyper Markov laws in the statistical analysis of decomposable graphical models. *Annals of Statistics*, 21:1272–1317, 1993.

11. K. Deb. Multi-objective genetic algorithms: Problem difficulties and construction of test problems. *Evolutionary Computation*, 7(3):205–230, 1999.

12. K. Deb, S. Agrawal, A. Pratab, and T. Meyarivan. A fast elitist non-dominated sorting genetic algorithm for multi-objective optimization: NSGA-II. In M. Schoenauer, K. Deb, G. Rudolph, X. Yao, E. Lutton, J. J. Merelo, and H.-P. Schwefel, editors, *Parallel Problem Solving from Nature - PPSN VI*, pp. 849–858. Springer, 2000.

13. K. Deb, A. Pratap, and T. Meyarivan. Constrained test problems for multi-objective evolutionary optimization. In E. Zitzler, K. Deb, L. Thiele, C. A. Coello Coello, and D. Corne, editors, *First International Conference on Evolutionary Multi-Criterion Optimization*, pp. 284–298, Berlin, 2001. Springer-Verlag.

14. A.P. Dempster, N.M. Laird, and D.B. Rubin. Maximum likelihood from incomplete data via the EM algorithm. *Journal of the Royal Statistic Society*, Series B 39:1–38, 1977.

15. M. Ehrgott and X. Gandibleux. An annotated bibliography of multi-objective combinatorial optimization. Technical Report 62/2000, Fachbereich Mathematik, Universität Kaiserslautern, Kaiserslautern, 2000.

16. C. M. Fonseca and P. J. Fleming. An overview of evolutionary algorithms in multiobjective optimization. *Evolutionary Computation*, 3(1):1–16, 1995.

17. D. E. Goldberg. *Genetic Algorithms in Search, Optimization, and Machine Learning*. Addison-Wesley, Reading, 1989.

18. G. Harik. Linkage learning via probabilistic modeling in the ECGA. IlliGAL Technical Report 99010, 1999.

19. G. Harik, F. Lobo, and D. E. Goldberg. The compact genetic algorithm. In *Proceedings of the 1998 IEEE International Conference on Evolutionary Computation*, pp. 523–528. IEEE Press, 1998.

20. J. A. Hartigan. *Clustering Algorithms*. John Wiley & Sons, Inc., 1975.

21. M.G. Kendall and A. Stuart. *The Advanced Theory Of Statistics, Volume 2, Inference And Relationship*. Charles Griffin & Company Limited, 1967.

22. J. Knowles and D. Corne. On metrics for comparing non-dominated sets. In *Proceedings of the 2002 Congress on Evolutionary Computation CEC 2002*, pp. 666–674, Piscataway, New Jersey, 2002. IEEE Press.

23. P. Larrañaga and J. A. Lozano. *Estimation of Distribution Algorithms. A New Tool for Evolutionary Computation*. Kluwer Academic Publishers, 2001.

24. M. Laumanns, E. Zitzler, and L. Thiele. On the effects of archiving, elitism, and density based selection in evolutionary multi-objective optimization. In E. Zitzler, K. Deb, L. Thiele, C. A. Coello Coello, and D. Corne, editors, *Proceedings of the First International Conference on Evolutionary Multi-Criterion Optimization - EMO 2001*, pp. 181–197. Springer-Verlag, 2001.

25. S. L. Lauritzen. *Graphical Models*. Clarendon Press, Oxford, 1996.

26. H. Mühlenbein and T. Mahnig. FDA - a scalable evolutionary algorithm for the optimization of additively decomposed functions. *Evolutionary Computation*, 7:353–376, 1999.

27. H. Mühlenbein and G. Paaß. From recombination of genes to the estimation of distributions I. binary parameters. In A. E. Eiben, T. Bäck, M. Schoenauer, and H.-P. Schwefel, editors, *Parallel Problem Solving from Nature - PPSN V*, pp. 178–187. Springer, 1998.

28. A. Ochoa, H. Mühlenbein, and M. Soto. A factorized distribution algorithm using single connected Bayesian networks. In M. Schoenauer et al., editor, *Parallel Problem Solving from Nature - PPSN VI*, pp. 787–796. Springer, 2000.

29. M. Pelikan and D. E. Goldberg. Escaping hierarchical traps with competent genetic algorithms. In L. Spector, E. D. Goodman, A. Wu, W. B. Langdon, H.-M. Voigt, M. Gen, S. Sen, M. Dorigo, S. Pezeshk, M. H. Garzon, and E. Burke, editors, *Proceedings of the GECCO-2001 Genetic and Evolutionary Computation Conference*, pp. 511–518. Morgan Kaufmann, 2001.

30. M. Pelikan, D. E. Goldberg, and F. Lobo. A survey of optimization by building and using probabilistic models. *Computational Optimization and Applications*, 21(1):5–20, 2002.

31. M. Pelikan, D. E. Goldberg, and K. Sastry. Bayesian optimization algorithm, decision graphs and Occam's razor. In L. Spector, E. D. Goodman, A. Wu, W. B. Langdon, H.-M. Voigt, M. Gen, S. Sen, M. Dorigo, S. Pezeshk, M. H. Garzon, and E. Burke, editors, *Proceedings of the GECCO-2001 Genetic and Evolutionary Computation Conference*, pp. 519–526. Morgan Kaufmann, 2001.

32. R. Santana, A. Ochoa, and M. R. Soto. The mixture of trees factorized distribution algorithm. In L. Spector, E. D. Goodman, A. Wu, W. B. Langdon, H.-M. Voigt, M. Gen, S. Sen, M. Dorigo, S. Pezeshk, M. H. Garzon, and E. Burke, editors, *Proceedings of the GECCO-2001 Genetic and Evolutionary Computation Conference*, pp. 543–550. Morgan Kaufmann, 2001.

33. M. Soto and A. Ochoa. A factorized distribution algorithm based on polytrees. In *Proceedings of the 2000 Congress on Evolutionary Computation CEC00*, pp. 232–237. IEEE Press, 2000.

34. D. A. Van Veldhuizen. *Multiobjective Evolutionary Algorithms: Classifications, Analyses, and New Innovations*. PhD thesis, Graduate School of Engineering of the Air Force Institute of Technology, WPAFB, Ohio, 1999.

35. E. Zitzler. *Evolutionary Algorithms for Multiobjective Optimization: Methods and Applications*. PhD thesis, Swiss Federal Institute of Technology (ETH), Zurich, Switzerland, 1999.

36. E. Zitzler, K. Deb, and L. Thiele. Comparison of multiobjective evolutionary algorithms: Empirical results. *Evolutionary Computation*, 8(2):173–195, 2000.

37. E. Zitzler, M. Laumanns, L. Thiele, C. M. Fonseca, and V. Grunert da Fonseca. Why quality assessment of multiobjective optimizers is difficult. In W. B. Langdon, E. Cantú-Paz, K. Mathias, R. Roy, D. Davis, R. Poli, K. Balakrishnan, V. Honavar, G. Rudolph, J. Wegener, L. Bull, M. A. Potter, A. C. Schultz, J. F. Miller, E. Burke, and N. Jonoska, editors, *Proceedings of the GECCO-2002 Genetic and Evolutionary Computation Conference*, pp. 666–674, San Francisco, California, 2002. Morgan Kaufmann.

38. E. Zitzler and L. Thiele. Multiobjective evolutionary algorithms: A comparative case study and the strength Pareto approach. *IEEE Transactions on Evolutionary Computation*, 3(4):257–271, 1999.

A Parallel Island Model for Estimation of Distribution Algorithms

Julio Madera[1], Enrique Alba[2], and Alberto Ochoa[3]

[1] Department of Computing, Camagüey University, Circunvalación Norte km. $5^{1/2}$, Camagüey, Cuba
jmadera@inf.reduc.edu.cu

[2] Department of Languages and Computer Science, Málaga University, Campus de Teatinos (3-2-12), 29071, Málaga, Spain
eat@lcc.uma.es

[3] Institute of Cybernetics, Mathematics and Physics, Calle 15 No. 551 e/ C y D, 10400, La Habana, Cuba
ochoa@icmf.inf.cu

Summary. In this work we address the parallelization of the kind of Evolutionary Algorithms (EAs) known as Estimation of Distribution Algorithms (EDAs). After an initial discussion on the types of potentially parallel schemes for EDAs, we proceed to design a distributed island version (dEDA), aimed at improving the numerical efficiency of the sequential algorithm in terms of the number of evaluations. After evaluating such a dEDA on several well-known discrete and continuous test problems, we conclude that our model clearly outperforms existing centralized approaches from a numerical point of view, as well as speeding up the search considerably, thanks to its suitability for physical parallelism.

1 Introduction

Estimation of Distribution Algorithms are a relatively recent type of optimization and learning techniques based on the concept of using a population of tentative solutions to iteratively approach the problem region where the optimum is located [21, 27]. EDAs are often listed as a kind of evolutionary algorithms in which an initial population of individuals, each one encoding a possible solution to the problem, is iteratively improved by the application of stochastic operators. Every individual encodes a solution that is weighted with respect to the others by assigning a fitness value according to the objective function being optimized.

Just as in other areas of learning and optimization, reducing the cost of the search process is a critical issue in EDAs. This cost is usually measured as the number of evaluations of the objective function. But reducing the wall-clock time is also very important in real world applications, in which time consuming operations lead to unaffordable computation times. A combined reduction (numerical-plus-physical)

J. Madera et al.: *A Parallel Island Model for Estimation of Distribution Algorithms*, StudFuzz **192**, 159–186 (2006)
www.springerlink.com

in the cost of the algorithm will permit researchers to address the many classes of complex problems that usually appear in academy and especially in industry.

Thus, our work is motivated by an apparently simple question: how could we reduce the number of total fitness function evaluations of an EA? Ideas for reducing the total number of function evaluations include, for example, the use of hybrid techniques combining global and local search. Also, decentralized algorithms such as distributed evolutionary algorithms and cellular genetic algorithms [2] can alleviate the problem of a large numerical effort. In fact, decentralized algorithms can be later parallelized to obtain a still higher degree of numerical and real time efficiency. This question relates to the term *Low Cost Evolutionary Algorithm* (LCEA) that was introduced in [30,31]. With this term we label a set of features that lead the considered strategy to:

- learn and use the probabilistic structure of the problem,
- learn "appropriate" evaluation functions,
- make partial evaluations of individuals, and
- use parallel and distributed techniques.

These directives are targeted to create efficient algorithms, in a similar way as Goldberg has recently defined *competent GA* [15]. We focus our research interest on the first and last points of the above list, namely: in learning the probabilistic structure of the problem and also in using parallelism.

The idea behind the first line of research is that detecting and using the most important interactions among the problem variables is a really important key to achieving an efficient sampling of the solution space. The aim of the last point is related to the use of parallel and distributed architectures to reduce the number of function evaluations (numerical point of view). Another motivation for parallelism is to reduce the computational cost of each EDA step (physical point of view).

The contributions of this paper are the discussion of distributed alternatives for EDAs and the proposal of a concrete distributed algorithm that outperforms its sequential version. Unlike many existing works, we deal in this study both with discrete and continuous benchmarks. The algorithm proposed here reduces drastically the number of function evaluations, as well as the run time.

The paper is structured as follows. Section 2 examines the state of the art in EDAs from the point of view of their parallelization. Section 3 reviews the parallel techniques available to design new EDA algorithms. Section 4 presents the parallel distributed version of the Univariate Marginal Distribution Algorithm (dUMDA) developed here. Then, we present the benchmark used in our analysis in Sect. 5. Later, in Sect. 6, we show some experimental results obtained from the analysis of two distributed versions of dUMDA over discrete and continuous domains. Finally, Sect. 7 contains some concluding remarks and discusses future work issues.

2 EDAs and Parallelism: State-of-the-Art

This section revisits the most relevant concepts concerning EDAs and parallelism needed to understand the present work. The general Estimation of Distribution Algorithm, or EDA, [25] can be defined conceptually as shown in Algorithm 5.

Algorithm 5 EDA

Set $t \leftarrow 1$;
Generate $N >> 0$ points randomly;
while termination criteria are not met **do**
    Select $M \leq N$ points according to a selection method;
    Estimate the distribution $p^s(x, t)$ of the selected set;
    Generate N new points according to the distribution $p^s(x, t)$;
    Set $t \leftarrow t + 1$;
end while

The chief step in this algorithm is to estimate $p^s(x, t)$ and to generate new points according to this distribution. This represents a clear difference with respect to other evolutionary algorithms that use recombination and/or mutation operators to compute a new population of tentative solutions. Since the results of the EDA depend on how the mentioned probability distribution is estimated, graphical models have became common tools capable of efficiently representing the probability distribution. Some authors [20, 26, 32, 36] have proposed Bayesian networks to represent the probability distribution for discrete domains, while Gaussian networks are usually employed for continuous domains [19]. The reader can find in [21] references to some popular implementations of the different EDA families, namely EBNA, EMNA, BOA, PADA, etc.

We will distinguish several levels at which an EDA could be parallelized:

- estimation of probability distribution level,
- sampling of new individuals level,
- population level,
- fitness evaluation level, and
- any combination of the above levels.

The first level, that is, learning of the model (either Bayesian or Gaussian) can be achieved in parallel. In general, learning Bayesian networks is an NP-hard problem [11], because learning a model requires exponentially increasing computational resources. Many algorithms for learning the probability distribution use a score+search procedure. These methods define a metric that measures the goodness of every candidate Bayesian network with respect to a database of cases. In addition, a search procedure to move efficiently through the space of possible networks is needed. In these algorithms, a single-link lookahead search is commonly adopted for efficiency. Xiang and Chu [12] have studied the parallel learning of belief networks. They propose an algorithm to decompose the learning task for parallel processing, which is based on a multi-link lookahead search [5]. They partitioned the processors into two types: one processor is designated as the search manager, and the other ones explore the network structure (see [12] for details). Following this idea, the authors of [22] propose an EDA that learns the probability distribution of the selected points in parallel. They concentrate on the property of decomposing the score BIC (Bayesian Information Criteria), i.e., the score can be calculated as the sum of the separate local BIC_i scores for the variables [35]. Similarly, in [6] the author extends this result applying two different parallelism techniques, shared memory with multithreads programming and messages passage between processes.

Mendiburu et. al. [24] implement an extension to the algorithm developed in [6, 22], that consists in realizing the generation of new individuals in a distributed form. Each slave receives from the manager process a variable order and the probabilities, to generate a portion of the population and send it to the master. As its ancestors, it suffers the problem of the communication highest costs.

Ocenásek and Schwarz proposed a different algorithm for computing the Bayesian network in parallel; this parallel algorithm is applied to BOA [32] and the resulting technique is called PBOA [29]. In this algorithm, the authors used explicit topological ordering of variables to keep the model acyclic.

One distributed version was implemented in [34]. A new semi naive-Bayes parallel algorithm was proposed. This algorithm is named Parallel Interval Estimation naive-Bayes (PIENB). The algorithm is based on the island model, where each island contains a different population evolving in isolation for a number of generations, and after a predetermined scheme of migration is applied, the islands interchange the best individuals with their newborn.

The second level is another hot topic, because few works have been proposed to achieve parallel sampling of the new individuals. In the Bivariate Marginal Distribution Algorithm [33], Pelikan proposes an algorithm for the generation of new individuals using the dependency graph. The algorithm is well suited for parallelization because the generation of different individuals is independent from each other. However, this algorithm has not been yet implemented by the authors. BMDA is mentioned just as an example, because many other EDAs show the same independency in the phase of generation of individuals.

The population-based parallel approach (third level) finds its inspiration in the observation that natural populations show a spatial structure. As a result, the so-called *demes* could be used to separately evolve a solution. Demes are semi-independent groups of individuals or subpopulations loosely coupled to other neighbor demes. This coupling takes the form of the migration or diffusion of some individuals from one deme to another. This technique admits an easy parallelization that has been largely investigated in the field of EAs (see [2, 10]), and the existing findings can be also applied to the parallelization of EDAs. In this paper, we deal with this approach; and in the following sections we will discuss how to use it in the distributed Univariate Marginal Distribution Algorithm (dUMDA) for discrete and continuous domains.

Parallelization at the fitness evaluation level (the fourth level) does not require any change in the standard EDA, since the fitness of an individual is independent of the rest of the population and can be computed in parallel in a farming style. Moreover, in many real-world problems, the calculation of the individual's fitness is by far the most time consuming step of the algorithm. This is also a necessary condition for the success of such a kind of parallelism, in order to keep the communication time small to the computation time. In this case, an obvious approach is to evaluate each individual fitness simultaneously on a different processor. A *master* process manages the population and hands out individuals to evaluate to a number of *slave* processes. After the evaluation, the master collects the results and applies the necessary operations to produce the next generation.

The last level, known as the hybrid level, is a combination of different parallelization methods. For example, we could have an algorithm that uses parallelization at the population level conjugated with a farming model for the evaluation of the individuals.

3 Parallel Evolutionary Algorithms

In this section we focus on the parallelization at the population level, what represents the main research line of this paper. First, we present the parallel architectures used to implement these algorithms, and then we analyze the so-called *coarse-grained* or *distributed* parallel EA.

3.1 Parallel Architectures

In this subsection we will present some widely used parallel architectures and their influence in the implementation of a parallel EA. Nowadays, the most popular type of parallel configuration is a distributed system, a kind of Multiple Instruction Multiple Data streams computer after the well-known Flynn's taxonomy. A distributed system is composed of separate computers (usually workstations or PCs) interconnected by a high-speed network. This world-wide trend of building parallel machines as networks of computers immediately suggests the application of coarse-grained parallelism, which is characterized by a high computation/communication ratio.

One distinguished parallel algorithm that successfully exploits these architectures efficiently is the coarse-grained parallel genetic algorithm (also called distributed, multi-population or island model) [37]. This kind of algorithm has multiple populations interconnected in a particular topology (ring, hyper-cube, etc.), performing sparse migrations of information (usually individuals) among its component islands. Our present study is based on this type of algorithm (see Fig. 1).

We should notice that the World Wide Web and the existing Wide Area Network technologies provide an important infrastructure for distributed computation, that could be used as a platform to implement coarse-grained parallel EAs. This is usually referred to as *grid computing* [14], a very interesting topic of research.

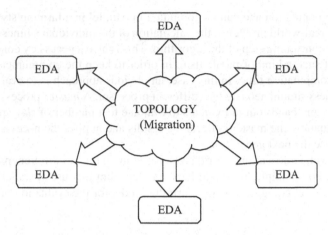

Fig. 1. Distributed Estimation of Distribution Algorithm (dEDA)

3.2 Coarse-Grained Parallel Evolutionary Algorithms

The coarse-grain computational EA model has been largely studied in the EA community, as well as in other branches of optimization and learning. The island model [13] features geographically separated sub-algorithms each one having its own subpopulation of a relatively large size. These subalgorithms may exchange information with a given frequency, e.g., by allowing some individuals to migrate from one island to another. The main idea of this approach is to periodically re-inject diversity into subpopulations which would otherwise converge prematurely. As could be expected, different islands will tend to explore different portions of the search space in parallel, and to provide independent solutions to the same problem [38]. Within each subpopulation, a standard sequential EA is usually executed between migration phases.

The algorithms having multiple populations must be tuned because they are controlled by several new parameters that affect their efficiency and precision. Among other things, we must decide the number and size of the subpopulations, the connection topology among them, the number of migrants (alternatively it can be defined as a migration rate), the frequency of the migrations, and the criteria for selecting the migrants and the replaced individuals when the new ones arrive. The importance of these parameters in the quality of the search and its efficiency has been largely studied [3, 17, 37], although the optimal values clearly depend on the problem being solved.

We are using for the present study a unidirectional ring topology, since it is easy to implement and analyze (see a discussion on this and other topologies in [10]).

3.3 Migration Policy in a Parallel Distributed Evolutionary Algorithm

The working principles of a distributed EA include a communication phase, which is governed by a migration policy. The migration policy determines how communication is carried out by the islands of the distributed EA, and it is defined by five parameters:

- **Number of migrants** (m). It is the number of individuals to exchange among the islands, $m \in \{0, 1, 2 \ldots\}$. The value 0 means in this case no interaction at all among the subpopulations (idle search). Alternatively, this parameter could be measured as a subpopulation percentage or rate.
- **Migration frequency** (r). Number of generations in isolation, $r \in \{0, 1, 2 \ldots\}$. Alternatively, it can be measured as the number of function evaluations before migration, which is more appropriate when comparing algorithms having a different step grain (in terms of the number of evaluations).
- **Policy for selecting migrants** (S). The migrant selection can be made according to any of the selection operators available in the literature (fitness proportional, tournament, etc.), e.g., $S = \{best, random\}$. The most used are truncation (select the best) and random.
- **Policy for migration replacement** (R). It is used for integrating the incoming individual in the target subpopulation, e.g., $R = \{worst, random\}$. It decides which individuals will be replaced by the incoming migrants.
- **Synchronization.** It is a flag indicating whether the algorithm islands are performing regular blocking input/output from/to another island, or whether individuals are integrated whenever they arrive, at any moment during the search.

In practice, many useful combinations of these techniques are possible. In our implementation, the algorithm can be tested with any combination of these parameters, although for the experiments we will keep some of them fixed.

4 Parallel Estimation of Distribution Algorithms Using Islands

After discussing the different parameters affecting the parallelization of EAs, we will now move to the EDA domain. As we pointed out, the distributed approach will be used here because it allows the exploitation of clusters of machines, which is the most popular parallel platform available in labs and departments. The resulting dEDA works as reported in the Algorithm 6.

The asynchronous dEDA algorithm can be seen as the combination of d $islands$ each one executing an EDA algorithm. This is graphically depicted in Fig. 1.

We directly stress the use of asynchronous dEDAs because they lead to faster executions than synchronous ones when the component subalgorithms run the same kind of algorithm on similar processors [4].

The main idea of our asynchronous algorithm is to execute in each island an EDA algorithm, and periodically (e.g., after the generation of each new individual) to verify whether the migration step has been reached. In that case, there will be an exchange of individuals with the neighbors according to the selected topology and the rest of migration parameters. The arriving individuals replace the selected individuals; e.g., worst or random individuals are replaced by the newcomers. In our case, we select the best individuals in the source island and replace the worst

---

Algorithm 6 dEDA

$Island_i$

  Set $t \leftarrow 1$;
  Generate $N >> 0$ points randomly;
  while termination criteria are not met **do**
    Select $M \leq N$ points according to a selection method;
    Estimate the distribution $p^s(x, t)$ of the selected set;
    Generate N new points according to the distribution $p^s(x, t)$;
    // *Communication between Islands -Migration-*
    Send and receive individuals asynchronously, according to the migr. parameters;
    Set $t \leftarrow t + 1$;
  end while

---

individuals of the target neighboring island. This choice is expected to induce a larger selection pressure that will hopefully accelerate the convergence of the algorithm as a whole [9].

In a more general conception, each island in a dEDA could execute a different EDA, resulting in a *heterogeneous* dEDA, which also represents a very interesting open research line. For example, we could have one dEDA of four islands where the first one executes UMDA, the second one MIMIC, the third one EBNA, and the last one executes PADA (see Fig. 2). Each algorithm, depending on the problem, has potential advantages and weaknesses that could be conjugated with the features of the other algorithms. To deal with the differences in the execution time of each algorithm, we suggest the distributed algorithm be implemented asynchronously, in order to better exploit the power of each algorithm. In a different heterogeneous scenario, each island could execute the same base algorithm, but with different parameters.

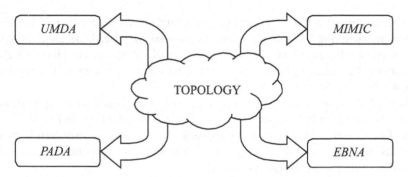

Fig. 2. Conceptual dEDA running different EDAs in each island (heterogeneity)

In this paper, our implementation is based on the execution of one asynchronous UMDA (for discrete and continuous domains) in each island, i.e., we use one homogeneous dEDA. We defer heterogeneous dEDAs for a future work, since they deserve careful consideration.

For the discrete version of dUMDAD the only difference with the above is the definition domain (discrete or continuous) of the fitness function, and the way in which the statistical model is estimated. In the discrete UMDA we estimate the (also discrete) distribution that better adjusts each variable in the selected population; in the continuous domain we estimate the mean and variance for each variable of the univariate normal distribution from the selected population of the algorithm (see Algorithm 7 for all the details).

Algorithm 7 dUMDA

$Island_i$ **executing UMDA**

Set $t \leftarrow 1$;

Generate $N >> 0$ points randomly according to the definition domain (discrete or continuous) of the fitness function;

while termination criteria are not met **do**

Select $M \leq N$ points according to a selection method;

$/*$

Estimate the distribution $p^s(x, t)$ of the selected set according to the domain

$*/$

if domain == discrete then

$/*$

Each univariate marginal distribution is estimated from marginal frequencies

$*/$

$$p^s(x, t) = \prod_{i=1}^{n} p^s(x_i, t);$$

else

$/*$

Since the univariate distributions are assumed to follow a normal distribution, the two parameters to be estimated for each variable are the mean, μ_i^t, and the standard deviation, σ_i^t

$*/$

$$p^s(x, t) = f(x, t, \mu^t, \sigma^t) = \prod_{i=1}^{n} f(x_i, t, \hat{\mu}_i^t, \hat{\sigma}_i^t);$$

and,

$$\hat{\mu}_i^t = \overline{X}_i^t = \frac{1}{N} \sum_{r=1}^{N} x_{i,r}^t \; ;$$

$$\hat{\sigma}_i^t = \sqrt{\frac{1}{N} \sum_{r=1}^{N} (x_{i,r}^t - \overline{X}_i^t)^2} \; ;$$

end if

Generate N new points according to the distribution $p^s(x, t)$ and the domain (continuous or discrete);

$/*$

Communication between Islands (*Migration*)

$*/$

Send and receive individuals asynchronously, according to the migr. parameters;

Set $t \leftarrow t + 1$;

end while

We now proceed to discuss the test benchmark we have selected to evaluate the performance of dUMDA.

5 Set of Test Functions

The set of test functions considered in our work is divided into two sub-groups: discrete problems and continuous problems. In the case of the discrete domain we study six functions, namely F_{OneMax}, $F_{Plateau}$, $F_{IsoPeak}$, $F_{Quadratic}$, F_{Muhl} and F_{Cuban1}. For the continuous problems we test our algorithms on four case of studies, namely F_{Sphere}, $F_{Griewangk}$, F_{Ackley} and F_{Water}. We selected such a benchmark to be representative and to facilitate the comparisons with other works. In all the cases, n represents the dimension of the problem.

5.1 Discrete Domain

Let us describe the functions used to test the proposed algorithms in discrete domains. For all functions $x_i \in \{0, 1\}$.

The OneMax Function

This function is defined as follows:

$$F_{OneMax}(\overrightarrow{x}) = \sum_{i=1}^{n} x_i$$

F_{OneMax} has $(n + 1)$ different fitness values, which are multinomially distributed. For *Additively Decomposed Functions* (ADFs) the multinomial distribution occurs fairly often [16] .

The objective is to maximize the function F_{OneMax}. The global optimum is located at the point $(1, 1, \ldots, 1)$.

Plateau Function

This problem was studied in [28]. It is also known as a 3-bit royal road problem. The solutions for this function consist of an n-dimensional vector, such that $n = 3 \times m$ (the genes are divided into groups of three). First, we define an auxiliary function g as:

$$g(x_1, x_2, x_3) = \begin{cases} 1, & if \quad x_1 = 1 \quad and \quad x_2 = 1 \quad and \quad x_3 = 1 \\ 0, & otherwise \end{cases}$$

Now, we can define the Plateau function as:

$$F_{Plateau}(\overrightarrow{x}) = \sum_{i=1}^{m} g(\overrightarrow{s_i})$$

Where $\overrightarrow{s_i} = (x_{3i-2}, x_{3i-1}, x_{3i})$. The goal is to maximize the function $F_{Plateau}$, and the global optimum is located at the point $(1, 1, \ldots, 1)$.

IsoPeak Function

This problem was investigated in [23]. The solutions for this function consist of an n-dimensional vector, such that $n = 2 \times m$ (the genes are divided into groups of two). First, we define two auxiliary functions $Iso1$ and $Iso2$ as:

$\overrightarrow{x}$	00	01	10	11
$Iso1$	m	0	0	$m-1$
$Iso2$	0	0	0	m

Now, we can define the IsoPeak function as:

$$F_{IsoPeak}(\overrightarrow{x}) = Iso2\,(x_1, x_2) + \sum_{i=2}^{m} Iso1\,(x_{2i-1}, x_{2i})$$

The goal is to maximize the function $F_{IsoPeak}$ and the global optimum is located at the point $(1, 1, 0, 0, \ldots, 0, 0)$.

Quadratic Function

This problem has been taken from [33]. The solution for this function is an n-dimensional vector, such that $n = 2 \times m$ (the genes are divided into groups of two). First, we define an auxiliary function g as:

$$g(u, v) = \begin{cases} 0.9, & if \quad u = 0 \quad and \quad v = 0 \\ 1.0, & if \quad u = 1 \quad and \quad v = 1 \\ 0.0, & otherwise \end{cases}$$

Now, we can define the Quadratic function as:

$$F_{Quadratic}(\overrightarrow{x}) = \sum_{i=1}^{m} g(x_{2i-1}, x_{2i})$$

The goal is to maximize the function $F_{Quadratic}$, and the global optimum is located at the point $(1, 1, \ldots, 1)$.

Cuban Function

This problem was proposed in [27]. The solution to this function is an n-dimensional vector, with $n = 4 \times m + 1$ and m odd. The definition is as follows:

$$F^3_{Cuban1}(\overrightarrow{x}) = \begin{cases} 0.595, & for \quad x = (0,0,0) \\ 0.200, & for \quad x = (0,0,1) \\ 0.595, & for \quad x = (0,1,0) \\ 0.100, & for \quad x = (0,1,1) \\ 1.000, & for \quad x = (1,0,0) \\ 0.050, & for \quad x = (1,0,1) \\ 0.090, & for \quad x = (1,1,0) \\ 0.150, & for \quad x = (1,1,1) \end{cases}$$

$$F^5_{Cuban1}(x,y,z,v,w) = \begin{cases} 4 \cdot F^3_{Cuban1}(x,y,z), & if \quad v = y \quad and \quad w = z \\ 0, & otherwise \end{cases}$$

Hence, we can define the F_{Cuban1} function as:

$$F_{Cuban1}(\overrightarrow{x}) = \sum_{i=1}^{m} F^5_{Cuban1}(x_{4i-3}, x_{4i-2}, x_{4i-1}, x_{4i}, x_{4i+1})$$

The goal is to maximize F_{Cuban1} and the global optimum is formed by alternating substrings 10000 and 00101. The first sub-string is the optima of the sub-function F^5_{Cuban1}, but the second one has only the third best value. The optimum is very difficult to reach even with local search [27].

Mühlenbein Function

This problem was proposed in [27]. The solution to this function is an n−dimensional vector, with $n = 5 \times m$. The definition follows:

$$F^5_{muhl}(\overrightarrow{x}) = \begin{cases} 3.0, & for \quad x = (0,0,0,0,1) \\ 2.0, & for \quad x = (0,0,0,1,1) \\ 1.0, & for \quad x = (0,0,1,1,1) \\ 3.5, & for \quad x = (1,1,1,1,1) \\ 4.0, & for \quad x = (0,0,0,0,0) \\ 0.0, & otherwise \end{cases}$$

Hence, we can define the F_{Muhl} function as:

$$F_{Muhl}(\overrightarrow{x}) = \sum_{i=1}^{m} F^5_{muhl}(\overrightarrow{s_i})$$

where $\overrightarrow{s_i} = (x_{5i-4}, x_{5i-3}, x_{5i-2}, x_{5i-1}, x_{5i})$.

The goal is to maximize F_{Muhl}, and the global optimum is located at point $(0, 0, \ldots, 0)$.

5.2 Continuous Problems

In this section we describe four functions broadly used in the literature for optimization [7, 20] to test the proposed algorithms in the continuous domain.

Sphere Model

This is a well-known minimization problem, used to provide a base line for comparison with other problems or algorithms. The variables x_i are defined in the interval $-600 \leq x_i \leq 600, i = 1, 2, \ldots, n$, and the fitness value for each individual is computed as follows:

$$F_{Sphere}(\overrightarrow{x}) = \sum_{i=1}^{n} x_i^2$$

The optimum fitness value $F_{Sphere}^*(\overrightarrow{x}) = 0$ is reached when all the variables equal 0.

Griewangk

This is a minimization problem. The variables x_i are defined in the interval $-600 \leq x_i \leq 600, i = 1, 2, \ldots, n$, and the fitness value for each individual is computed as follows:

$$F_{Griewangk}(\overrightarrow{x}) = 1 + \sum_{i=1}^{n} \frac{x_i^2}{4000} - \prod_{i=1}^{n} \cos\left(\frac{x_i}{\sqrt{i}}\right)$$

The optimum fitness value $F_{Griewangk}^*(\overrightarrow{x}) = 0$ is reached when all the variables equal 0.

Ackley

This minimization problem has an optimum value at $F_{Ackley}^*(\overrightarrow{x}) = 0$. This fitness value is obtained when all the variables are set to 0. The variables x_i are defined in the interval $-6.0 \leq x_i \leq 6.0, i = 1, 2, \ldots, n$. The definition of the fitness function for n dimensions is as follows:

$$F_{Ackley}(\overrightarrow{x}) = -20 \cdot \exp\left(-0.2 \cdot \sqrt{\frac{1}{n} \cdot \sum_{i=1}^{n} x_i^2}\right) - \exp\left(\frac{1}{n} \cdot \sum_{i=1}^{n} \cos(2 \cdot \pi \cdot x_i)\right)$$

Water Function

This minimization problem was proposed by [8] for two variables, having an optimum value at $F^*_{Water}(\overrightarrow{x}) = 0$. We extend this problem to n variables. The minimum fitness value is obtained when all the variables are equal to 0. The variables x_i are defined in the interval $-1.5 \leq x_i \leq 1.5, i = 1, 2, \ldots, n$. The definition of the fitness function for n dimensions is as follows:

$$F_{Water}(\overrightarrow{x}) = \sum_{i=1}^{n/2} a \cdot x_{2i-1}^2 + b \cdot x_{2i}^2 - c \cdot \cos(\alpha \cdot x_{2i-1}) - d \cdot \cos(\gamma \cdot x_{2i}) + c + d$$

Where $a = 1.0, b = 50.0, c = 3.0, d = 4.0, \alpha = 3 \cdot \pi$, and $\gamma = 4 \cdot \pi$ have been used for all test runs. This function is hard for many optimization algorithms as well as for very specialized search methods [18]. We show how dUMDAC can optimize this problem efficiently.

6 Computational Experiments

In this section, we report and discuss the results of applying two kinds of EDA on the previously explained test functions. In the case of the discrete domain UMDAD (UMDA for discrete problems) versus dUMDAD are tested. In the continuous domain, we will test the UMDAC (UMDA for continuous problems) versus dUMDAC.

The methodology that we follow is first to find the population size for a UMDA that solves the proposed problem with at least 95% out success. It is to be able to perform further comparisons for the success rate and number of function evaluations. We analyze different parameterizations of dUMDA to study the influence of the migration in the results. Our goal is also to look for the existence of a dUMDA that could show a much smaller number of evaluations of the objective function than the panmictic (single population) one. After this, a separate study on the run time is addressed in a different subsection to report the physical efficiency in a cluster of workstations.

We initially focus on the behavior of the algorithm for the F_{OneMax} problem exhaustively, and then we analyze the hypothesis that dUMDA is more efficient than the single-populated UMDA for the rest of problems. We did not introduce any special bias, and then did not search for highly tuned parameters of the decentralized version.

All the algorithms, discrete and continuous, use truncation selection with ratio 0.3 (30% of the population) without elitism (the new generation completely replaces the old one). For each problem the population size N is shown in the result tables. The algorithms stop after finding the optimum (hit) or after reaching 10^5 evaluations. The number of evaluations is averaged over 100 independent runs. For the distributed versions, the number of function evaluations is the sum of the evaluations carried out by each island. All the results (fitness evaluations and speedup) are average values over the successful runs. In the continuous case, the first generation was created by using a normal distribution and estimating μ and σ in the definition interval (a, b) for the problem variables, $\mu = \frac{a+b}{2}$ and $\sigma = \frac{b-a}{6}$. This definition allows out 99% of the generated points to be in the interval and around the center. When a value is out of the definition interval it is sampled again until a valid point is generated.

6.1 dUMDA Can Decrease the Numerical Effort

We begin the experimental section by trying to answer the following question: can our dUMDAD perform with a larger numerical efficiency with respect to UMDAD? We empirically explore the answer to this question by analyzing the behavior of these algorithms in a set of problems.

The F_{OneMax} Problem

We begin by analyzing the results related to the F_{OneMax} function. Table 1 shows the success percentage (hits) and the number of evaluations of UMDAD for four different population sizes. In the experiments, the F_{OneMax} function is defined over a vector of 1000 variables. The execution stops if the optimum has not been reached after 40000 fitness evaluations.

Table 1. Success percentage and number of evaluations (mean plus standard deviation) obtained with UMDAD for different population sizes (N) when solving the F_{OneMax} problem with 1000 variables

N	Success Percentage	Number of Evals.
400	97%	16437.11 ± 224.22
200	3%	8600 ± 200
100	0%	-
50	0%	-

Note in Table 1 that UMDAD only converges to the optimal solution with a success percentage above 95% for a population size of 400. Our goal is to reduce the reference cost attained by UMDAD with 400 individuals (97% success), by using dUMDAD algorithms of two, four and eight islands (with subpopulations of 200, 100 and 50 individuals, respectively).

Let us begin by analyzing the case of 2 islands. Figure 3 shows the percentage of success for different values of the parameters in a ring topology of two islands for the F_{OneMax} problem. We plot one line per r value, where r is the migration frequency. The number of migrants (m) varies from 5 to 60, which is approximately the size of the selected population. As we can observe, the percentage of success stays relatively high for most of the combinations of the migration parameters. With the exception of $r = 1$ (high coupling), all of them stay over 90% of success for any of the tested m

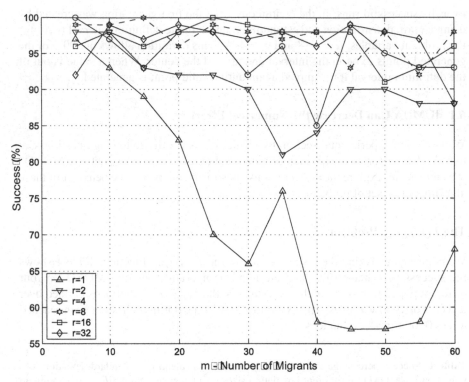

Fig. 3. Percentage of success versus number of migrants for different migration frequencies (r) and migration rates (m) in two islands connected in a ring topology for the F_{OneMax} problem

between 5 and 30. For $m \leq 20$ practically every r surpasses 95% with the exception of $r = 1$ and $r = 2$. All this means that accurate results can be obtained by enlarging the isolation time (r) or/and by enlarging the set of exchanged individuals (with the mentioned exceptions of highly coupled subalgorithms having $r = 1$ and $r = 2$). This result confirms similar experiments with different EAs reported in the past [3].

Table 2 shows in each cell the percentage of hits (%) and the average number of fitness evaluations with their standard deviation. Almost every case with $r < 8$ (specially with $m = 10$), the average number of evaluations is clearly reduced with respect to the centralized algorithm having a population size of 400 individuals. An example is for $r = 1$ and $m = 5$, where the percentage of success is as high as 97% with a saving of around 1620 evaluations.

In the case with four islands (see Table 3) we can conclude that the best (m, r) configuration is a dUMDAD using $r = 1$ and $m = 5$. This claim comes out if we inspect all the combinations where the success percentage is 95% and then select the one that executes the smaller number of fitness evaluations. However, it is clear that the algorithm is quite sensitive to the used parameterizations when $r = 1$ is used.

Table 2. Success percentage plus mean number of evaluations to converge to the optimum with the two island dUMDAD when solving the F_{OneMax} problem. Results are shown for different combinations of the number of migrants (m) and migration frequencies (r)

	$m = 5$	$m = 10$
$r = 1$	97%	93%
	14816.5 ± 237.4	13882.6 ± 226.1
$r = 2$	98%	98%
	15737.9 ± 226.8	15109.6 ± 257.9
$r = 4$	100%	97%
	16356 ± 204.1	15904.1 ± 238.8
$r = 8$	99%	99%
	16868.7 ± 280.1	16448.4 ± 230.5
$r = 16$	96%	98%
	17229.2 ± 493.3	16914.2 ± 281.3
$r = 32$	92%	99%
	18639.1 ± 2841.4	17442.4 ± 307.3

The Rest of the Discrete Problems

In the case of the other problems we exhaustively test two functions, F_{Muhl} and $F_{Plateau}$. The frequency of the migration (r) variate from 1 to 32 and the number of migrants (m) ranges from 1 to 50; we show the percentage of success in the Tables 4 and 5. Fig. 4 shows the number of function evaluation for the F_{Muhl} and $F_{Plateau}$ problems. We only show the values when the percentage of success is larger than 95%. This is due to the fact that we must select the configuration (r, m) that less function evaluations makes of all those that obtain over 95% of hits. In general, we observe that a high coupling makes the dUMDAD resemble the single-population behavior in terms of success rate but with a more reduced effort. In academic problems like F_{OneMax}, F_{Muhl} and $F_{Plateau}$ a high coupling is the better strategy. In other more complex problems larger isolation would probably be more efficient. In fact, for a physically parallel implementation high coupling is a undesirable strategy due to its higher communication overhead. We similarly set (m, r) for the other problems, with the results shown in Table 6, where we show first the percentage of hits and second the average fitness evaluations with their associated standard deviation.

The figures in Table 6 allow us to clearly conclude the higher efficiency of dUM-DAD. We must notice that all the results are statistically significant for the t-student test (p-value well below 0.05, for a 95% significance level). This algorithm is either more efficient for the same success rate than UMDAD (this holds for the first 4 out of 5 problems) or it works out a clearly higher success rate than UMDAD (see the last

Table 3. Success percentage plus mean number of evaluations to converge to the optimum with the four island dUMDAD when solving F_{OneMax} problem. Results are shown for different combinations of the number of migrants (m) and migration frequencies (r)

	$r = 1$	$r = 2$	$r = 4$	$r = 8$
$m = 1$	99%	99%	98%	97%
	16921.2 ± 293.9	18371.4 ± 411.1	20639.1 ± 944.5	25266.6 ± 2263.2
$m = 5$	95%	99%	97%	96%
	14265.2 ± 264.8	15653.1 ± 267.1	16900.0 ± 338.4	18160.0 ± 486.1
$m = 10$	76%	95%	95%	97%
	13236.8 ± 237.1	14863.8 ± 264.3	16234.1 ± 283.8	17316.6 ± 332.9
$m = 15$	62%	77%	90%	97%
	12954.8 ± 209.3	14736.0 ± 271.4	16094.3 ± 225.8	17204.1 ± 335.8
$m = 20$	58%	81%	94%	97%
	12979.3 ± 261.4	14680.0 ± 214.8	16038.7 ± 257.5	17091.6 ± 340.7
$m = 25$	53%	80%	89%	93%
	12988.6 ± 230.1	14673.4 ± 252.5	16050.0 ± 269.9	17191.3 ± 355.6
$m = 30$	51%	81%	92%	89%
	12933.3 ± 206.5	14670.0 ± 244.6	16128.8 ± 298.7	17290.9 ± 362.5

Table 4. Percentage of success for the F_{Muhl} problem

	$r = 1$	$r = 2$	$r = 4$	$r = 8$	$r = 16$	$r = 32$
$m = 1$	44%	18%	16%	7%	11%	10%
$m = 5$	95%	77%	48%	26%	21%	10%
$m = 10$	88%	96%	73%	43%	20%	17%
$m = 15$	93%	93%	87%	64%	30%	14%
$m = 20$	81%	88%	93%	77%	26%	7%
$m = 25$	84%	88%	92%	83%	42%	13%
$m = 30$	86%	90%	95%	94%	44%	19%
$m = 35$	85%	88%	91%	92%	53%	20%
$m = 40$	80%	84%	90%	95%	69%	24%
$m = 45$	72%	89%	92%	93%	79%	31%
$m = 50$	77%	86%	93%	96%	95%	85%

Table 5. Percentage of success for the $F_{Plateau}$ problem

	$r = 1$	$r = 2$	$r = 4$	$r = 8$	$r = 16$	$r = 32$
$m = 1$	100%	99%	75%	5%	0%	0%
$m = 5$	96%	98%	100%	100%	91%	55%
$m = 10$	97%	97%	99%	99%	98%	100%
$m = 15$	94%	95%	99%	97%	100%	98%
$m = 20$	97%	100%	99%	98%	99%	97%
$m = 25$	87%	99%	99%	100%	99%	100%
$m = 30$	95%	97%	97%	98%	98%	98%
$m = 35$	95%	97%	98%	98%	99%	99%
$m = 40$	94%	97%	98%	99%	94%	88%
$m = 45$	91%	96%	97%	98%	74%	37%
$m = 50$	91%	96%	96%	96%	68%	27%

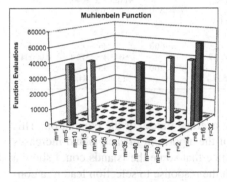 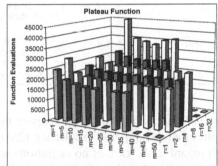

Fig. 4. Numerical effort for F_{Muhl} (*left*) and $F_{Plateau}$ (*right*) problems

function). In the first column of Table 6 we list the problem to be solved. The second column (dUMDAD header) shows for each problem, the (r, m) pair where dUM-DAD overcomes the centralized version plus the result of the run in total isolation for comparison (no migration between islands $(r = 0, m = 0)$). The third column shows the results with the above parameters (r, m). The fourth and fifth columns present the parameters and results, respectively, of running UMDAD in the problem. In all cases the results columns show the percentage of convergence, and the average number of fitness evaluations.

Why dUMDA Reduce the Numerical Effort?

In the following, we explain the reasons for these reductions (for the case of truncation selection, two islands, and $r = 1$). The initial motivation resides in that, in

Table 6. Success percentage plus mean number of evaluations to converge to the optimum with the four island dUMDAD when solving all the discrete problems. Results are shown for different combinations of the number of migrants (m), migration frequencies (r) and population sizes (N)

Functions	dUMDAD with 4-islands			UMDAD		p-value
$F_{IsoPeak}, n = 64$	$r = 1, m = 40$	98%	$N = 3200$	97%		
		45954.6 ± 2739.7		55158.7 ± 4424.7		0.0
	$r = 0, m = 0$	46%	$N = 800$	10%		
		54191.3 ± 3193.0		13360.0 ± 386.4		0.0
$F_{Plateau}, n = 600$	$r = 1, m = 30$	95%	$N = 600$	100%		
		17640 ± 627.4		22152.0 ± 550.5		0.0
	$r = 0, m = 0$	0%	$N = 150$	0%		
		-		-		-
$F_{Quadratic}, n = 66$	$r = 1, m = 12$	95%	$N = 2000$	96%		
		33452.6 ± 2448.4		34583.3 ± 2426.4		0.0015
	$r = 0, m = 0$	18%	$N = 500$	0%		
		34000.00 ± 0.00		-		-
$F_{Muhl}, n = 200$	$r = 1, m = 5$	95%	$N = 1400$	96%		
		39505.4 ± 1503.1		41183.3 ± 1646.3		0.0
	$r = 0, m = 0$	1%	$N = 350$	0%		
		43400.0 ± 0.0		-		-
$F_{Cuban1}, n = 21$	$r = 2, m = 10$	92%	$N = 800$	46%		
		5408.7 ± 521.9		5043.4 ± 372.1		0.00004
	$r = 0, m = 0$	53%	$N = 200$	45%		
		8256.6 ± 1535.4		1355.5 ± 84.1		0.0

each generation, the migration increases the average fitness of the population. This implies that the response to selection [25] of the algorithm in each island increases with regard to the case of no migration (notice that some the islands could show a larger increase than the others). The gains in the response to selection lead to a convergence acceleration in the dEDA, but the required balance between exploitation and exploration causes this to reach a limit. This analysis is valid and extensible to any other fitness function.

Let $R(t)$ be the response to selection of generation t before the migration, then:

$$R(t) = \overline{f}(t + 1) - \overline{f}(t) \tag{1}$$

where $\overline{f}(t)$ and $\overline{f}(t + 1)$ are the means of the fitness function evaluations in the population t and $t + 1$.

Let $\overline{f}_b(t + 1), \overline{f}_w(t + 1), \overline{f}_r(t + 1)$ be the respective averages of the values of the fitness function for the subsets of the best M, the worst M and the $(N - 2 \cdot M)$ remaining individuals of a population with size N. In our analysis M is equal to m, i.e., the number of individuals exchanged among the islands. Then:

$$\overline{f}(t + 1) = \frac{M \cdot \overline{f}_b(t + 1) + (N - 2 \cdot M) \cdot \overline{f}_r(t + 1) + M \cdot \overline{f}_w(t + 1)}{N} \tag{2}$$

We can assume that the values $\overline{f}_b(t+1)$ of each island are similar. This assumption is correct and largely used in the literature, because the islands are homogenous and the subsets of the best M (in each island) are an approximation of an equally fitted individuals set. Now we can write for the population average (after the migration of the best M takes place, and replace the worst M in the neighbor islands):

$$\overline{f}_{mig}(t+1) = \frac{2 \cdot M \cdot \overline{f}_b(t+1) + (N - 2 \cdot M) \cdot \overline{f}_r(t+1)}{N} \tag{3}$$

Therefore, the migration produces an increase in the response to selection equal to:

$$\Delta R_{mig}(t) = R_{mig}(t) - R(t) = \overline{f}_{mig}(t+1) - \overline{f}(t+1) = \frac{M \cdot \left(\overline{f}_b(t+1) - \overline{f}_w(t+1)\right)}{N} \tag{4}$$

Note that if M increases (larger coupling), then $\overline{f}_b(t+1)$ tends to decrease, while $\overline{f}_w(t+1)$ tends to increase. Therefore, $\left(\overline{f}_b(t+1) - \overline{f}_w(t+1)\right)$ decreases when M increases. This provokes a convergence acceleration of the algorithm. This is seen in the experimental results for F_{OneMax} (see Tables 2 and 3): when M increases first, the results improve, but later the percentage of success decreases considerably. It is important to remember that the levels of response to the selection of the algorithm without migration are not enough to obtain the convergence to an optimum with high probability. Finally, note that the magnitude $\left(\overline{f}_b(t+1) - \overline{f}_w(t+1)\right)$ depends on the problem, and thus for each problem, different (m, r) pairs could minimize the total number of fitness evaluations that the dUMDA makes.

On the other hand, if the frequency of migration r is decreased, then we will decrease also the response to selection. In this case, we would observe that the convergence decelerates, decreasing also drastically the success percentage, but with an enhanced exploration that is suitable for many complex problems. In some way, existing studies like that of [9] and [38] lead to the same conclusions by different means.

Figure 5 shows the change in the response to selection (left) and the average fitness value for the F_{OneMax} function. We must point out that the UMDAD response to selection is very similar to that of the dUMDAD with $r = 1$ and $m = 1$. In fact this is common sense, since such a high coupling resembles a centralized behavior. The right figure shows the increment of the average fitness evaluations per generation for the tested bounding cases of dUMDAD and UMDAD. This corroborates that dUMDAD with $r = 1$ and $m = 1$ can show the same selection pressure as the UMDAD. Another important result is the fact that dUMDA with high isolation ($r = \{32, 0\}$ and $m = \{1, 0\}$) represents a too slow search for F_{OneMax} and the algorithm does not converge to the solution. The most efficient behavior is detected when the algorithm uses the parameters shown in the preceding tables. This is an indication that the exploration and exploitation phases change when different parameters are used.

Figure 6 shows the algorithm behavior for another two functions ($F_{Plateau}$ left and $F_{IsoPeak}$ right). Again, the results observed in F_{OneMax} are confirmed in these problems, with special clarity in the graphic of the $F_{Plateau}$ function.

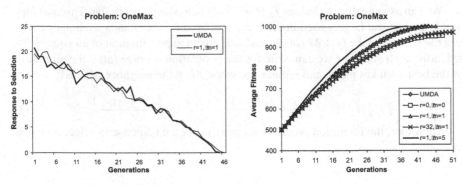

Fig. 5. Response to selection (*left*) and average fitness evaluations for the F_{OneMax} problem

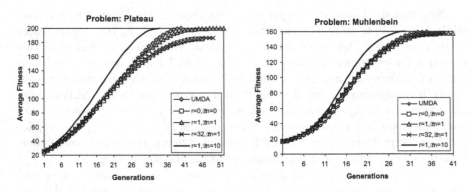

Fig. 6. Change in the average fitness for $F_{Plateau}$ (*left*) and F_{Muhl} (*right*)

Continuous Problems

After finishing the analysis with discrete problems, we encompass in this section a similar study on continuous ones. Applications of parallel or distributed EDAs to continuous domains in optimization are rare in literature, because in fact there are not many works accounting for parallelism in EDAs in general.

In the case of the continuous problems we test exhaustively two functions, F_{Sphere} and F_{Water}. The frequency of the migration (r) variate from 1 to 32 and the number of migrants (m) ranges from 1 to 50, we show the percentage of success in the Tables 7 and 8. Fig. 7 show the number of function evaluation for the F_{Sphere} and F_{water} problems, where we only show the values when the percentage of success is larger than 95%. In general, as in discrete domain, we observe that a high coupling makes dUMDAC resemble the single-population behavior in terms of success rate but with a more reduced effort. In all the results a high coupling is the better strategy. In other more complex problems larger isolation would probably be more efficient.

Table 7. Percentage of success for the F_{Sphere} problem

	$r = 1$	$r = 2$	$r = 4$	$r = 8$	$r = 16$	$r = 32$
$m = 1$	100%	100%	100%	100%	95%	77%
$m = 5$	100%	100%	100%	100%	100%	100%
$m = 10$	100%	100%	100%	100%	100%	100%
$m = 15$	100%	100%	100%	100%	100%	100%
$m = 20$	100%	100%	100%	100%	100%	100%
$m = 25$	100%	100%	100%	100%	100%	100%
$m = 30$	100%	100%	100%	100%	100%	100%
$m = 35$	100%	100%	100%	100%	100%	100%
$m = 40$	100%	100%	100%	100%	100%	100%
$m = 45$	100%	100%	100%	100%	100%	100%
$m = 50$	100%	100%	100%	100%	100%	100%

Table 8. Percentage of success for the F_{Water} problem

	$r = 1$	$r = 2$	$r = 4$	$r = 8$	$r = 16$	$r = 32$
$m = 1$	100%	100%	100%	100%	100%	100%
$m = 5$	100%	100%	100%	100%	100%	100%
$m = 10$	100%	100%	100%	100%	100%	100%
$m = 15$	100%	100%	100%	100%	100%	100%
$m = 20$	100%	100%	100%	100%	100%	100%
$m = 25$	100%	100%	100%	100%	100%	100%
$m = 30$	100%	100%	100%	100%	100%	100%
$m = 35$	100%	100%	100%	100%	100%	100%
$m = 40$	100%	100%	100%	100%	100%	100%
$m = 45$	100%	100%	100%	100%	100%	100%
$m = 50$	100%	100%	100%	100%	100%	100%

Table 9 shows the results of executing dUMDAC over four functions defined in the continuous domain. Notice that in all the cases we confirm the preceding results found on discrete problems, i.e., the dUMDAC algorithm improves the results of UMDAC. In the continuous function set the results are still more relevant: even a more clear reduction of the number of evaluations than the one noticed for the discrete problems, can be appreciated. The distributed version, independently of whether we are interested in using a parallel execution platform or not, makes a much smaller number of function evaluations.

Notice also that when dUMDAC has no migration and UMDAC is executed with the same population size of one island the percentage of convergence is very low or

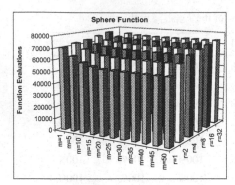
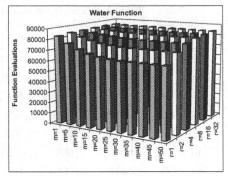

Fig. 7. Numerical effort for F_{Sphere} (*left*) and F_{Water} (*right*) problems

Table 9. Success percentage plus mean number of evaluations to converge to the optimum with the four island dUMDAD when solving all the continuous problems. Results are shown for different combinations of the number of migrants (m), migration frequencies (r) and population sizes (N)

Functions	dUMDAC with 4-islands			UMDAC		p-value
$F_{Sphere}, n = 100$	$r = 1, m = 45$	100%	$N = 600$	100%		
		55002.0 ± 330.8		75630.0 ± 421.0		0.0
	$r = 0, m = 0$	45%	$N = 150$	16%		
		78000.0 ± 3167.1		20820.0 ± 2097.1		-
$F_{Ackley}, n = 100$	$r = 1, m = 20$	97%	$N = 400$	100%		
		45430.9 ± 315.3		62832.0 ± 358.1		0.0
	$r = 0, m = 0$	0%	$N = 100$	0%		
		-		-		-
$F_{Griewangk}, n = 100$	$r = 1, m = 20$	99%	$N = 400$	100%		
		29301.0 ± 281.9		40336.0 ± 315.1		0.0
	$r = 0, m = 0$	0%	$N = 100$	0%		
		-		-		-
$F_{Water}, n = 100$	$r = 1, m = 45$	100%	$N = 800$	99%		
		64856.0 ± 498.9		128945.4 ± 2941.3		0.0
	$r = 0, m = 0$	0%	$N = 200$	0%		
		-		-		-

null. This table again corroborates the results presented in discrete domains: complete isolation is not an interesting technique from a numerically efficiency point of view. Also, like for the discrete cases, all the $p - values$ point out a high confidence in the claims, since they all resulted significant.

We want to comment apart, the case of the difficult F_{Water} problem, in which the decentralized version reduces the number of functions evaluation drastically (even more intensely than for the other problems). We selected this problem since it motivated a considerably large number of papers in the past trying to solve it with multi-start, parallel and advanced search methods. Our dUMDAC seems also an efficient alternative to traditional and enhanced mathematical algorithms.

6.2 Run Time Analysis

In this subsection, we show how dUMDA (discrete and continuous) can improve the results in terms of the execution time used to solve the same problem that UMDA. The speedup can be defined as follow:

$$Speedup = \frac{sequential\ time}{parallel\ time}$$

To evaluate the speedup we use the taxonomy proposed in [1]. This taxonomy divides the speedup analysis into two types, *strong speedup, - type I -* and *weak speedup, - type II -*. Strong speedup is intended to compare the parallel algorithm with the best-so-far sequential algorithm. The second measure compares the parallel algorithm with its own sequential version that is more pragmatic in most works. Before the algorithm runs, the stop condition is set (find an optimum) and then the speedup is measured (type II.A). In this work we select a weak speedup (type II.A.2), called *Orthodox*. This type of analysis runs the same algorithm (dUMDA) with four islands over 1, 2, and 4 processors to provide a meaningful comparison, i.e. the algorithm run in one processor is not the panmictic UMDA, but the dUMDA itself (in parallel studies it is not fair to compare time against a different algorithm, since any result could be obtained). We demonstrate that the execution time decreases significantly as more processors are used. All the tests were executed on four Pentium 4 at 2.4 GHz and 512 MB of RAM running Linux, inter-connected with a Gigabit Ethernet network.

Table 10 shows the speedup in the discrete and continuous problems. Note that in the discrete case, specially for problems needing large population sizes ($F_{IsoPeak}$ and $F_{Quadratic}$ functions) the speedup is super-linear. The reason can be found in the stochastic nature of the algorithms, that asynchronously perform the search in a different way depending on the number of processors.

For continuous problems all the results are super-linear. In this case we have the same reasons as for the discrete results, with additional reductions in times since the use of float values, because the operations are encoded in more complex expressions and larger data structures in memory are needed (that are faster to deal with in parallel when split among the processors).

7 Conclusions and Future Work

This paper has presented an asynchronous distributed implementation of an EDA algorithm that we call dUMDA, although the basic parallel model could be generalized to other EDA subtypes. The algorithm distributes the population among the available processors and make migrations among them. We have performed a very complete set of tests to analyze the numerical and wall-clock time behavior on a network of computers, and we did so for continuous and discrete test functions. All these considerations were made before engaging in the study to ensure that our conclusions are really of valuable interest for other researchers and free of bias.

Table 10. Speedup results on discrete and continuous problems

Functions	1-Processor	2-Processors	4-Processors
Discrete problems			
F_{OneMax}	1.00	1.65	2.28
$F_{IsoPeak}$	1.00	10.55	13.75
$F_{Plateau}$	1.00	2.71	3.13
$F_{Quadratic}$	1.00	6.40	7.51
F_{Muhl}	1.00	3.19	3.62
Continuous problems			
F_{Sphere}	1.00	4.94	6.43
F_{Ackley}	1.00	5.86	7.96
$F_{Griewangk}$	1.00	5.57	7.47
F_{Water}	1.00	3.35	6.82

Results show that dUMDA is able to solve problems of considerable complexity (hard problems and large population sizes), and that its capabilities can be improved with an adequate configuration of the migration policy (tuning the (m, r) pairs) with important reductions of the number of function evaluations; this can lead to obtain remarkable results also in the speedup.

Globally stated, the dUMDA algorithm is a step forward to low cost algorithms, and it shows that the decentralization of EDAs can provide fast algorithms that increase the numerical efficiency and reduce the run time.

According to these results, we will extend this analysis to study other types of EDAs (PADA, EBNA, MIMIC) to include multi-population behavior, because these models are easy to implement and exploit. We also will extend this work by applying the algorithms to more complex problems, as training neural networks, and other complex industrial problems. Also, a different next step could be the analysis of parallelization at the learning level, and the way it could be conjugated with the algorithms proposed in this paper. Of course, we will check for new evidences of the results shown in this paper with a larger set of machines in the near future.

Acknowledgements

This work has been funded by MCYT and FEDER under contract TIC2002-04498-C05-02 (the TRACER project) http://tracer.lcc.uma.es.

References

1. E. Alba. Parallel evolutionary algorithms can achieve superlinear performance. *Information Processing Letters*, 82(1):7–13, 2002.

2. E. Alba and J. M. Troya. A survey of parallel distributed genetic algorithms. *Complexity*, 4(4):31–52, 1999.

3. E. Alba and J. M. Troya. Influence of the migration policy in parallel distributed gas with structured and panmictic populations. *Applied Intelligence*, 12(3):163–181, 2000.

4. E. Alba and J. M. Troya. Analyzing synchronous and asynchronous parallel distributed genetic algorithms. *Future Generation Computer Systems*, 17(4):451–465, 2001.

5. S. Baluja. Y. xiang and s. k. wong and n. cercone, n. *A Microscopic Study of Minimum Entropy Search in Learning Decomposable Markov Networks*, 26(1):65–92, 1996.

6. E. Bengoetxea. Inexact graph matching using estimation of distribution algorithms. Technical report, Ecole Nationale Supérieure des Télécommunications, Paris, France, 2002.

7. E. Bengoetxea, T. Mikelez, J. A. Lozano, and P. Larrañaga. Experimental results in function optimization with EDAs in continuous domain. In P. Larrañaga and J. A. Lozano, editors, *Estimation of Distribution Algorithms. A New Tool for Evolutionary Computation*. Kluwer Academic Publishers, 2002.

8. I.O. Bohachevsky, M.E. Johnson, and M.L. Stein. Generalized simulated annealing for function optimization. *Technometrics*, 28(3):209–217, 1986.

9. E. Cantú-Paz. *Efficient and Accurate Parallel Genetic Algorithms*. Kluwer Academic Press, 2000.

10. E. Cantú-Paz and D. E. Goldberg. Predicting speedups of idealized bounding cases of parallel genetic algorithms. In T. Bäck, editor, *Proceedings of the Seventh International Conference on GAs*, pp. 113–120. Morgan Kaufmann, 1997.

11. D. M. Chickering, D. Geiger, and D. Heckerman. Learning Bayesian networks: Search methods and experimental results. In *Preliminary Papers of the Fifth International Workshop on Artificial Intelligence and Statistics*, pp. 112–128, 1995.

12. T. Chu and Y. Xiang. Exploring parallelism in learning belief networks. In *Proceedings of Thirteenth Conference on Uncertainty in Artificial Intelligence*, pp. 90–98, 1997.

13. J. P. Cohoon, S. U. Hedge, W. N. Martin, and D. Richards. Punctuated Equilibria: A Parallel Genetic Algorithm. In J. J. Grefenstette, editor, *Proceedings of the Second International Conference on GAs*, pp. 148–154. Lawrence Erlbaum Associates, 1987.

14. I. T. Foster and C. Kesselman. *Computational Grids*. Morgan Kaufmann, 1998.

15. D. E Goldberg. *The Design of Innovation*. Kluwer Academic Publishers, 2002.

16. D. E. Goldberg, K. Deb, H. Kargupta, and G. Harik. Rapid, accurate optimization of difficult problems using fast messy genetic algorithms. In S. Forrest, editor, *Proceedings of the Fifth International Conference on GAs*, pp. 56–64. Morgan Kaufmann, 1993.

17. J. J. Grefenstette. Parallel adaptative algorithms for function optimization. Technical Report CS-81-19, Vanderbilt University, 1981.

18. S. Höfinger, T. Schindler, and A. Aszodi. Parallel global optimization of high-dimensional problems. In *Lecture Notes in Computer Science*, pp. 148–155, 2002. 2474.

19. J.Whittaker. *Graphical Models in Applied Multivariate Statistics*. John Wiley & Sons, Inc., 1990.

20. P. Larrañaga, R. Etxeberria, J. A. Lozano, and J. M. Peña. Optimization by learning and simulation of Bayesian and Gaussian networks. Technical Report KZZA-IK-4-99, Department of Computer Science and Artificial Intelligence, University of the Basque Country, 1999.

21. P. Larrañaga and J. A. Lozano. *Estimation of Distribution Algorithms. A New Tool for Evolutionary Computation*. Kluwer Academic Publishers, 2002.

22. J.A. Lozano, R. Sagarna, and P. Larrañaga. Parallel Estimation of Distribution Algorithms. In P. Larrañaga and J. A. Lozano, editors, *Estimation of Distribution Algorithms. A New Tool for Evolutionary Computation*. Kluwer Academic Publishers, 2002.

23. T. Mahnig and H. Mühlenbein. Comparing the adaptive Boltzmann selection schedule SDS to truncation selection. In *III Symposium on Artificial Intelligence. CIMAF01. Special Session on Distributions and Evolutionary Optimization*, pp. 121–128, 2001.

24. A. Mendiburu, J. Miguel-Alonso, and J.A. Lozano. Implementation and performance evaluation of a parallelization of estimation of Bayesian networks algorithms. Technical Report EHU-KAT-IK-XX-04, Department of Computer Architecture and Technology, University of the Basque Country, 2004.

25. H. Mühlenbein. The equation for response to selection and its use for prediction. *Evolutionary Computation*, 5:303–346, 1998.

26. H. Mühlenbein and T. Mahnig. Evolutionary optimization using graphical models. *New Generation of Computer Systems*, 18(2):157–166, 2000.

27. H. Mühlenbein, T. Mahnig, and A. Ochoa. Schemata, distributions and graphical models in evolutionary optimization. *Journal of Heuristics*, 5:215–247, 1999.

28. H. Mühlenbein and D. Schlierkamp-Voosen. The science of breeding and its application to the breeder genetic algorithm (bga). *Evolutionary Computation*, 1:335–360, 1993.

29. J. Ocenásek and J. Schwarz. The parallel bayesian optimization algorithm. In *Proceedings of the European Symposium on Computational Inteligence*, pp. 61–67, 2002.

30. A. Ochoa. How to deal with costly fitness functions in evolutionary computation. In *Proceedings of the 13th ISPE/IEE International Conference on CAD/CAM. Robotics & Factories of the Future*, pp. 788–793, 1997.

31. A. Ochoa and M. Soto. Partial evaluation of genetic algorithms. In *Proceedings of the X International Conference on Industrial and Engineering Applications of AI and Expert Systems*, pp. 217–222, 1997.

32. M. Pelikan, D. E. Goldberg, and E. Cantú-Paz. BOA: The Bayesian optimization algorithm. In W. Banzhaf, J. Daida, A. E. Eiben, M. H. Garzon, V. Honavar, M. Jakiela, and R. E. Smith, editors, *Proceedings of the Genetic and Evolutionary Computation Conference GECCO-99*, volume 1, pp. 525–532. Morgan Kaufmann Publishers, San Francisco, CA, 1999. Orlando, FL.

33. M. Pelikan and H. Mühlenbein. The bivariate marginal distribution algorithm. *Advances in Soft Computing-Engineering Design and Manufacturing*, pp. 521–535, 1999.

34. V. Robles. Clasificación supervisada basada en redes bayesianas. aplicación en biología computacional. Doctoral Dissertation, Universidad Politécnica de Madrid, Madrid, Spain, 2003.

35. G. Schwarz. Estimating the dimension of a model. *Annals of Statistics*, 7(2):461–464, 1978.

36. M. Soto, A. Ochoa, S. Acid, and L. M. de Campos. Introducing the polytree aproximation of distribution algorithms. In *Second Symposium on Artificial Intelligence and Adaptive Systems. CIMAF 99*, pp. 360–367, 1999.

37. R. Tanese. Parallel genetic algorithms for a hypercube. In J. J. Grefenstette, editor, *Proceedings of the Second International Conference on GAs*, pp. 177–183. Lawrence Erlbaum Associates, 1987.

38. D. L. Whitley. An executable model of a simple genetic algorithm. In D. L. Whitley, editor, *Proceedings of the Second Workshop on Foundations of Genetic Algorithms*, pp. 45–62. Morgan Kaufmann, 1992.

GA-EDA: A New Hybrid Cooperative Search Evolutionary Algorithm

Victor Robles[1], Jose M. Peña[1], Pedro Larrañaga[2], María S. Pérez[1], Vanessa Herves[1]

[1] Department of Computer Architecture and Technology, Universidad Politécnica de Madrid, Madrid, Spain
{vrobles,jmpena,mperez,vherves}@fi.upm.es

[2] Department of Computer Science and Artificial Intelligence, University of the Basque Country, San Sebastián, Spain
ccplamup@si.ehu.es

Summary. Hybrid metaheuristics have received considerable interest in recent years. A wide variety of hybrid approaches have been proposed in the literature. In this paper a new hybrid approach, named GA-EDA, is presented. This new hybrid algorithm is based on genetic and estimation of distribution algorithms. The original objective is to benefit from both approaches and attempt to achieve improved results in exploring the search space. In order to perform an evaluation of this new approach, a selection of synthetic optimization problems have been proposed, together with some real-world cases. Experimental results show the competitiveness of our new approach.

1 Introduction

Over the last years, interest in hybrid metaheuristics has risen considerably among researchers. The best results found for many practical or academic optimization problems are obtained by hybrid algorithms. Combination of algorithms such as descent local search [32], simulated annealing [21], tabu search [12] and evolutionary algorithms have provided very powerful search algorithms.

Two competing goals govern the design of a metaheuristic [39]: exploration and exploitation. Exploration is needed to ensure every part of the search space is searched thoroughly in order to provide a reliable estimate of the global optimum. Exploitation is important since the refinement of the current solution will often produce a better solution. Population-based heuristics (where genetic algorithms [18] and estimation of distribution algorithms [23] are found) are powerful in the exploration of the search space, and weak in the exploitation of the solutions found.

With the development of our new approach, GA-EDA, a hybrid algorithm based on genetic algorithms (GAs) and estimation of distribution algorithms (EDAs), we aim to improve the exploration power of both techniques.

V. Robles et al.: *GA-EDA: A New Hybrid Cooperative Search Evolutionary Algorithm*, StudFuzz **192**, 187–219 (2006)
www.springerlink.com

This hybrid algorithm has been tested on combinatorial optimization problems (with *discrete* variables) as well as *real-valued* variable problems. Results of several experiments show that the combination of these algorithms is extremely promising and competitive.

This paper is organized in the following way: First, we will focus on different taxonomies of hybrid algorithms found in the literature; in Sect. 3, the new GA-EDA approach is proposed with a complete performance study presented in Sect. 4. Finally we close with our conclusions and further, future work.

2 Taxonomy of Hybrid Algorithms

The goal of the general taxonomies is to provide a mechanism to allow comparison of hybrid algorithms in a qualitative way. Additionally, taxonomies are useful to indicate areas in need of future work, as well as assist in classifying new hybrid approaches. In this section we include a survey of the current, most important hybrid taxonomies.

In [4] three different forms of hybridization are described:

- *Component Exchange Among Metaheuristics.*
  One of the most popular ways of hybridization concerns the use of trajectory methods, such as local search, Tabu Search or Simulated Annealing, in population-based algorithms. Most of the successful applications of Evolutionary Computation (EC) make use of local search algorithms. The reason for the success comes from the strengths of trajectory methods and population-based methods, finding a proper balance between diversification (exploration) and intensification (exploitation).
  The power of population-based methods is based on the concept of recombining solutions to obtain new ones. In EC algorithms, explicit recombinations are implemented by one or more recombination operations. In EDAs recombination is implicit because new solutions are generated using a distribution over the search space which is a function of earlier populations. This allows making guided steps in the search space which are usually larger than the steps done by trajectory methods.
  The strength of trajectory methods is found in the way they explore a promising region of the search space. A promising area in the search space is searched in a more structured way than in population-based methods. In this way, the danger of being close to good solutions but "missing" them is not as high as in population-based methods.
  In summary, population-based methods are better at identifying promising areas in the search space, whereas trajectory methods are better at exploring promising areas in the search space. Thus, metaheuristic hybrids that manage to combine the advantages of population-based methods with the strength of trajectory methods are often very successful.

Some examples of this trend are: GASAT [14] which incorporates local search within the genetic framework for solving the satisfiability problem or [45] a hybrid algorithm based on the combination of EDA with Guided Local Search for Quadratic Assignment Problems.

- *Cooperative Search.*
  A loose form of hybridization is provided by cooperative search [1, 8, 17, 38, 42, 43], which consists of a search performed by possibly different algorithms that exchange information about states, models, entire sub-problems, solutions or other search space characteristics. Typically, cooperative search algorithms consist of the parallel execution of search algorithms with a varying level of communication. The algorithms can be different or they can be instances of the same algorithm working on different models or running with different parameters settings.
  Presently, cooperative search receives more attention which, among other reasons, is due to the increasing research on parallel implementations of metaheuristics [3, 6, 24, 34, 35]. The aim of research on parallelization of metaheuristics is twofold. First, metaheuristics should be redesigned to make them suitable for parallel implementation in order to exploit intrinsic parallelism. Second, an effective combination of metaheuristics has to be found, both to combine different characteristics and strengths, and to design efficient communication mechanisms.

- *Integrating Metaheuristics and Systematic Methods.*
  This approach has recently produced very effective algorithms especially when applied to real-world problems. Discussions on similarities, differences and possible integration of metaheuristics and systematic search can be found in [11, 12, 15]. A very successful example of such an integration is the combination of metaheuristics and Constraint Programming [10].

Our hybrid GA-EDA algorithm, which is a completely new approach, can be classified in the second form; cooperative search, of Blum and Roli's classification.

Another excellent taxonomy can be found in [39]. In this hierarchical classification, at the first level, *low-level* and *high-level* hybridizations are distinguished. In low-level algorithms, a given function of a metaheuristic is replaced by another metaheuristic. In high-level algorithms, the different metaheuristics are self-contained; we have no direct relationship to the internal workings of a metaheuristic.

At the second level, *relay* and *co-evolutionary* hybridizations are distinguished. In relay hybridization, a set of metaheuristics is applied one after another, each using the output of the previous as its input, acting in a pipeline fashion.

Four classes are derived from this hierarchical taxonomy:

- *LRH (Low-level Relay Hybrid).*
  Algorithms in which a given metaheuristic is embedded into a single-solution metaheuristic. A few examples from the literature belong to this class. For instance in [28] a LRH hybrid which combines simulated annealing with local search to solve the travelling salesman problem, is introduced.
- *LCH (Low-level Co-evolutionary Hybrid).*

Algorithms in which population based heuristics have been coupled with local search heuristics such as hill-climbing, Simulated Annealing and Tabu Search. The local search algorithms will try to optimize locally, while the population based algorithms will try to optimize globally. It is exactly the same form as previously defined *component exchange among metaheuristics*.

- *HRH (High-level Relay Hybrid).*

  In HRH hybrid self-contained metaheuristics are executed in a sequence. For example, a HRH hybridization may use a greedy heuristic to generate a good initial population for an EC algorithm. Once the high performance regions are located, it may be useful to apply local search heuristics to these regions; thus, in this example, we have three pipelined algorithms. Many authors have used the idea of HRH hybridization for EC. In [25] the authors introduce simulated annealing to improve the population obtained by a GA. In [27] the proposed algorithm starts from simulated annealing and uses GAs to enrich the solutions found.

- *HCH (High-level Co-evolutionary Hybrid).*

  This schema is similar to the previously defined *cooperative search*. It involves several self-contained algorithms performing a search in parallel, and cooperating to find an optimum. Intuitively, HCH will ultimately perform at least as well as one algorithm alone, and more often perform better. Each algorithm provides information to the others to help them. An example of HCH based on parallel EDAs is the island model [34, 35].

In Talbi's taxonomy GA-EDA is heterogeneous because different metaheuristics are used; global because the algorithm search the whole state space, and general because both, GAs and EDAs, solve the same target optimization problem (HCH algorithm).

3 Hybrid GA-EDA Algorithm

Hybrid GA-EDA are new algorithms based on both techniques [33]. The original objective is to get benefits from both approaches. The main difference from these two evolutionary strategies is how new individuals are generated. These new individuals generated on each generation are called *offspring*. On the one hand, GAs use crossover and mutation operators as a mechanism to create new individuals from the best individuals of the previous generation. On the other, EDAs builds a probabilistic model with the best individuals and then sample the model to generate new ones.

3.1 Introduction

Our new approach generates two groups of offspring individuals, one generated by the GA mechanism and the other by EDA one. $Population_{p+1}$ is composed of the best overall individuals from (i) the past population ($Population_p$), (ii) the GA-evolved offspring, and (iii) EDA-evolved offspring.

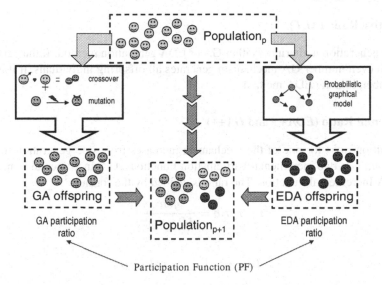

Fig. 1. Hybrid Evolutionary Algorithm Schema

The individuals are selected based on their fitness function. This evolutionary schema is quite similar to Steady State GA in which individuals from one population, with better fitness than new individual from the offspring, survive in the next one. In this case we have two offspring pools. Figure 1 shows how this model works.

3.2 Participation Functions

In this approach an additional parameter appears, this parameter has been called *Participation Function*. Participation Function provides a ratio of how many individuals are generated by each mechanism. In other words, the size of GA and EDA offspring sets. The size of these sets also represents how each of these mechanisms participates on the evolution of the population. These ratios are only a proportion for the number of new individuals each method generates, it is not a proportion of individuals in the next population, which is defined by the quality of each particular individual. If a method were better than the other in terms of how it combines the individuals, there would be more individuals from this offspring set than from the other.

The following alternatives for Participation Functions are introduced:

Constant Ratio (*x% EDA / y% GA*)

The percentage of individuals generated by each method is constant during all the generations.

Alternative Ratio (*ALT*)

On each generation it alternates either GA or EDA generation method. If the generation is an even number GA mechanism generates all offspring individuals, if it is an odd number, it is the EDA method.

Incremental Ratio (*EDA++* and *GA++*)

The partition ratio for one of the mechanism increases from one generation to the other. There are two incremental Participation Functions, GA Incremental Function and EDA Incremental Function. The ratio is defined by the formula[3]:

$$i - ratio = \frac{gen}{M + gen} \tag{1}$$

where,

$$i - ratio_{GA} = 1 - i - ratio_{EDA} \tag{2}$$

Dynamic Ratio (*DYNAMIC*)

The previous functions do not take into account the quality of the evolution methods they are merging. There is no simple method that outperforms the other in all the cases and a static Participation Function could lead toward the same problem. A constant ratio function, like *50% EDA / 50% GA* could balance the benefits and the problems from each approach, but if, for one problem, GAs do not provide good individuals, the former Participation Function would waste computational resources dealing with genetic operators and fitness evaluations for individuals that would not improve the overall population.

As a difference with the previous Participation Functions that are static and deterministic, we also propose a dynamic adaptative Participation Function. The idea is to have a mechanism that increases the participation ratio for the method which happens to generate better individuals. This function evaluates each generation considering the possibility to change the participation criterion as defined by the ratio array.

This function performs according to the algorithm in Fig. 2.

In Fig. 2 *avg_score* represents an array of the average fitness score of the top 25% of the individual generated by each of the offspring methods. *base* is the average fitness of the first generation. *dif* represents the relative difference in terms of improvement that the best method has compared with the other. $ADJUST$ is a constant that defines the trade-off between these two methods when one of them performs better than the other (5% in our experimentation).

This algorithm starts with 50%/50% ratio distribution between the two methods. On each generation the best offspring individuals from each method are compared

---

[3] *gen* is the number of the current generation and M, called the Mid-point, represents at which generation the ratio is *50%/50%*. Participation Function is 0 at the first generation and never reaches 1

```
diff=(MAX(avg_score[GA],avg_score[EDA])-base) /
     (MIN(avg_score[GA],avg_score[EDA])-base);
if (avg_score[GA]>avg_score[EDA]) {
   ratio_inc=ratio[EDA] * ADJUST * dif;
   ratio[GA] += ratio_inc;
   ratio[EDA] = 1.0 - part[GA];
}
else if (avg_score[GA]<avg_score[EDA]) {
   ratio_inc=ratio[GA] * ADJUST * dif;
   ratio[EDA] += ratio_inc;
   ratio[EDA] = 1.0 - part[GA];
}
```

Fig. 2. Pseudocode of Dynamic Participation Function

and the wining method gets a 5% of the opposite method ratio (scaled by the amount of relative difference between the methods, dif variable). This mechanism provides a contest-based DYNAMIC function in which methods are competing to get higher ratios as they generate better individuals.

4 Binary-encoded Problems

Part of the experiments have been performed considering six different binary-encoded problems:

❶ The *MaxBit* problem.
❷ Two deceptive tramp functions.
❸ A Feature Subset Selection wrapper approach for a classification problem.
❹ The *Holland Royal Road* function.
❺ One Satisfiability (SAT) problem.

On the figures, which represent the experiments, it is shown the results using five different constant ratio functions: CONST 0.00 ($0\%GA$ / $100\%EDA$, pure EDA algorithm), CONST 0.25 ($25\%GA$ / $75\%EDA$), CONST 0.50 ($50\%GA$ / $50\%EDA$), CONST 0.75 ($75\%GA$ / $25\%EDA$) and CONST 1.00 ($100\%GA$ / $0\%EDA$, pure GA). The best of these five constant Participation Functions is included also in the second figure of the experiment, as well as the four variable Participation Functions: ALT (Alternative Function), GA++ (Incremental GA function), EDA++ (Incremental EDA function), and DYNAMIC (Dynamic Participation Function).

The proposed hybrid algorithm is composed of the simplest versions of both GA and EDA components. In this sense a single bit-string chromosome (for binary-encoded problems) and real string (for continuous problem) have been used to code all the problems. GA uses *Roulette Wheel* selector, one-point crossover, flip mutation (with probability 0.01) and uniform initializer. EDA uses the Univariate Marginal Distribution Algorithm (UMDA) [30] in discrete problems and the continuous version (UMDA$_c$) [22] in continuous problems. The overall algorithms generate an offspring twice the size of the population. Depending on the ratios provided by the Participation Function, this offspring is then divided between the two methods. The composition of the new population is defined by a deterministic method, selecting the best fitness scores from the previous population and both offspring sets. The stopping criteria is quite straightforward, we stop when the difference of the sum of the fitness values of all individuals in two successive generations is smaller than a predefined value.

The experiments have been executed ten times and the average of these executions are presented. Several population sizes have been tested, but only the most representative size has been included. All the experiments have been performed in an 8-nodes cluster of bi-processors with Intel Xeon 2.4Ghz with 1GB of RAM and Gigabit network running Linux 2.4.

In most cases we have applied the Mann-Whitney statistical test to compare the results achieved by the algortihms. The fitness values of the best solutions found in the search are used for this purpose.

It is important to highlight that the use of different individual representations to the ones here used, can guide to very different results.

4.1 The MaxBit Problem

Definition

We try to obtain the maximum of the function defined as:

$$f_{M256}(\boldsymbol{x}) = \frac{\sum_{i=1}^{256} x_i}{n}$$

$$x_i \in \{0, 1\}$$

$$f_{M256}(\boldsymbol{x}^*) = max(f_{M256}(\boldsymbol{x}))$$

This problem is a typical benchmark function to evaluate the performance of evolutionary algorithms and the global maximum is found in 1.

Results

We have done this experiment using a population size of 100 individuals. Figure 3a shows that the performance of the pure genetic algorithm is very poor, while EDA outperforms all the constant Participation Functions, although these other functions also reach the optimum value.

Variable Participation Functions (see Fig. 3b) also succeed in finding the maximum but with few more iterations to converge. Dynamic Participation Function is the second best approach.

This problem shows a lineal independence among the genes of each of the individuals. EDA profits from this characteristic better than any other Participation Function. It should be considered that this feature is not quite realistic when considering real-world problems.

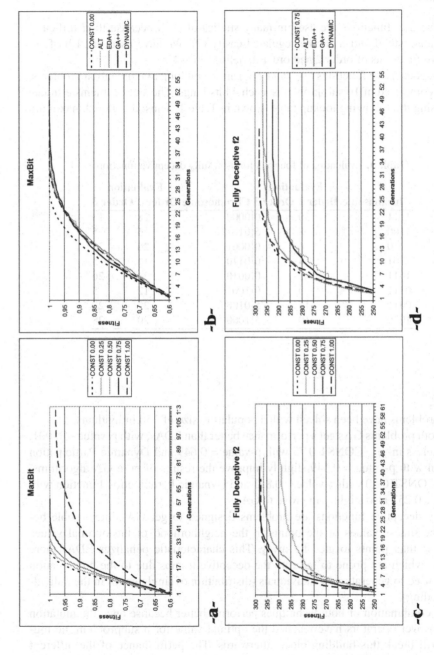

Fig. 3. MaxBit problem solved with **-a-** constant Participation Functions and **-b-** variable Participation Functions and Fully Deceptive f2 solved with **-c-** constant Participation Functions and **-d-** variable Participation Functions

4.2 4-bit Fully Deceptive Function

Definition

Deceptive trap functions are used in many studies of GAs because their difficulty is well understood and it can be regulated easily [7]. We have used the 4-bit fully deceptive functions of order 2 and order 3, defined in [44].

These deceptive functions (f_{D_2} and f_{D_3}) are 40 bit long maximization problems, and are comprised of 10 sub-problems, each 4 bits longs. The sub-problems evaluate 4 bits using the following lookup table shown in Table 1. Thus, the global maximum is 300.

Table 1. Evaluation of four bits for 4-bit fully deceptive function

Chromosome	Evaluation Order 2	Order 3	Chromosome	Evaluation Order 2	Order 3
1111	30	30	0000	28	10
1100	8	5	0101	16	5
1110	6	0	0001	26	25
1101	4	0	0110	14	5
1011	2	0	0010	24	26
0111	0	0	1001	12	5
0011	18	5	0100	22	27
1010	10	5	1000	20	28

Results

These problems have been solved with a population size of 250 individuals.

In both problems GAs get a performance better than EDAs, with p-value < 0.001. Nevertheless in f_{D_2} CONST 0.75 with p-value = 0.649 and Dynamic Participation Function with p-value = 0.649 slightly improve the results of pure GA algorithms. In f_{D_3} CONST 0.50 with p-value = 0.811 and Dynamic Participation Function with p-value = 0.257 also slightly improve to GAs.

Fully deceptive functions are problems designed to get GAs into trouble because the fitness values of the points in the neighborhood of the optimal values are worse than points located far away. This characteristic penalizes EDAs more severely, which are prone to fall into the deceptive tramps due to the combination method used to generate new individuals (distribution of single genes in the individual encoding).

The combination of both techniques performs better because once the mutation and crossover operators have reached the optimal value for a subproblem, no mutation will break this building block afterwards. The performance of the different constant or variable Participation Functions is not the same based on the particular characteristics of the deceptive tramp. DYNAMIC seems to perfectly adapt the participation ratio in order to balance GA and EDA recombination techniques to deal with these problems.

4.3 Feature Subset Selection

Definition

Feature Subset Selection (FSS) [20] is a well-known task in the Machine Learning, Data Mining, Pattern Recognition and Text Learning fields. FSS formulates as follows: Given a set of candidate features, select the best subset under some learning algorithm. As the learning algorithm, we are going to use naïve Bayes [9,13]. A good review of FSS algorithm can be found in [26]. To test the FSS problem we will use the chess dataset from the UCI repository [31], which has a total of 36 features and 699 instances.

Results

We have done this experiment using a population size of 1000 individuals. Figure 4c shows that pure GAs are a better option than EDAs for the FSS problem, with p-value = 0.004. The other constant Participation Functions do not reach results as good as GAs. For instance, with respect to CONST 0.25, the best constant Participation Function, the Mann-Whitney p-value is 0.197.

On the other hand, Fig. 4d also shows that variable Participation Functions are close to the results of GAs, being DYNAMIC the best of these functions, with p-value = 0.819. Nevertheless, the results achieved by GAs are the best for this problem.

It is important to consider that the number of generations is quite low and the complexity of the problem is not very significant. A detailed study of more complex FSS scenarios should be addressed to confirm the performance of the different algorithms on this problem. The morphology of the problem, the dataset, its features and the relationships among them, is very relevant to evaluate the performance of the algorithms in them.

4.4 240 bit Holland Royal Road - JHRR

Definition

The Holland Royal Road functions were introduced in [29]. They were designed as functions that would be simple for a genetic algorithm to optimize, but difficult for a hillclimber. In [19], Holland presented a revised class of Royal Road functions that were designed to create insurmountable difficulties for a wider class of hillclimbers, and yet still admissible to optimization by a GA.

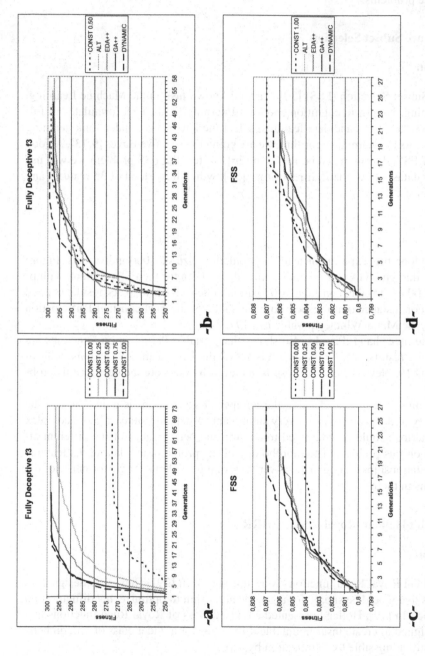

Fig. 4. Fully Deceptive f3 solved with **-a-** constant Participation Functions and **-b-** variable Participation Functions and Feature Subset Selection solved with **-c-** constant Participation Functions and **-d-** variable Participation Functions

The Holland Royal Road function takes a binary string as input and produces a real value. The function is used to define a search task in which one wants to locate strings that produce high function values. The string is composed of $2k$ non-overlapping continuous regions, each of length $b+g$. With Holland's defaults, $k = 4$, $b = 8$, $g = 7$, there are 16 regions of length 15, giving an overall string length of 240. Each region is divided into two non-overlapping pieces. The first, of length b, is called the block, and the second, of length g, is called the gap. In the fitness calculation, only the bits in the block part of each region are considered. The fitness calculation proceeds in two steps: the PART calculation, that considers each block individually and, the BONUS calculation, created to reward completed blocks and some combinations of completed blocks.

Results

We have done this experiment using a population size of 500 individuals. Holland Royal Road problem, as shown by Fig. 5a, is a very complex scenario for EDAs. This problem was designed to highlight and compare the benefits of GAs to hill climbers or other optimizers that are neighborhood-oriented search methods. Although GAs are well-suited for this problem, a combination of 25% EDAs and 75% GAs gets better results, with Mann-Whitney p-value = 0.0353. As one of the benefits mentioned in the introduction of this technique, hybrid algorithms improves the results by using two different exploratory techniques which increase the probability to find the optimal values as the range of possible movements is more complete.

In this case, ALT Participation Function outperforms all the other functions, including CONST 0.75 with p-value = 0.306. Figure 5b presents how GA++ also performs better than CONST 0.75 with less number of generations and a p-value = 0.3267. ALT gets more iterations to converge ($\sim 10\%$ more) which means that GA and EDA offsprings change a little more when they are near to the optimum value and then the exploration is more exhaustive.

4.5 SAT problem

Definition

The goal of the satisfiability (SAT) problem [36] is to attempt to find an assignment of truth values to the literals of a given Boolean formula, in its conjunctive normal form, that satisfies it. In theory, SAT is one of the basic core NP-complete problems. In practice, it has become increasingly popular in different research fields, given that several problems can be easily encoded into propositional logic formula such as planning, formal verification, knowledge representation and so on. In GAs and EDAs the SAT problem can be represented using binary strings of length n in which the i-th bit represents the true value of the i-th propositional variable in the formula. The fitness function used is the fraction of clauses satisfied.

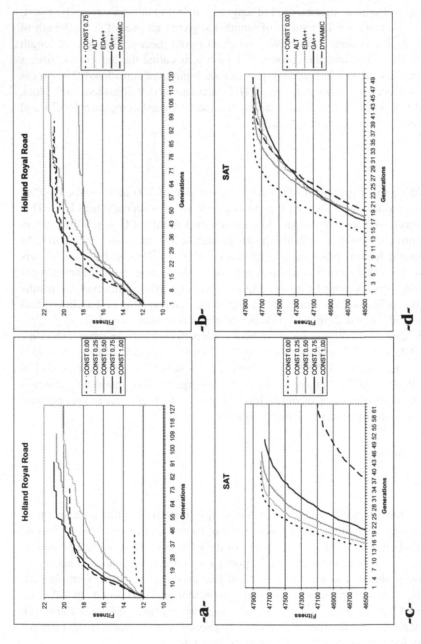

Fig. 5. Holland Royal Road solved with -**a**- constant Participation Functions and -**b**- variable Participation Functions and SAT problem solved with -**c**- constant Participation Functions and -**d**- variable Participation Functions

$$f_{SAT}(x) = \frac{1}{C} \sum_{c=1}^{C} s(x_c)$$

$$s(x_c) = \begin{cases} 1 & \text{if the clause } c \text{ is satisfied} \\ 0 & \text{in other case} \end{cases}$$

In previous equation C denotes the number of clauses that the formula has. To test the developed algorithm, the SAT instances *4blocksb.cnf* was used since they are widely-known and easily available from the SATLIB benchmark[4]. *4blocksb.cnf* contains 24758 clauses, 410 propositional variables and is satisfiable.

Results

SAT problem is one of the best scenarios for EDAs (Fig. 5c) by getting the top results using less number of generations. The difference, when compared with GAs, are very significant with a p-value < 0.001. This problem has been solved with a population size of 1000 individuals.

A very interesting issue is that as the constant ratio varies the progression of the algorithm seems to be the same with a gap between each of the graphs. This means that the lack of accuracy shown by the GA components is located on the earliest stages of the evolutionary process (the first iterations). The evolution curves are quite similar after these first generations.

None of the constant or variable Participation Functions, see Figs. 5c and 5d, reach the same fitness value as pure EDAs. Although their performance is not as bad as GAs, their p-values are in all the cases smaller than 0.001.

Dynamic Participation Function goes quite slowly on the first generation. This could drive to a new definition of this Dynamic Participation Function, with more aggressive behavior in early generations and more conservative changes later.

5 Continuous Problems

The other part of the experiments have been performed considering ten continuous problems:

① Five well-known continuous optimization problems: *Branin* RCOS function, *Griewank* function, *Rastrigin* function, *Rosenbrock* function and *Schwefel's* problem. [16,41]
② A new synthetic problem has been also defined (proportional Participation Function).
③ A continuous version of the *MaxBit* problem.
④ A real-coded solution for three different TSP problems.

---

[4] http://www.satlib.org/benchm.html

5.1 Branin RCOS Function

Definition

Results

This problem is a two-variable continuous problem with three global minimum and no local minimum. The problem is defined as follows [5]:

$$f_B(x_1, x_2) = \left(x_2 - \frac{5}{4\pi^2} x_1^2 + \frac{5}{\pi} x_1 - 6 \right)^2 + 10 \left(1 - \frac{1}{8\pi} \right) \cos(x_1) + 10$$

$$-5 < x_1 < 10$$

$$0 < x_2 < 15$$

The global optimum for this problem is 0.397887 that is reached in the points $(x_1, x_2) = (-\pi, 12.275), (\pi, 2.275), (9.42478, 2.475)$.

This problem is considered easy not only because of the number of variables, but the small chance to miss the basin of the global minimum in a global optimization procedure. This is due to the probability of reaching the global optimum using local optimization methods, started with a small number of random points, is quite high.

This problem was solved using a population size of 150 individuals.

Branin is a very simple problem where in few generations all the algorithms converge. Figure 6 shows CONST 0.25 is the best function, and GAs a very poor option to solve this problem. CONST 0.25 aheads EDA with p-value = 0.063 and CONST 0.50 with p-value = 0.339.

In this problem GA++ performs similarly to EDA. This is due to the reduced number of generations which represents the participation share of GAs which do not increase too fast to recover the majority of individuals generated by the EDA mechanism. See Fig. 6b.

The simplicity of this function biases the performance of the algorithm towards the trend addressed by the very first generations. Few modifications are achieved after these generations. For the DYNAMIC function, this could be a good justification to define a more radical variations of the first iterations of the algorithm.

5.2 Griewank Function

Definition

This problem has ten variables with a unique global optimum with many ($O(10^3)$) local minima nearby.

$$f_{G10}(\boldsymbol{x}) = 1 + \frac{1}{d} \sum_{i=1}^{n} x_i^2 - \prod_{i=1}^{n} \cos \left(\frac{x_i}{\sqrt{i}} \right)$$

$$d = 4000; n = 10$$

$$-500 < x_i < 500$$

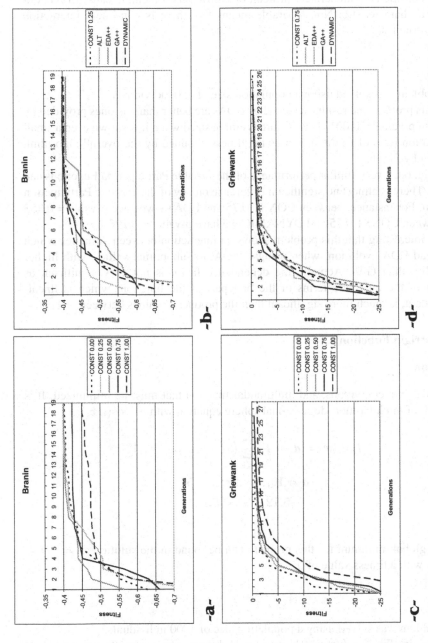

Fig. 6. Branin function solved with -**a**- constant Participation Functions and -**b**- variable Participation Functions and Griewank function solved with -**c**- constant Participation Functions and -**d**- variable Participation Functions

The global minimum for this problem can be found in the solution $x_i = 0, i = 1, \ldots, n$ with a fitness value of 0.

It is considered a moderately difficult optimization problem, because of its non-separable characteristic. Non-separable means that there is non-linear interaction among variables.

Results

This problem was solved using a population size of 250 individuals.

In this problem the results achieved by EDAs are better than the ones provided by GAs with p-value < 0.001. This feature is emphasized when it is shown even a small participation ratio of EDAs increases the fitness obtained by the overall algorithm, as shown by Fig. 6c.

In Fig. 6d, a very similar performance of the variable Participation Functions can be seen. There is almost no significant difference on any of the variable Participation Function. For instance, between CONST 0.75 and EDA++ we have p-value = 0.853 and between CONST 0.75 and DYNAMIC we have p-value = 0.795.

It is interesting that this problem shows an interaction between variables which could lead EDA evolution, when using UMDA, to suboptimal values, with higher probability than GAs. Actually the dependence factor is not very significant as the sum of the quadratic terms of the first part of the function, which is lineal-independent, is much more significant than the product of values between $[-1, 1]$.

5.3 Rastrigin Function

Definition

It is a scalable, continuous, and multimodal function that must be minimized. It is the result of modulating n-dimensional sphere equation with $a \cdot \cos(\omega x_i)$.

$$f_{Ra5}(\boldsymbol{x}) = a \cdot n + \sum_{i=1}^{n} \left(x_i^2 - a \cdot \cos(\omega x_i) \right)$$

$$a = 10; \omega = 2\pi; n = 5$$

$$-5.12 < x_i < 5.12$$

The global minimum for this problem can be found in the solution $x_i = 0, i = 1, \ldots, n$ with a fitness value of 0.

Results

This problem was solved using a population size of 1000 individuals.

Rastrigin (Fig. 7a) function has no lineal dependency among the variables, but the performance of EDAs is very poor. Near the optimum value there are many local optimum and EDAs seems to be very sensitive to this characteristic. The best constant ratio function is CONST 0.50, with a p-value < 0.001 respect to GA. This means that pure GA could be improved by the help of EDAs even if this method is not very well-suited by itself.

In Fig. 7b, we can see that DYNAMIC is able to provide the best participation ratio to outperform CONST 0.50 with a p-value = 0.006 and the other variable Participation Functions.

5.4 Rosenbrock Function

Definition

It is a continuous, non-separable, and unimodal function. It has the global minimum located in a steep parabolic valley with a flat bottom [37]. This issue represents a big challenge to the optimization process.

$$f_{Ro10}(\boldsymbol{x}) = \sum_{i=1}^{n-1} \left(100 \cdot (x_{i+1} - x_i^2)^2 + (x_i - 1)^2 \right)$$

$$n = 10$$

$$-500 < x_i < 500$$

The global minimum for this problem can be found in the solution $x_i = 0, i = 1, \ldots, n$ with a fitness value of 0.

Results

This problem was solved using a population size of 1000 individuals.

In Fig. 7c, EDAs perform much better than the other approaches, and GAs are far worse than any Participation Function with at least a small EDA ratio.

Variable Participation Functions on this problem are not better than pure EDAs. It is very significant, as shown by Figure 7d, that performance of GA++ is much better than EDA++. This is due to the small number of generations that represents a more intensive participation of EDAs.

DYNAMIC does not perform very well. This can be explained in the same terms previously used. In problems with few generations our Dynamic Participation Functions has no time to balance the participation ratios of the algorithms.

5.5 Schwefel's Problem

Definition

It is also a continuous unimodal function. Its difficulty also concerns the fact of searching along the coordinate axes only gives a poor rate of convergence because

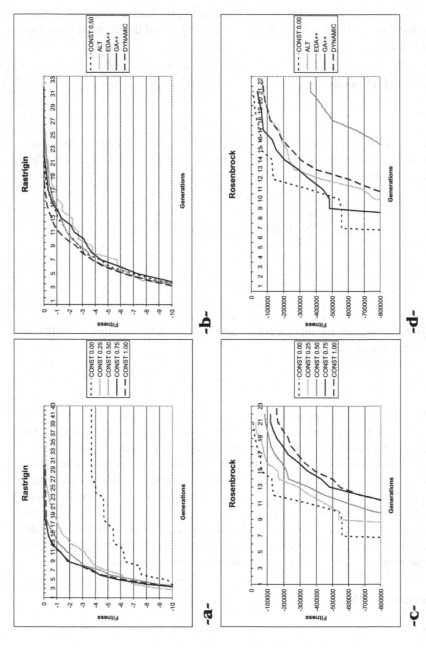

Fig. 7. Rastrigin function solved with **-a-** constant Participation Functions and **-b-** variable Participation Functions and Rosenbrock function solved with **-c-** constant Participation Functions and **-d-** variable Participation Functions

function gradient is not oriented along the axes. As in the previous case global optimum is surrounded by several local optimum in the neighborhood.

$$f_{S10}(\boldsymbol{x}) = \sum_{i=1}^{n} x_i \cdot \sin(\sqrt{|x_i|})$$

$$n = 10$$

$$-500 < x_i < 500$$

$$f_{S10}(\boldsymbol{x^*}) = \min(f_{S10}(\boldsymbol{x}))$$

The global minimum for this problem can be found in the solution $x_i = 420.9687, i = 1, \ldots, n$ with a fitness value of 0.

Results

This problem has been solved with a population of 250 individuals.

Schwefel's problem is very difficult due to the large number of suboptimal points, especially those near the global optimum. This feature drives EDAs to a very poor performance, also due to the non-lineal relationships among the variables.

In Fig. 8a GAs is the best approach, much better than any other of the constant ratio functions, although they are not able to find the optimal value in all of the cases.

All the results achieved by other than pure GA algorithms are not able to improve the results reached after the first generations. Even GA++, which increments the ratio of GA-based individuals, are very poor on this problem (see Fig. 8b). Many studies have proved that significantly high mutation rates could help improve the results of this problem.

5.6 Proportion Problem

Definition

This new function represents a model of similar real-world problems that deal with the search of the correct proportions that should make it true that:

$$\sum_{i=1}^{n} x_i = 1 \tag{3}$$

The fitness function is:

$$f_{P128}(\boldsymbol{x}) = \frac{1}{n} \sum_{i=1}^{n} \left(1 - |x_i - x_i^*|^{1/p}\right)$$

$$n = 128; p = 2; x_i^* = \frac{i}{\frac{n \cdot (n+1)}{2}}$$

$$0 \leq x_i \leq 1$$

$$f_{P128}(\boldsymbol{x^*}) = \max(f_{P128}(\boldsymbol{x}))$$

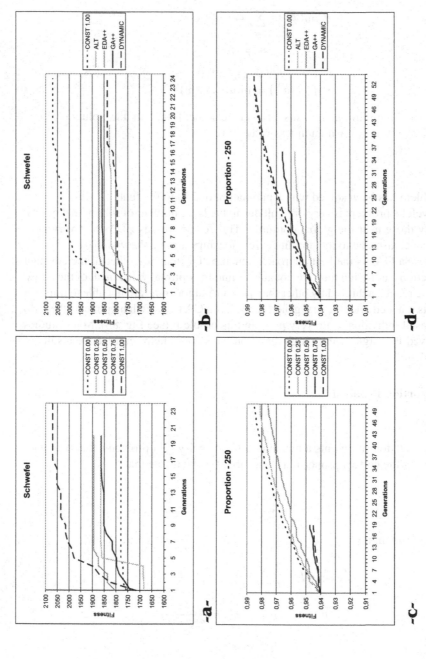

Fig. 8. Schwefel function solved with -a- constant Participation Functions and -b- variable Participation Functions and Proportion problem solved with -c- constant Participation Functions and -d- variable Participation Functions

The objective of the function is to find the right proportion (represented by x^*), with p-order distance function.

This problem is not difficult, as the distance function is separable and lineal. In order to be compliant with the restriction expressed by Eq. 3 a Lamarckian correction is performed to the individual represented by x, instead of dropping malformed elements.

$$x_i^c = \frac{x_i}{\sum_{i=1}^n x_i} \tag{4}$$

The individual x^c substitutes the individual x in the population, before fitness calculation is performed.

Results

This problem has been solved with a population of 250 individuals.

EDAs deal with this problem much better than any other constant functions. Between EDAs and CONST 0.25, the best constant function, the p-value is p-value $<$ 0.001.

DYNAMIC slightly outperforms EDAs (with p-value = 0.185) and seems to adapt perfectly to the characteristic of this problem (see Fig. 8d). EDA++ converges prematurely due to the heavy ratio of the GA-based individuals.

5.7 The MaxBit Continuous Problem

Definition

This problem is a redefinition of the binary MaxBit problem previously presented. The aim is to maximize:

$$f_{M12}(x) = \frac{\sum_{i=1}^n x_i}{n}$$
$$x_i \in \{0,1\}; n = 12$$

In the continuous domain this problem is more complex, as the optimum value of the function is located on the boundary of the search space.

Results

This problem has been solved with a population of 250 individuals.

Figures 9a and 9b contain the obtained results. All the constant and variable Participation Functions perform in a very similar way reaching the global optimum in almost the same number of generations. However, pure GAs, which converge after more generation, only reach a suboptimal value.

MaxBit performance, as we can see, is very similar in both cases with continuous and with bit-string individuals.

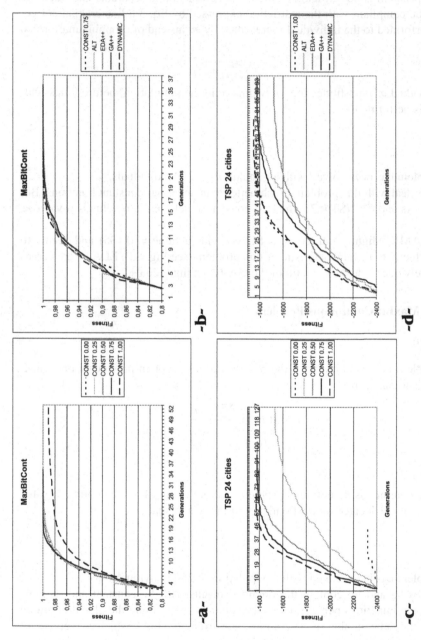

Fig. 9. MaxBit continuous problems solved with **-a-** constant Participation Functions and **-b-** variable Participation Functions and TSP problem with 24 cities solved with **-c-** constant Participation Functions and **-d-** variable Participation Functions

5.8 TSP Continuous

Definition

The Travelling Salesman Problem (TSP) objective is to find the shortest route for a travelling salesman who, starting from his home city, has to visit every city on a given list precisely once and them return to his home city. The main difficult of this problem is the immense number of possible tours: $(n-1)!/2$ for n cities.

The TSP is a relatively old problem. It was documented as early as 1759 by Euler, however not using that name, whose interest was in solving the knights' tour problem in chess. A correct solution would have a knight visit each of the 64 squares of a chessboard exactly once on its tour. The term "traveling salesman" was first used in 1932, in a German book written by a veteran traveling salesman. The RAND CORPORATION introduced the TSP in 1948. The corporation's reputation helped to make the TSP a well-known and popular problem.

Although there are different alternatives to encode this problem, in this paper individuals of population are represented by using vectors with real numbers. Thus, we need a method to translate these real vectors to a valid tour for the TSP. In the following table we see one of these translations.

In Table 2 we can see a 6-city example. In the original vector the generated real numbers are between 3 and -3. The obtained tour will be an integer vector in which each of the elements is the index after the values of the original vector are sorted. Thus, the calculus of the fitness function of individuals is more complex to compute.

Table 2. Translation of an individual to a correct tour

Original vector:	1.34	2.14	0.17	0.05	−1.23	2.18
Resulting tour:	4	5	3	2	1	6

The following files have been used in the empirical study: The well known Gröstel24, Gröstel48 and Gröstel120. These are files that can be obtained via web or ftp in many sites. They represent the distances between 24, 48 and 120 imaginary cities. They are often used in TSP problems to know the fitness of the algorithm we use, and can be defined as a classical experiment in the TSP.

Results for Gröstel24 Problem

This problem has been solved with a population of 1000 individuals.

In the TSP problem of 24 cities (see Fig. 9c) the best algorithms are pure GA, CONSTANT 0.75 and CONSTANT 0.50 (p-values > 0.9 between them). The worst algorithm is EDA which presents a very poor performance.

In the variable Participation Functions (see Fig. 9d) obtained results are excellent, being the DYNAMIC approach being better than GA with a p-value = 0.161.

Although it has a very good beginning because of the number of GA individuals created in the first generations, the EDA++ approach presents bad results.

Results for Gröstel48 Problem

This problem has been solved with a population of 1500 individuals.

The results obtained for 48 cities with the constant Participation Functions (see Fig. 10a) are very similar to the previous ones, CONSTANT 0.75 and pure GAs being the best constant approaches, without statistical significance difference between them.

However, in this case, in the variable Participation Functions (Fig. 10b), GA++ is similar to the DYNAMIC approach, with p-value = 0.722, but worse than CONST 0.75, with p-value = 0.147.

Results for Gröstel120 Problem

This problem has been solved with a population of 1500 individuals.

TSP with 120 cities is a very hard problem for heuristic optimization approaches such as GAs and EDAs without the help of local optimization techniques. However, the obtained results are quite similar to the previous ones, GA being the best approach with constant Participation Function, and GA++ the best with variable Participation Function, without significant difference respect to GA (p-value = 0.7393).

6 Conclusion and Further Work

In this chapter we have proposed a new hybrid algorithm based on genetic and estimation of distribution algorithms. This new algorithm has been tested on a set of different problems. Although the hybrid algorithm proposed is composed by the simplest versions of both GA and EDA components, the experimentation shows it is really promising and competitive. In most of the experiments we reach the best of the values found by GAs or EDAs or we even improve them. There is still a lot of further future work. Here are some possibilities: Extend the implementation to support more sophisticated individual representations, make new Participation Functions based on statistical tests, implement a parallel version based on the island model or use more complex GAs and EDAs in the hybrid solution.

6.1 Evolution of the Dynamic Participation Function

One interesting issue is to survey the evolution of the dynamic Participation Function in the series of different experiments. This function, as we have seen, adjusts the participation ratio depending on the quality of the individuals each of the methods is providing. Indirectly, this measure could be used to evaluate the quality of each of the methods across the continuous generations of an algorithm.

As we see in Fig. 11 the evolution of the dynamic functions are able to guide the hybrid algorithms towards the best option, either GA, EDA or other constant ratio Participation Functions. For example, in TSP, GAs outperforms clearly all the other

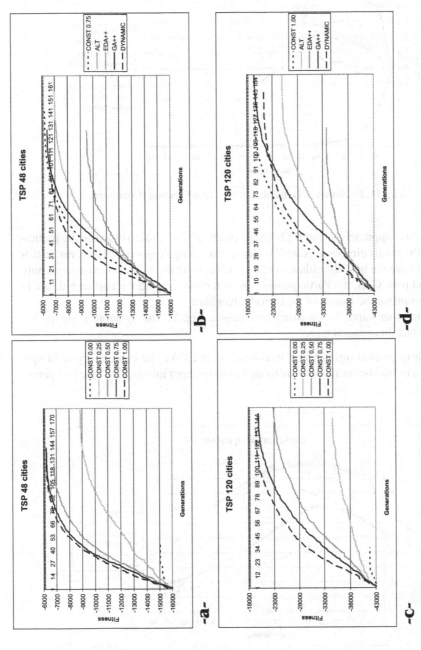

Fig. 10. TSP problems solved with -a- constant Participation Functions and -b- variable Participation Functions and TSP problem with 24 cities solved with -c- constant Participation Functions and -d- variable Participation Functions

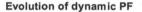

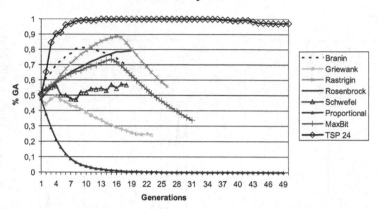

Fig. 11. Evolution of Dynamic Participation Function (Continuous)

constant ratio approaches, and DYNAMIC (with similar results) changes the participation ratio in this direction. A similar case is shown by Proportion. However, in this case EDAs are the best algorithm. In other problems, with a best option between pure EDAs and pure GAs, the Participation Function moves to find the correct balance in order to improve the results of the overall algorithm.

On the other hand, for bit-string problems in Fig. 12, the dynamic Participation Function has this general trend:

- On early generations GA performs better than EDAs, the exploratory technique is able to find better individuals using the same input information (the last generation).

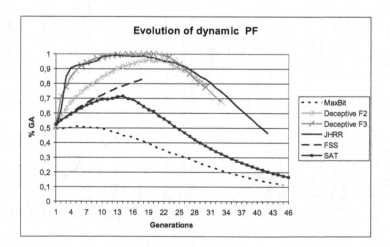

Fig. 12. Evolution of Dynamic Participation Function (Bit-string)

- When the algorithm is close to the optimum, EDAs generate the best solutions. That is probably due to the mutation ratio, which is very useful to avoid local optimum, but once the environment of the global solution is reached it drives towards malformed individuals, far from the local optimum.

This trend is also shown, with small variations, by the continuous problems shown before, in the case neither pure GAs nor pure EDAs are clear options. Although special abnormalities are present, for example Rosenbrock is best solved by EDAs, but DYNAMIC trend increases the participation ratio of GAs instead.

These trends of the fitness, provided by each of the methods, could be useful in order to tune up either genetic and estimation of distribution algorithms by themselves. Updating mutation rate is one of the issues already considered by works such as [2, 40].

6.2 Experiments Summary

On Tables 3 and 4, the summary of the results obtained by these experiments show that hybrid algorithms, in most of the cases, are a better option than the pure EDA or GA algorithms by themselves. Although there are specific problems in which EDA (SAT, Proportion and Rosenbrock) or GA (FSS and Schwefel) are the best options, hybrid algorithms show a competitive behavior. The opposite is not as common, as there are experiments (like JHRR, Branin, Griewank, Rastrigin, and MaxBit Continuous) in which neither EDAs nor GAs present good results compared to most of the hybrid approaches.

Table 3. Result Summary Table

	EDA		CONST 0.25		CONST 0.50		CONST 0.75		GA	
Problem	Mean	Gen	Mean	Gen	Mean	Gen	Mean	Gen	Mean	Gen
Max Bit	1	45	1	48	1	51	1	59	0,9793	115
Deceptive f2	280,0	28	289,8	62	296,4	45	298,2	42	297,6	38
Deceptive f3	272,4	68	296,5	74	298,6	51	298,3	41	298,2	40
FSS	0,8041	20	0,8059	22	0,8057	20	0,8057	20	0,8070	27
JHRR	12,85	43	20,09	130	20,75	111	20,99	92	19,37	74
SAT	47803,6	42	47800,1	44	47790,9	47	47752,7	51	47096,5	63
Branin	−0,4035	19	−0,3999	19	−0,4006	19	−0,4235	19	−0,4513	19
Griewank	626,38	24	626,45	25	626,51	25	626,42	25	625,45	32
Rastrigin	−3,69683	43	−0,06368	34	−0,00054	32	−0,00473	30	−0,10823	30
Rosenbrock	−12403	21	−29197	21	−81222	22	−118522	22	−157107	23
Schwefel	1778,36	19	1863,85	20	1894,66	20	1862,60	20	2068,70	24
Proportion	0,9851	51	0,9810	51	0,9756	51	0,9473	19	0,9449	19
MaxBit Cont	0,9999	37	0,9999	37	1	36	1	35	0,9909	53
TSP 24	−2324,7	45	−1531,3	128	−1381,3	104	−1378,9	95	−1372,2	69
TSP 48	−15037,2	45	−8873	181	−6814,2	126	−6222,8	142	−6227,3	131
TSP 120	−41809,1	35	−34666,9	145	−22859,9	154	−19983,7	149	−19640,5	134

Table 4. Result Summary Table

Problem	ALT Mean	Gen	GA++ Mean	Gen	EDA++ Mean	Gen	DYNAMIC Mean	Gen
Max Bit	1	52	1	56	1	51	1	50
Deceptive f2	296,8	55	292,4	49	296,6	48	298,2	42
Deceptive f3	298,0	55	297,7	52	297,3	58	299,2	45
FSS	0,8057	21	0,8060	22	0,8059	21	0,8065	21
JHRR	21,48	121	21,33	82	18,40	105	20,88	97
SAT	47783,0	48	47745,9	47	47799,4	46	47798,7	50
Branin	−0,4006	19	−0,4001	19	−0,4085	19	−0,4033	19
Griewank	626,42	25	626,47	25	626,49	26	626,42	25
Rastrigin	−0,00159	33	−0,00027	32	−0,00472	31	−0,00005	30
Rosenbrock	−70688	22	−36544	21	−366490	22	−73786	22
Schwefel	1826,47	19	1863,35	20	1874,48	20	1836,26	24
Proportion	0,9574	36	0,9659	36	0,9426	19	0,9851	54
MaxBit Cont	1	36	1	36	0,9999	37	1	35
TSP 24	−1371,6	93	−1378	95	−1522,5	95	−1351,3	78
TSP 48	−6962,8	169	−6413,1	128	−9484,5	127	−6641,9	113
TSP 120	−24875,1	159	−19457,4	159	−33759,5	121	−21327,8	147

In order to compare the results, for the experiments carried out, the relative position (ranked-based) has been computed. This ranking has been developed using fitness-driven criteria. The best fitness is #1, next one #2, and so on. Using this method, the average ranking has also been computed:

$$avg\_rank(PF) = \frac{\sum_{i=1}^{N} rank(PF, i)}{N} \tag{5}$$

being $rank(PF, P)$ the relative ranking of Participation Function PF in the problem P.

DYNAMIC is the best Participation Function, as can be seen on Table 5. Another interesting result is that CONST 0.50 also behaves quite well. Among the worst results are both pure EDAs and pure GAs. Of course, the set of experiments is not representative of all the possible optimization problems, but have been selected to cover a wide spectrum of possible real-world scenarios. In the performance of EDA++ and GA++, we should consider that several experiments do not last for many generations, thus the influence of the first generations biases the results achieved by these approaches.

The experimentation and research perspective of hybrid methods is very promising, and several issues are still open in terms of alternatives of the presented Participation Functions, using of more complex EDAs approaches, hybridization with local heuristics, three or multi hybrid algorithms (using more than one GA or one EDA algorithm), and parallel definition of the hybrid algorithms.

Table 5. Average ranking of the Participation Functions

Participation Function	avg_rank
EDA	6,625
CONST 0.25	4,9375
CONST 0.50	3,8125
CONST 0.75	4,4375
GA	5,4375
ALT	4,25
GA++	3,5
EDA++	5,6875
DYNAMIC	3,0625

References

1. V. Bachelet and E. Talbi. Cosearch: A co-evolutionary metaheuritics. In *Proceedings of Congress on Evolutionary Computation – CEC2000*, pp. 1550–1557, 2000.
2. T. Back and M. Schutz. Intelligent mutation rate control in canonical genetic algorithms. In *International Syposium on Methodologies for Intelligent Systems*, 1996.
3. T.C. Belding. The distributed genetic algorithm revisited. In *Proceedings of the Sixth International Conference on Genetic Algorithms*, pp. 114–121, 1995.
4. C. Blum and A. Roli. Metaheuristics in combinatorial optimization: Overview and conceptual comparison. *ACM Computing Surveys*, 35(3):268–308, September 2003.
5. F.K. Branin. A widely convergent method for finding multiple solutions of simultaneous nonlinear equations. *IBM Journal of Research and Development*, pp. 504–522, 1972.
6. E. Cantú-Paz. *Efficient and accurate parallel genetic algorithms*. Kluwer Academic Publisher, 2001.
7. K. Deb and D.E. Golberg. Analyzing deception in trap functions. In L.D. Withley, editor, *Foundation of Genetic Algorithms 2*, pp. 93–108, San Mateo, CA, 1993. Morgan Kaufmann.
8. J. Denzinger and T. Offerman. On cooperation between evolutionary algorithms and other search paradigms. In *Proceedings of Congress on Evolutionary Computation – CEC1999*, pp. 2317–2324, 1999.
9. R. Duda and P. Hart. *Pattern Classification and Scene Analysis*. John Wiley and Sons, 1973.
10. F. Foccaci, F. Laburthe, and A. Lodi. Local search and constraint programming. In F. Glover and G. Kochenberger, editors, *Handbook of Metaheuristics*, volume 57 of *International Series in Operations Research and Management Science*, Kluwer Academic Publishers, Norwell, MA, 2003.
11. E.C. Freuder, R. Dechter, M.L. Ginsberg, B. Selman, and E.P.K. Tsang. Systematic versus stochastic constraint satisfaction. In *Proceedings of the 14th International Joint Conference on Artificial Intelligence, IJCAI'1995*, pp. 2027–2032. Morgan-Kaufmann, 1995.
12. F. Glover and M. Laguna. *Tabu Search*. Kluwer Academic Publishers, 1997.
13. D.J. Hand and K. Yu. Idiot's Bayes – not so stupid after all? *International Statistical Review*, 69(3):385–398, 2001.
14. J. Hao, F. Lardeux, and F. Saubion. A hybrid genetic algorithm for the satisfiability problem. In *Proceedings of the First International Workshop on Heuristics*, Beijing, 2002.

15. W.D. Harvey. *Nonsystematic backtracking search*. PhD thesis, University of Oregon, 1995.

16. F. Herrera and M. Lozano. Gradual distributed real-coded genetic algorithms. *IEEE Transactions on Evolutionary Computation*, 4(1), 2000.

17. T. Hogg and C. Williams. Solving the really hard problems with cooperative search. In *Proceedings of AAAI'1993*, pp. 213–235. AAAI Press, 1993.

18. J.H. Holland. *Adaption in natural and artificial systems*. The University of Michigan Press, Ann Harbor, MI, 1975.

19. J.H Holland. Royal road functions. *Internet Genetic Algorithms Digest*, 7(22), 1993.

20. I. Inza, P. Larrañaga, and B. Sierra R. Etxeberria. Feature subset selection by Bayesian networks based optimization. *Artificial Intelligence*, 123(1–2):157–184, 2000.

21. S. Kirkpatrick, C.D. Gelatt, and M.P. Vecchi. Optimization by simulated annealing. *Science*, 220(4598):671–680, May 1983.

22. P. Larrañaga, R. Etxeberria, J. A. Lozano, and J. M. Peña. Optimization in continuous domains by learning and simulation of Gaussian networks. In A. S. Wu, editor, *Proceedings of the 2000 Genetic and Evolutionary Computation Conference Workshop Program*, pp. 201–204, 2000.

23. P. Larrañaga and J.A. Lozano. *Estimation of Distribution Algorithms. A New Tool for Evolutionary Computation*. Kluwer Academic Publisher, 2002.

24. D. Levine. *A Parallel Genetic Algorithm for the Set Partitioning Problem*. PhD thesis, Illinois Institute of Technology, Mathematics and Computer Science Division, Argonne National Laboratory, 1994.

25. F.T. Lin, , C.Y. Kao, and C.C. Hsu. Incorporating genetic algorithms into simulated annealing. *Proceedings of the Fourth International Symposium on Artificial Intelligence*, pp. 290–297, 1991.

26. H. Liu and H. Motoda. *Feature Selection for Knowledge Discovery and Data Mining*. Kluwer Academic Publisher, 1998.

27. S.W. Mahfoud and D.E. Golberg. Parallel recombinative simulated annealing: A genetic algorithm. *Parallel computing*, 21:1–28, 1995.

28. O.C. Martin and S.W. Otto. Combining simulated annealing with local search heuristics. *Annals of Operations Research*, 63:57–75, 1996.

29. M. Mitchell, S. Forrest, and J.H. Holland. The royal road for genetic algorithms: Fitness landscapes and GA performance. In *Proceedings of the First European Conference on Artificial Life*, Cambridge, 1992. MIT Press/Bradford Books.

30. H. Mühlenbein. The equation for response to selection and its use for prediction. *Evolutionary Computation*, 5:303–346, 1998.

31. P.M. Murphy and D.W. Aha. UCI repository of machine learning databases. http://www.ics.uci.edu/~mlearn/, 1995.

32. C.H. Papadimitriou and K. Steiglitz. *Combinatorial Optimization: Algorithms and Complexity*. Prentice-Hall, 1982.

33. J.M. Peña, V. Robles, P. Larrañaga, V. Herves, F. Rosales, and M.S. Pérez. GA-EDA: Hybrid evolutionary algorithm using genetic and estimation of distribution algorithms. In *Innovations in Applied Artificial Intelligence: Proceeding of IEA/AIE 2004*, 2004.

34. V. Robles. *Clasificación supervisada basada en redes Bayesianas. Aplicación en biología computacional*. PhD thesis, Universidad Politécnica de Madrid, Departamento de Arquitectura de Computadores, Junio, 2003.

35. V. Robles, M.S. Pérez, V. Herves, J.M. Peña, and P. Larrañaga. Parallel stochastic search for protein secondary structure prediction. In *Lecture Notes in Computer Science 2724*, pp. 1162-1170, Czestochowa, Poland. 5th International Conference on Parallel Processing and Applied Mathematics, 2003.

36. E. Rodriguez-Tello and J. Torres-Jimenez. ERA: An algorithm for reducing the epistasis of SAT problems. Number 2724 in Lecture Notes in Computer Science, pp. 1283–1294. GECCO 2003, Genetic and Evolutionary Computation Conference, Springer Verlag, 2003.

37. H.H. Rosenbrock. An automatic method for finding the greatest or least value of a function. *Computational Journal*, pp. 175–184, 1960.

38. L. Sondergeld and S. Voß. Cooperative intelligent search using adaptive memory techniques. In I. Osman S. Voß, S. Martello and C. Roucairol, editors, *In MetaHeuristics: Advances and Trends in Local Search Paradigms for Optimization*, chapter 21, pp. 297–312. Kluwer Academic Publishers, 1999.

39. E-G. Talbi. A taxonomy of hybrid metaheuristics. *Journal of Heuristics*, 8(5):541–564, 2002.

40. D. Thierens. Adaptive mutation rate control schemes in genetic algorithms. In *Proceedings of the 2002 IEEE World Congress on Computational Intelligence: Congress on Evolutionary Computation*, 2002.

41. A. Törn, M. M. Ali, and S. Viitanen. Stochastic global optimization: Problem classes and solution techniques. *Journal of Global Optimization*, 14:437-47, 1999.

42. M. Toulouse, T. Crainic, and B. Sansó. An experimental study of the systemic behavior of cooperative search algorithms. In I. Osman S. Voß, S. Martello and C. Roucairol, editors, *In Meta-Heuristics: Advances and Trends in Local Search Paradigms for Optimization*, chapter 26, pp. 373–392. Kluwer Academic Publishers, 1999.

43. M. Toulouse, K. Thulasiraman, and F. Glover. Multi-level cooperative search: A new paradigm for combinatorial optimization and application to graph partitioning. In *Proceedings of the 5th International Euro-Par Conference on Parallel Processing. Lecture Notes in Computer Science. Springer-Verlag*, pp. 533–542, New York, 1999.

44. D. Withley and T. Starkweather. Genitor II: a distributed genetic algorithm. *Journal of Experimental and Theoretical Artificial Intelligence*, 2:189–214, 1990.

45. Q. Zhang, J. Sun, E. Tsang, and J. Ford. Combination of guided local search and estimation of distribution algorithm for quadratic assignment problems. In *Proceedings of the Bird of a Feather Workshops, Genetic and Evolutionary Computation Conference, in GECCO 2003*, pp. 42–48, 2003.

Bayesian Classifiers in Optimization:
An EDA-like Approach

Teresa Miquélez[1], Endika Bengoetxea[1] and Pedro Larrañaga[2]

[1] Department of Computer Architecture and Technology, University of the Basque Country,
P.O. Box 649, 20080 San Sebastian, Spain
{teresa,endika}@si.ehu.es
[2] Department of Computer Science and Artificial Intelligence, University of the Basque
Country, P.O. Box 649, 20080 San Sebastian, Spain
ccplamup@si.ehu.es

Summary. This chapter introduces a new Evolutionary Computation method which applies Bayesian classifiers in the construction of a probabilistic graphical model that will be used to evolve a population of individuals. On the other hand, the *satisfiability* problem (SAT) is a central problem in the theory of computation as a representative example of NP-complete problems. We have verified the performance of this new method for the SAT problem. We compare three different solution representations suggested in the literature. Finally, we apply local search methods for this problem.

1 Introduction

This chapter introduces Evolutionary Bayesian Classifier-based Optimization Algorithms (EBCOAs) as a new approach in Evolutionary Computation. The originality of these algorithms comes from the fact that they evolve a generation of individuals by constructing Bayesian classifier models that take into account deeper differences rather than simply a subset of the better individuals of the previous population. The main difference between this approach and Estimation of Distribution Algorithms (EDAs) is the fact that the probabilistic graphical model in discrete EDAs is a Bayesian network, while in EBCOAs we construct a Bayesian classifier that includes an extra C node that represents the different classes to which each individual of a population is classified. EBCOAs take into account the differences between the individuals in the population that make them be more or less fit regarding their fitness value, and apply this knowledge to create a new population by enhancing the characteristics of the fitter ones and tries to avoid such of the less fit ones. In order to better understand the motivation for EBCOAs, this chapter analyzes the issues that allows Estimation of Distribution Algorithms (EDAs) to converge to the best solution of a problem, as well as several new methods for improving the way in which this convergence is done. The aim of this idea is to avoid a too fast convergence that could lead to fall in local optima.

T. Miquélez et al.: *Bayesian Classifiers in Optimization: An EDA-like Approach*, StudFuzz **192**, 221–242 (2006)
www.springerlink.com

In order to analyze the potential of this new method, we tried its performance with a typical NP-hard optimization algorithm such as the SAT problem. This optimization problem is regarded as a very interesting one in the field of computer science due to the big number of problems that can be formulated and solved using SAT instances.

The rest of the chapter is structured in the following way. Section 2 is devoted to the introduction to EBCOAs and the different Bayesian classifiers that can be applied. The SAT problem is defined in Sect. 3. Section 4 shows the experimental results obtained with this new method, and the conclusions of the chapter as well as ideas for future work can be found in Sect. 5.

2 The Evolutionary Bayesian Classifier-based Optimization Algorithm Approach

This section describes the EBCOAs [22]. Similarly as EDAs, EBCOAs combine both probabilistic reasoning and evolutionary computing. The main characteristic of EBCOAs that distinguish them from EDAs is that the learning of the probabilistic graphical model is based on using Bayesian classifiers.

2.1 Motivation

In many Evolutionary Computation techniques such as Genetic Algorithms (GAs) and EDAs only the best individuals of the generation are taken into account to proceed to apply crossover and mutation techniques–in GAs – or to learn the probabilistic graphical model – in EDAs. In these approaches, the aim is to take into account the characteristics of the N fittest individuals of the population. However, in most of the cases the fitness value differences among the individuals are not taken into account but for the purpose of deciding whether the phenotype of an individual is relevant for generating the next generation. Therefore, in the case of most EDAs, the best and worst individuals within the selected population of the lth generation are considered to have the same relevance for the learning of the probabilistic graphical model.

However, in many optimization problems the fitness differences between the selected individuals are also important in order to ensure an adequate convergence of the search process. This is essential in the case of EDAs in order to ensure convergence. The literature shows different possibilities for taking into account the fitness of each of the selected individuals in the learning step of EDAs:

- **Making fitter individuals to influence the learning of the probabilistic graphical model regarding their fitness value:** this approach assigns a weight to each individual to have more influence in the learning step in EDAs regarding the respective fitness value. An example of this idea was proposed in BSC [26].
- **Applying a proportional selection method:** An example of this approach is the use of a Boltzman distribution based selection [24].

- **Considering the fitness value as an additional node in the probabilistic graphical model:** the fact of including such a new variable together with variables $X_1, \ldots, X_n$ makes a direct influence on the learning and sampling steps of EDAs. Unfortunately, this fitness value variable is typically continuous and it is therefore difficult to apply learning algorithms that are able to handle at the same time discrete and continuous variables. In addition, these learning procedures are computationally more complex and require considerable CPU time.

- **Transform the learning of the probabilistic graphical model into a supervised classification problem:** this approach is the one proposed in EBCOAs, in which all the individuals of a population are classified in different classes, and then these are used to build Bayesian classifiers that have the form of a Bayesian network. These Bayesian networks have the characteristic of including the class-variable in the probabilistic graphical model as the parent of the rest of the variables $X_1, \ldots, X_n$. The main idea of this approach is to guide the search taking into account both the genotypes of the fittest and the less fit individuals.

EBCOAs are not the only approach in the literature that apply classification techniques in optimization. The most relevant statistical approach that follows a similar idea is the Learnable Evolution Model (LEM) [20] –in which an original machine learning method based on inductive rules is applied– although other approaches that apply decision trees [18] and hill climbing methods [2] can also be found. Examples on the use of classification paradigms in optimization for the continuous domain can also be found in the literature, such as the use of Gaussian modeling [1].

2.2 Main Steps of the EBCOA Approach

EBCOAs can be regarded as an Evolutionary Computation approach very similar to EDAs in the sense that there are steps such as the learning of a probabilistic graphical model and the posterior sampling of this model in order to obtain the individuals of the new population. These two steps are present in both paradigms, although the most relevant differences are precisely in the type of Bayesian network to build: in EDAs the learning algorithms applied are general purpose Bayesian network induction algorithms while EBCOAs build Bayesian classifiers using the information provided by the fitness function. In order to better compare the main differences between these two paradigms, Figs. 1 and 2 illustrate the EDA and EBCOA approaches respectively. These two figures evidence this difference, in which it appears clearly that the probabilistic graphical model learned in EBCOAs contains the additional class variable which is the parent of the rest, and is denoted as C. Note also the difference in both approaches regarding the selection of individuals –in EDAs– or the division in classes following a supervised classification approach made in EBCOAs. Figures 3 and 4 show the pseudocode of these two different paradigms for a clearer explanation of these main differences.

We denote by $X = (X_1, \ldots, X_n)$ an n–dimensional random variable, and we represent by $x = (x_1, \ldots, x_n)$ one of its possible instantiations –that is, one of the possible individuals. The probability distribution of X will be denoted as $p(x)$, and

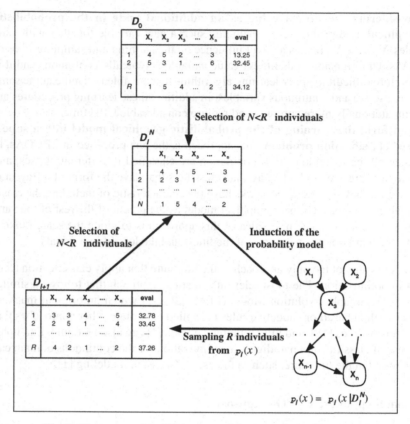

Fig. 1. Illustration of the EDA approach in the optimization process

the conditional probability distribution of the variable X_i given the value x_j of the variable X_j will be written as $p(x_i|x_j)$.

Let D_l be the population (database) of the l-th generation, formed by R individuals. This population has to evolve to the $(l+1)$-th one. In EBCOA, instead of having a selection of the fittest step as in EDAs, the population D_l is firstly divided in $|E|$ different classes, where following a supervised classification approach the variable E is defined so that to take the values $\{1, 2, \ldots, |E|\}$. We denote by D_l^E the database D_l after being divided in $|E|$ classes, in which each of the individuals in the population is assigned to a class of the variable E. In many cases we will be interested in enhancing the characteristics of the fittest and least fit classes, and therefore it is very likely not to use all the different classes for the learning. Therefore, we select $|C| \leq |E|$ classes and we ignore the rest of them for learning the Bayesian classifier. We denote by D_l^C the subset of D_l^E that will be used for the learning, and similarly we denote by C the variable that assigns a class c –with $1 \leq c \leq |C|$– to each of the individuals in D_l^C.

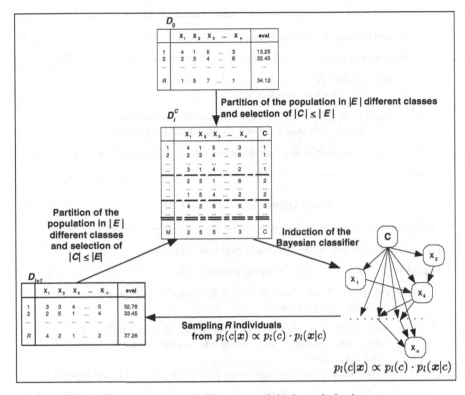

Fig. 2. Illustration of the EBCOA approach in the optimization process

The hardest task in EBCOA and the most critical one regarding the convergence aspect, is the estimation of $p_l(\boldsymbol{x}, c)$. This probability is estimated from the Bayesian classifier that is learned every generation. As EBCOAs are based on Bayesian classifiers, the Bayesian network structure S that is induced as a result of the learning step contains the variables $X_1, \ldots, X_n$ as in EDAs, but also the newly defined variable C, and it will always be the parent of all the other variables in S. The next section introduces the main characteristics of the different methods for building Bayesian classifiers. See [22] for more details.

2.3 Bayesian Classifiers

The problem of *supervised classification* consists of assigning one of the $|C|$ classes of a variable C to a vector $\boldsymbol{x} = (x_1, \ldots, x_n)$. The true class is denoted by c and it takes values in $\{1, 2, \ldots, |C|\}$. Following this definition, a classifier can be seen as a function $\gamma : (x_1, \ldots, x_n) \rightarrow \{1, 2, \ldots, |C|\}$ that assigns a class label to each observation.

The loss function is defined as the cost of misclassifying an observation. A loss function $0/1$ is defined as a loss function in which the cost of misclassifying an element is always 1. In this case, it has been demonstrated that the optimum Bayesian

$D_0 \leftarrow$ Generate R individuals (the initial population) at random

Repeat for $l = 0, 1, \ldots$ until satisfying a stopping criterion

    $D_l^N \leftarrow$ Select $N < R$ individuals of D_l following
        a certain selection method

    $p_l(\boldsymbol{x}) = p(\boldsymbol{x}|D_l^N) \leftarrow$ Estimate the distribution of probability
        for an individual to be among the selected individuals

    $D_{l+1} \leftarrow$ Sample R new individuals (the new population) from $p_l(\boldsymbol{x})$

Fig. 3. Generic pseudocode of EDA

$D_0 \leftarrow$ Generate R individuals (the initial population) randomly

Repeat for $l = 0, 1, 2 \ldots$ until a stopping criterion is met

    $D_l^E \leftarrow$ Divide the R individuals in $E < R$ different classes from D_l
        according to a criterion

    $D_l^C \leftarrow$ Select the $C \leq E$ classes of D_l^E that will be used for building the
        Bayesian classifier, usually taking into account at least the best
        and worst classes.
        The individuals of the classes not included in $D_l^C \subset D_l^E$ are ignored

    $p_l(c|\boldsymbol{x}) \propto p_l(c) \cdot p_l(\boldsymbol{x}|c) \leftarrow$ Estimate the probability distribution of an individual
        in D_l^C of being part of any of the different possible C classes

    $D_{l+1} \leftarrow$ Sample R individuals (the new population) from $p_l(\boldsymbol{x}|c)$

Fig. 4. Generic pseudocode for the EBCOA approach

classifier that minimizes the total misclassification error cost is obtained by assigning to the observation $\boldsymbol{x} = (x_1, \ldots, x_n)$ the class with the highest a posteriori probability [5].

This optimum Bayesian classifier is expressed more formally as follows:

$$\gamma(\boldsymbol{x}) = \arg \max_c p(c|\boldsymbol{x}) \tag{1}$$

In informative or generative classifiers such as the ones that are revised in this section, $p(c|x_1, \ldots, x_n)$ is obtained indirectly by applying the Bayes rule:

$$p(c|x_1, \ldots, x_n) \propto p(c, x_1, \ldots, x_n) \propto p(c)p(x_1, \ldots, x_n|c)$$

This section revises Bayesian classifiers that have been proposed specifically for classification problems.

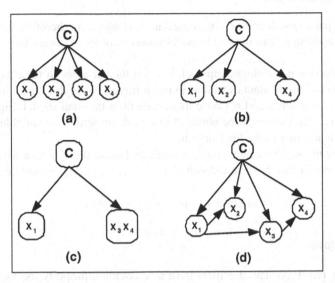

Fig. 5. Example of graphical structures of different Bayesian classifiers for a problem of four variables. The Bayesian classifiers presented are examples of (**a**) Naive Bayes, (**b**) Selective Naive Bayes, (**c**) Seminaive Bayes, and (**d**) Tree Augmented Naive Bayes

Naive Bayes

The naive Bayes approach [21] is the most simple one presented in this section. The Bayesian network structure that is applied is always fixed: all the variables $X_1, \ldots, X_n$ are considered to be conditionally independent given the value of the class value C. Figure 5(a) shows the structure that would be obtained in a problem with four variables.

In the naive Bayes classifier, when classifying an example $\boldsymbol{x}$, this will be assigned to the class c which has a higher a posteriori probability. This a posteriori probability is computed as follows:

$$p(c \mid \boldsymbol{x}) \propto p(c, \boldsymbol{x}) = p(c) \prod_{i=1}^{n} p(x_i|c) \tag{2}$$

The a priori probability of the class, $p(c)$, and the conditional probabilities $p(x_i|c)$ are estimated from the database of cases.

Selective Naive Bayes

The main restriction of naive Bayes is that this model forces the classifier to take into account all the variables. This aspect appears to be a drawback for some classification problems, since some of the observed variables could be irrelevant or redundant for classification purposes. Furthermore, it is known [13, 17] that the behavior of the naive Bayes paradigm degrades with redundant variables, and therefore the motivation for this approach is to remove those variables in order to obtain more efficient classifiers.

In the selective naive Bayes approach [14, 16] the variables in the classifier are considered to be independent as well as in naive Bayes, but in this case some of the variables can be ignored and not have them present in the final model. Figure 5(b) shows the structure that could be obtained in a problem with four variables, where one of them is missing in the final structure.

Following the selective naive Bayes model, and using the selective naive Bayes classifier shown in Fig. 5(a), an individual $x = (x_1, x_2, x_3, x_4)$ would be assigned to the class

$$c^* = \arg\max_c p(c)p(x_1|c)p(x_2|c)p(x_4|c) \tag{3}$$

Seminaive Bayes

The previous two Bayesian classifiers have as a common property the fact that all the variables in the structure are considered to be conditionally independent. That is, all the variables can have uniquely the class variable C as a parent. The seminaive Bayes approach [15] is able to take into account dependencies between the variables $X_1, \cdots, X_n$ as it allows groups of variables to be considered as a single node in the Bayesian network. Figure 5(c) illustrates an example of a seminaive Bayesian classifier in a problem with four variables, showing that the Bayesian network structure treats those grouped variables as a single node regarding the factorization of the probability distribution. The grouping of variables as a single node means that all the dependencies between them are considered implicitly for classification purposes. On the other hand, and similarly as in selective naive Bayes, in seminaive Bayes it is also allowed that some variables are not included in the final classifier (Fig. 5(c) shows an example of this).

In [25] we can find a greedy approach to build seminaive Bayes classifiers, in which redundant as well as dependent variables are detected. When dependent variables are found, a new variable is created as the cartesian product of these. Two greedy algorithms are presented, the first of them on a forward direction called *FSSJ (Forward Sequential Selection and Joining)*, and the second on the opposite backward direction named *BSEJ (Backward Sequential Elimination and Joining)*. The pseudocode of *FSSJ* is shown in Fig. 6. The algorithm *BSEJ* follows an analogous approach, and can be interesting in optimization problems in which the objective function depends on all or nearly all the variables. Figure 5(c) shows an example of a possible structure for the Bayesian classifier that could be obtained as a result of the seminaive Bayes approach.

Therefore, applying the seminaive Bayes model and using the final classifier obtained in Fig. 5(c), an individual $x = (x_1, x_2, x_3, x_4)$ would be assigned to the following class:

$$c^* = \arg\max_c p(c)p(x_1|c)p(x_3, x_4|c) \tag{4}$$

Initialize the set of variables to be used to the null set

Classify all the examples as being of the class with higher $p(c)$

Repeat in every iteration: choose the best option between

    (a) Consider each variable that is not in the model as a new one to be
        included in it. Each variable should be added as conditionally
        independent of the variables in the model given the class

    (b) Consider grouping each variable not present in the model with a variable
        that is already in it

    Evaluate each possible option by means of the estimation of the percentage
        of cases well classified

Until no improvement can be obtained

Fig. 6. Pseudocode of the *FSSJ* algorithm for seminaive Bayes models

Tree Augmented Naive Bayes

The tree augmented naive Bayes [7] is another Bayesian network classifier in which dependencies between the variables $X_1, \cdots, X_n$ are also taken into account, conditioned to the class variable C, by using a tree structure. In seminaive Bayes a wrapper approach[3] is applied to search for a good structure. In the tree augmented naive Bayes algorithm the method follows a procedure analogous to the filter approaches where only pairwise dependencies are considered.

The tree augmented naive Bayes structure is built in a two phase procedure illustrated in Fig. 7. Firstly, the dependencies between the different variables $X_1, \ldots, X_n$ are learned by means of a score based on the information theory. The weight of a branch (X_i, X_j) is given by $I(X_i, X_j | C)$, which is the mutual information measure conditioned to the class variable. These conditional mutual information values are used to build a tree structure. Secondly, the structure is augmented to the naive Bayes paradigm.

Following the tree augmented naive Bayes model building pseudocode presented in Fig. 7, if we obtain the Bayesian classifier structure shown in Fig. 5(d) an individual $x = (x_1, x_2, x_3, x_4)$ will be assigned to the class

$$c^* = \arg \max_c p(c)p(x_1|c)p(x_2|c, x_1)p(x_3|c, x_1)p(x_4|c, x_3) \tag{5}$$

2.4 Description of the Main Steps of EBCOAs

Evolutionary Bayesian Classifier-based Optimization Algorithms (EBCOAs) is an approach that combines Bayesian classifiers such as the ones presented in the previous section and Evolutionary Computation to solve optimization problems. The

---

[3] The wrapper approach consists of applying the induction algorithm itself as a part of the evaluation function. On the other hand, the filter approach looks only at the intrinsic characteristics of the data, such as probabilistic or distance scores, or the mutual information

---

Calculate $I(X_i, X_j \mid C) = \sum_{i=1}^{n} \sum_{j=1}^{m} \sum_{r=1}^{w} p(x_i, y_j, c_r) \log \frac{p(x_i, y_j \mid c_r)}{p(x_i \mid c_r) p(y_j \mid c_r)}$
   with $i < j, j = 2, \ldots, n$

Build an undirected complete graph, where the nodes correspond to the predictor
   variables: $X_1, \ldots, X_n$. Assign the weight $I(X_i, X_j \mid C)$ to the edge connecting
   variables X_i and X_j

Assign the largest two branches to the tree to be constructed

Repeat in every iteration:
   Examine the next largest branch and add it to the tree unless it forms a loop.
   In the latter case discard it and examine the next largest branch
Until $n - 1$ branches have been added to the structure

Transform the undirected graph in a directed one, by choosing a random
   variable as the root
Build the tree augmented naive Bayes structure adding a node labelled as C, and
   later add one arc from C to each of the predictor variables X_i $(i = 1, \ldots, n)$

Fig. 7. Pseudocode of the *tree augmented naive Bayes* algorithm

main idea is that, having a population of solutions for the optimization problem, this population will be evolved to a new generation formed by a next population of fitter individuals. The main difference between EDAs and EBCOAs is that in EDAs the evolution to the next population is performed by learning a probabilistic graphical model using uniquely the information of the best individuals, ignoring simply the worst ones, whilst in EBCOAs these worst individuals are also taken into account. The EBCOA approach is based on constructing a Bayesian classifier that will represent the main characteristics between the fittest and the least fit individuals. This approach, illustrated in Fig. 2, contains the following steps:

- Firstly, the initial population D_0 of R individuals is generated. This initial population is generated usually by assuming an uniform distribution on each variable, similarly as in EDAs. Each of the created individuals is evaluated by means of the fitness function.

- Secondly, each of the individuals in D_l are given a label $e \in E$ to classify them regarding their respective fitness value. This is the supervised classification step, in which each of the R individuals is assigned an e label. As a result of this, the class variable E is created in the new database denoted by D_l^E.

- Thirdly, D_l^C is created by selecting from D_l^E only the $|C| \leq |E|$ classes that will be used for building the Bayesian classifier. In EBCOAs we take into account uniquely the best and worst classes of individuals. The rest of the classes in D_l^E could be discarded to facilitate the learning by enhancing the differences between the most distant classes. The individuals which are in $D_l^E \setminus D_l^C$ are simply ignored.

- A Bayesian classifier is built using the database D_l^C and applying Bayesian classifier model construction techniques such as the ones described in the previous section. This classifier estimates the probability distribution $p_l(c|\boldsymbol{x}) \propto p_l(c)p_l(\boldsymbol{x}|c)$ which represents the probability of any individual $\boldsymbol{x}$ to be classified in any of the different possible $|C|$ classes.

- Finally, the new population D_{l+1} constituted by the R new individuals is obtained by carrying out the sampling of the probability distribution $p_l(c)p_l(\boldsymbol{x}|c)$. This step can be performed very similarly as in EDAs.

Steps 2, 3, 4 and 5 are repeated until a stopping criterion is satisfied. Examples of stopping conditions are: achieving a fixed number of populations or a fixed number of different evaluated individuals, uniformity in the generated population, and the fact of not obtaining an individual with a better fitness value after a certain number of generations.

The performance of EBCOAs is mainly responsibility of the step of learning the Bayesian classifier. In order to compare possible classifiers to be applied in EBCOAs, the most strict criterion to be used would be the use of an honest validation using the initial database of cases. However, this procedure is very expensive in computation time and very often this is approximated using a filter approach. It is important to note that in EBCOAs the most important criterion when choosing a Bayesian classifier is not to apply the one that best represents a strictly correct classifier, since the convergence speed and computation time are also important aspects to take into account. Indeed, the best Bayesian classifiers are usually the most time consuming ones. A balance between these two performance criteria is required since this learning step (i.e. the classifier building step) is going to be applied every generation.

3 The Satisfiability Problem

The SAT is one of the most known problems in computational theory because it results in a generic model in which many different decision making problems can be represented. The SAT problem is known to be NP-hard.

This particular problem has been analysed for many years and diverse methods have been applied for its resolution. Evolutionary computation methods have also been applied to it, mainly GAs, but also EDAs [11].

3.1 Definition of the Optimization Problem

The SAT problem is proposed in a formula in conjunctive normal form composed by a set of clauses. Each clause is formed by a set of literals, and a literal is a variable or its negation. A possible interpretation for the problem is a function that assigns a boolean value to each of the variables. An instantiation is said to be satisfiable if it satisfies every single clause. In order to satisfy a clause, it is necessary that at least one of the literals of the clause is satisfied by the instantiation.

The SAT problem is based on a set of Boolean variables $\mathbf{Z} = (Z_1, \ldots, Z_v)$, and a Boolean function $g : B^v \rightarrow B, B = \{0, 1\}$. The formula g is in conjunctive normal form $g(\mathbf{z}) = c_1(\mathbf{z}) \wedge \ldots \wedge c_u(\mathbf{z})$, where u is the number of clauses of the problem and each clause c_i is a disjunction of literals. A literal is a variable or its negation. A literal z_i is said to be satisfied if the value 1 is assigned to z_i, while the literal $\bar{z}_i$ is satisfied if the value 0 is assigned. A SAT instance is called *satisfiable* if there exists such an **z**, and otherwise it is called unsatisfiable. Here the aim is to analyze and check the existence of an assignation of $\mathbf{z} = (z_1, \ldots, z_v) \in B^v$ such that $g(\mathbf{z}) = 1$.

For example,

$$g(\mathbf{z}) = (z_1 \vee z_3 \vee \bar{z}_4) \wedge (z_2 \vee \bar{z}_3 \vee z_4) \wedge (\bar{z}_1 \vee z_2 \vee \bar{z}_3) \tag{6}$$

This instance is considered satisfiable if all its clauses are satisfied, and each of the single clauses is satisfiable if at least one of the literals is according to Z. For instance,

$$\mathbf{z} = (z_1, z_2, -, z_4) \tag{7}$$

satisfies the instance, where the value $-$ in Z_3 means that this variable can take the positive or negated value.

One might find easy to find a solution to this type of problems, but when we increase the number of variables and clauses the problem becomes very difficult to be solved.

The sub-type of SAT problems called k-SAT is defined as the one that contains in each clause exactly k different literals. While the 2-SAT class is solved in polynomial time, k-SAT appears to be NP-hard for $k \geq 3$. In 1971, Cook [3] showed that this problem is NP-hard and that the SAT problem can be understood as a representative for solving other NP-hard problems.

The optimization version of the SAT problem is known as MAXSAT, and it consists on finding an assignation that maximizes the number of clauses satisfied for a particular SAT problem. If the original set of clauses is insatisfiable, MAXSAT is supposed to search for the assignation that satisfies the highest number of clauses possible.

3.2 Related Research Work

SAT is a generic method for problem solving that has been analysed for many years in the literature using very different approaches. This section concentrates on the work oriented to solve SAT by means of heuristic methods, and more precisely on Evolutionary Computation methods. In 1989, in [4] the authors apply GAs to the SAT problem to transform other NP-hard problems into SAT. This initial work concludes that GAs cannot perform better for the SAT, than other problem-specific algorithms, although GAs are referred to as a robust, efficient, and promising method.

More recent works apply different ways of representing solutions, techniques to adapt the definition of the fitness function, variations in crossover and mutation operators, or local optimization techniques. The aim of all these techniques in most of the cases is to improve the results obtained by GAs. As an example of these, [10] proposes to concentrate on the importance of the clause regarding the problem and proposes a new individual representation to guide the search following this idea.

In [9] this option and its corresponding fitness function with improvements is compared to using other type of representations, which also enhances the relevance of the clause, although the paper concentrates on the need to satisfy a single literal in each clause so that it becomes satisfied.

Finally, we would like to mention the work [8] in which the authors show an interesting revision of evolutionary algorithms for the SAT problem.

3.3 Representation Types and Their Associated Fitness Function

When solving any optimization problem using Evolutionary Computation techniques we need to choose a means of representing each of the possible solutions and to define a fitness function that measures how good each solution is for this problem. These two components are critical for an appropriated convergence of the algorithm. We can find in the literature many different options to represent a solution for the SAT problem. Each of these representations has a different appropriated fitness function. Next, we will analyse three different individual representations and their respective fitness functions: the first of them focuses on the individual literals that form the problem, while the other two are concentrated on the clauses and on how to satisfy them.

Bit-String Representation

This option is one the most applied in the literature, since it is the most natural way of representing the SAT problem. It consists on representing each possible solution as a string of bits of length v: $(z_0, \ldots, z_v)$, where v is the number of variables of the problem, and each variable Z_i is associated to a bit.

In this case, the aim is to satisfy the maximum number of clauses. Taking into account that the MAXSAT fitness function consists on counting the number of clauses that are satisfied, and the idea is to maximize this value, in the MAXSAT formulation the fitness value is equivalent to the number of satisfied clauses directly in the following way:

$$g_{MAXSAT}(\mathbf{z}) = c_1(\mathbf{z}) + \ldots + c_u(\mathbf{z}) \tag{8}$$

where c_i represents the true value of the ith clause.

> EXAMPLE 1:
> Given Equation 6, a possible solution would be $\mathbf{z}_a = (1, 0, 0, 0)$ where $g_{MAXSAT}(\mathbf{z}_a) = 3$, since all the clauses are satisfied with this particular assignation.
> On the other hand, if we consider another individual $\mathbf{z}_b = (1, 0, 1, 0)$, we have that $g_{MAXSAT}(\mathbf{z}_b) = 1$, since the second and third clauses are not satisfied.

Path Representation

This representation was suggested in [9]. It is based on the idea that for satisfying each particular clause it is enough with satisfying a single literal. Therefore, in order to create a possible satisfiable solution, it also would be possible to select a clause each time and generate a continuous path among clauses by choosing a literal in each single clause. That is, in this case a possible solution has as many variables as clauses, and each variable takes a value in $1, \ldots, k$ (k-SAT) which expressed the literal that is chosen in a particular clause.

The obvious problem created using this idea is that the same literal that appears in different clauses with different boolean values can be selected in two positions of the path, creating inconsistencies.

> EXAMPLE 2:
> Given Equation 6, the path (1-1-3), which means $z_1 = 1, z_2 = 1$, satisfies
>   the SAT problem and induces the next complete assignations to literals:
> $\mathbf{z} = (1, 1, 0, 0), \mathbf{z} = (1, 1, 0, 1), \mathbf{z} = (1, 1, 1, 0), \mathbf{z} = (1, 1, 1, 1)$.
> On the other hand, the path (1-3-1), $z_1 = 1, z_4 = 1, z_1 = 0$, contains an
>   inconsistency for z_1.

A reasonable fitness function for this representation is the one that measures the number of inconsistencies, with the aim of minimizing this function.

If the path with a smaller number of inconsistencies has at least a single one, that particular SAT problem is not satisfiable. Otherwise, the fact of not having a single inconsistency in the path means that it exists at least a sequence of assigning variables that satisfies the SAT problem.

Clause Representation

A new individual representation also based on the clause-variables is proposed in [10], in which the effect of each of the variables within the clause is stressed. In other words, the main idea is to concentrate initially on finding an assignment of values to satisfy individual clauses, and next to search for the particular assignment of values in the literals of them to satisfy the problem globally.

Each clause with k literals can be regarded as to take $2^k - 1$ possible combinations of value assignments to its literals that satisfy the clause, while only one would not satisfy it. In the particular case of a 3-SAT problem, there exist eight different possible values for each clause, and only one of would not satisfy the clause. Therefore, in this representation the main idea is to choose any of the $2^k - 1$ combinations of literal values in the clause that would satisfy it.

Following this idea, we can obtain an assignment that would satisfy all the clauses. However, this representation is likely to propose different values (according to whether the variable is negated or not) for the same literal which is present in different clauses. This case would result in an inconsistency in a similar way than in the previous path representation. Therefore, the idea is to choose any other clause-value so that the total number of inconsistencies among clauses is minimized.

Just as a simple example, for a clause $(z_2 \vee \bar{z}_3 \vee z_4)$, the assignation of variables not satisfied by this clause would be $(z_2, z_3, z_4) = (0, 1, 0)$. Next, we can see the illustration of how to use this representation in an overall 3-SAT problem.

EXAMPLE 3:
Given Equation 6 the forbidden assignations are: (0 0 1), (0 1 0),
(1 1 0) respectively, which correspond to the representation (1 2 6). A possible solution is (0 1 1), that is (0 0 0), (0 0 1), (0 0 1),
$000 \rightarrow z_1 = 0, z_3 = 0, z_4 = 0,$
$001 \rightarrow z_2 = 0, z_3 = 0, z_4 = 1,$
$001 \rightarrow z_1 = 0, z_3 = 0, z_2 = 1,$
however, this solution presents inconsistencies in z_2 and z_4.
On the other hand, the solution (2 6 3), that is (0 1 0), (1 1 0),
(0 1 1), does not present inconsistencies.
$010 \rightarrow z_1 = 0, z_3 = 1, z_4 = 0,$
$110 \rightarrow z_2 = 1, z_3 = 1, z_4 = 0,$
$011 \rightarrow z_1 = 0, z_3 = 1, z_2 = 1,$
therefore there exists a global assignation of variables (0 1 1 0) that satisfies the problem.

Similarly as with the previous representation, the fitness function proposed for this representation measures the amount of inconsistencies, and the aim of optimization algorithms is to minimize it.

When the clause representation is used, if the solution with the less possible number of inconsistencies has at least one inconsistency, the SAT is not satisfiable. Otherwise, the SAT problem is satisfiable.

3.4 Local Optimization

We propose in this section two types of local optimization techniques to converge to a satisfactory solution. The local optimization types presented in this section are applied in this case to EBCOAs, although they can also be used in other evolutionary computation methods. These techniques are suggested to be applied after the search process arrives to a determined generation number.

Firstly, we introduce a local optimization technique that flips each of the values in the individual after it has been generated, evaluating the fitness of each of the flips. If the individual is not improved, the original individual is included in the next generation, and otherwise, the fittest flipped version is included in the population. We call this method *local optimization with flip*.

Secondly, we present a method that detects the most difficult clauses of the problem to be satisfied, assigning them a higher weight in an adapted fitness function. This method is called *optimization through adaptation of weights*.

These local optimization procedures try to improve the search for a satisfiable version. Initially, the fitness function guides the search in a very efficient way, but due to the ambiguity on assigning the same fitness value to very different individuals that are given rise by many individual representations and fitness functions, this guidance is not precise enough to guide the algorithm towards finding a satisfiable solution in the search space. We suggest to apply this local optimization procedure once the search has found a number of different individuals that have the same best fitness value, since the algorithm will have no appropriated means to continue the search. The key decision here is to choose the best time to start applying the local optimization. The criterion that we propose is to start applying the local optimization when all the individuals of the highest (fittest) class have all the same fitness value. Other options are also possible, for instance to start to apply it once a concrete generation number has been reached.

Local Optimization with Flip

The local optimization that we propose in this section is based on the FlipGA method proposed in [19] for GAs. The main idea of FlipGA consists on applying the "Flip" heuristic to each individual after the crossover and mutation operations. The heuristic explores the genes randomly: each gen is "flipped", and this modification is accepted only if the improvement is bigger or equal than zero (that is, the number of satisfied clauses after the modification is the same or higher).

We propose in this section a procedure to apply local optimization as follows:

1. Each generated individual will be evaluated and its variables that worsen the individual – i.e. they make a clause not to be satisfied or create inconsistencies – will be marked.
2. Next, the marked variables are taken individually, and each time we modify a single variable and re-evaluate the individual. If a better fitness value is obtained, this modification is kept for this variable, otherwise we discard it.

Depending on the type of individual representation chosen, the local optimization will be applied in a different way.

Bit-string representation: The variables of the individuals represent directly the literals of the problem, while in the path and clause representations the variables of the individual represent the different clauses and the value of the literals are implicit.

Path representation: When a variable creates an inconsistency this means that for the same literal a clause has assigned a value while in another the value is the opposite. When this is identified, the literal that creates an inconsistency is marked in order to choose another literal of the clause and try to satisfy it. This helps at finding a satisfiable solution.

Clause representation: When a variable of the individual that creates an inconsistency, this means that the value of at least a literal that has been assigned to satisfy the clause is said to take a different value in two different clauses. This variable of the individual is marked and the local optimization will try to choose another combination of values of the clause to satisfy it and avoid the inconsistency.

Optimization Through Adaptation of Weights

The improvement presented in this section proposes to consider the different clauses differently according to the difficulty to satisfy each of them. This idea has been proposed for the first time in [6], and they apply the stepwise adaption of weights principle (SAW) combined with the fitness function.

$$g_{SAW}(\mathbf{z}) = w_1 \cdot c_1(\mathbf{z}) + \ldots + w_u \cdot c_u(\mathbf{z}) \tag{9}$$

The weights $w_i \in N$ are adapted to identify the most difficult clauses to satisfy in each step of the search. Initially, all the weights are initialized to 1 ($w_i = 1$), which is equivalent to apply a MAXSAT function. After some time, the weights are adapted according to $w_i \leftarrow w_i + 1 - q_i(z^*)$, where $q_i(z^*)$ is the actual fitness. This adaptation increases only the weights that correspond to clauses that have not been satisfied at that time. This forces implicitly the evolutionary search focusing it on these *difficult* clauses, and therefore guiding the search process applying these weights.

Here one of the tasks to perform is to decide when to start applying the optimization to the search. Similarly as in the previous case, we consider that the search process must start per se without applying the optimization procedure until some concrete conditions are detected. A possible proposal is to start applying this procedure when reaching a concrete generation number, or also when all the individuals in the fittest class have the same fitness value. Another decision to take is the number of generations after which we will adapt periodically the vector of weights.

4 Experimental Results

4.1 Comparison of the Different EBCOAs

In order to perform some experiments we have chosen *3*-SAT instances randomly from the collection of problems offered publicly by SATLIB [12]. The files selected contain definitions of different SAT problems for which we know that a satisfiable solution exists. We have used 5 different problem instances with 20 literals and 91 clauses each. For each instance we run each EBCOA presented in Sect. 2.3 (EBCOA$_{NaiveBayes}$, EBCOA$_{SelectiveNB}$, EBCOA$_{FSSJ}$, EBCOA$_{BSEJ}$ and EBCOA$_{TAN}$) 10 times under the same conditions. For each algorithm we have also applied three exposed possible individual representations and their respective fitness functions as described in Sect. 3.3.

Table 1 shows the results of these experiments for each individual representation type and EBCOA algorithm, without the addition of any local optimization at all. The column *Porc.* shows the percentage of runs in which a satisfiable solution has been found. The *Dif.* column represents the mean number of clauses that are not satisfied in the last generation of the runs. As we can see, this column corresponds to the percentage of runs in which a possible solution was found for the SAT problem: the value in *Dif.* is bigger when the percentage of runs in which a satisfiable solution

Table 1. Experimental results obtained with 5 different SAT problems containing 20 literals and 91 clauses. No local optimization technique has been applied in any of these results

	Bits			Path			Clause		
	Porc.	Dif.	Eval.	Porc.	Dif.	Eval.	Porc.	Dif.	Eval.
EBCOA$_{NaiveBayes}$	0.86	0.18	2222.84	0.06	1.72	54443.32	0.00	39.12	49832.62
EBCOA$_{SelectiveNB}$	0.82	0.20	2256.10	0.02	1.78	53788.66	0.00	41.78	49856.22
EBCOA$_{FSSJ}$	0.98	0.02	2472.10	0.00	2.76	50253.64	0.00	53.56	49753.52
EBCOA$_{BSEJ}$	0.92	0.08	1700.60	0.02	1.74	52330.42	0.00	49.50	49881.02
EBCOA$_{TAN}$	0.52	0.66	5285.12	0.00	4.60	52767.50	0.00	26.02	49639.46

has been found. On the other hand, the *Eval.* column represents the number of different individuals that have been evaluated, that is, the number of different individuals that have been analyzed in the search process. The stopping criterion chosen for all the cases is the fact of finding a satisfiable solution or to reach a maximum of 500 generations.

As we can see in Table 1 the bit-string representation is the one that shows the best performance. The other two representations obtain considerably worse results due to having a bigger search space and more different individuals with the same fitness value. The EBCOA that performed best is the one that learns a seminaive Bayes classifier, with which we obtain a satisfiable version in more than 90% of the executions. In the case of EBCOA$_{FSSJ}$ we obtain even a result of 98%. EBCOA$_{TAN}$ is the one that obtains a satisfiable solution in only 50% of the executions.

4.2 The Result of Applying Local Optimization

In order to analyse the effect of local optimization in EBCOAs, we have performed a set of experiments to compare the performance of not applying local optimization, applying the optimization described in Sect. 3.4, and the optimization through adaptation of weights of Sect. 3.4. Table 2 shows the results obtained in our experiments.

On the other hand, when local optimization with flip is applied more balanced results are obtained, and the differences between EBCOAs are reduced. This is the result on a better guidance for the different EBCOAs over the search space. EBCOA$_{TAN}$ improves its performance to the 76% of the executions, and EBCOA$_{FSSJ}$ obtains a lower percentage of success, although it is still far behind the 90%.

When the weight adaptation optimization is applied, a satisfiable solution is obtained in the 100% of the cases when applying an EBCOA based on a Selective Naive Bayes classifier. For the rest of EBCOAs the success rate is also quite high (98% for EBCOA$_{NaiveBayes}$ and EBCOA$_{FSSJ}$) except for EBCOA$_{TAN}$ which remains in a 72%.

It is also important to note that the effect of the individual representation is very important in EBCOAs. Other studies in the literature also show that the path representation has a worse performance than bit-string representation even when applied to GAs [9]; however, under the conditions of our experiments using EBCOAs the

Table 2. Experimental results obtained with 5 different SAT problems containing 20 literals and 91 clauses

	Results applying local optimization with flip								
	Bits			Path			Clause		
	Porc.	Dif.	Eval.	Porc.	Dif.	Eval.	Porc.	Dif.	Eval.
EBCOA$_{NaiveBayes}$	0.92	0.08	1326.24	0.16	1.40	51113.16	0.00	12.52	49656.04
EBCOA$_{SelectiveNB}$	0.80	0.22	2274.98	0.12	1.48	50822.18	0.00	12.90	49654.14
EBCOA$_{FSSJ}$	0.92	0.08	2137.82	0.00	2.20	49956.84	0.00	29.28	49655.04
EBCOA$_{BSEJ}$	0.94	0.08	1806.74	0.06	2.02	48897.10	0.00	15.04	49600.00
EBCOA$_{TAN}$	0.76	0.28	3437.08	0.00	4.10	50130.92	0.00	70.44	49625.38

	Results applying weight adaptation optimization								
	Bits			Path			Clause		
	Porc.	Dif.	Eval.	Porc.	Dif.	Eval.	Porc.	Dif.	Eval.
EBCOA$_{NaiveBayes}$	0.98	0.04	2155.10	0.58	0.76	31726.54	0.00	12.58	49600.00
EBCOA$_{SelectiveNB}$	1.00	0.00	4912.80	0.52	0.88	34284.70	0.00	11.42	49600.00
EBCOA$_{FSSJ}$	0.98	0.04	3105.50	0.00	5.86	49600.00	0.00	33.86	49600.00
EBCOA$_{BSEJ}$	0.94	0.08	5420.42	0.64	0.76	32599.72	0.00	12.38	49600.00
EBCOA$_{TAN}$	0.72	0.38	14698.92	0.00	6.42	49600.00	0.00	47.86	49600.00

performance is quite poor. EBCOAs are able to find a satisfiable version when using this representation, and only if we apply weight adaptation optimization we manage to find satisfiable solutions in more than 50% of the cases – but not for all the EBCOA types. Results are even worse for the case of clause representation, since no EBCOA did manage to find a satisfiable solution in any of the executions.

4.3 Performance of EBCOAs Depending on the Complexity of the SAT Problem

It is a common practise to define the complexity of a SAT problem by using a u/v ratio, where u is the number of clauses and v is the number of variables. In all our experiments, we use instances with ratio $u/v \approx 4, 3$, which have been reported by [23] to be the hardest instances. With this ratio, we performed an experiment to analyse the performance of EBCOAs when increasing the number of literals of the SAT problem. For this, we have applied other problems from SATLIB [12] containing SAT with 50 literals and 218 clauses. We chose again 5 different instances of these SAT problems and we run 10 times each EBCOA using the different local optimization possibilities. This time we have applied uniquely the bit-string representation. The results obtained are presented in Table 3.

As we can see in this table, the tendency shown in the previous experiments is kept for this case, although the increase in complexity results in a much lower performance than the case with 20 literals.

Table 3. Results for a SAT problem with 50 literals and 218 clauses. These results have been obtained using a bit-string representation

Results without local optimization

	Porc.	Dif.	Eval.
EBCOA$_{NaiveBayes}$	0.18	1.62	15933.72
EBCOA$_{SelectiveNB}$	0.08	1.78	16633.78
EBCOA$_{FSSJ}$	0.34	1.14	98698.86
EBCOA$_{BSEJ}$	0.14	1.90	17962.96
EBCOA$_{TAN}$	0.00	4.06	25691.80

Results applying local optimization with flip

	Porc.	Dif.	Eval.
EBCOA$_{NaiveBayes}$	0.10	1.62	13620.32
EBCOA$_{SelectiveNB}$	0.20	1.36	14232.72
EBCOA$_{FSSJ}$	0.46	0.94	90296.14
EBCOA$_{BSEJ}$	0.26	1.44	14749.36
EBCOA$_{TAN}$	0.00	3.76	20471.18

Results applying weight adaptation optimization

	Porc.	Dif.	Eval.
EBCOA$_{NaiveBayes}$	0.50	1.06	69199.38
EBCOA$_{SelectiveNB}$	0.52	0.94	66921.32
EBCOA$_{FSSJ}$	0.62	0.74	72924.72
EBCOA$_{BSEJ}$	0.46	1.14	71453.36
EBCOA$_{TAN}$	0.10	2.82	110144.16

5 Conclusions

This chapter presents a new evolutionary computation approach called EBCOA, and its performance has been studied for a classical NP-hard problem, the satisfiability one. We have analysed the behavior of EBCOAs for three different individual representations together with the appropriated fitness function for each case. In our experimental results we demonstrate that the bit-string representation obtains the best results than the ones based on the clause structures. This is due to the fact that the bit-string representation reduces the search space since the variables of the individual are binary.

We have also applied two different local optimization options based on the specific characteristics of the SAT problem. In this sense, in one of them we differentiate the clauses which are more difficult to be satisfied and a weight is assigned to each clause regarding this aspect. The results obtained in this local optimization method are the most promising.

EBCOAs are a very new research paradigm that has still a lot of work to be done. Possible future working trends that we are currently facing are the use of other more complex Bayesian classifiers, the definition of other mechanisms to adapt the learning step in EBCOAs to take into account the a priori information of the optimization problem, and the use of other local optimization techniques designed specifically for EBCOAs.

Acknowledgements

This chapter has been partially supported by the University of the Basque Country with the project 9/UPV-EHU 00140.226-15334/2003, as well as by the Basque Government with projects ETORTEK BIOLAN, and ETORTEK GEMMODIS. The authors would like to thank these institutions for their kind support.

References

1. S. Akaho. Statistical learning in optimization: Gaussian modeling for population search. In *Proc. of 5th International Conference on Neural Information Processing*, pp. 675–678, 1998.
2. J.A. Boyan and A.W. Moore. Learning evaluation functions for global optimization and boolean satisfiability. In *Proceedings of the Fifteenth National Conference on Artificial Intelligence AAAI-98*, 1998.
3. S. Cook. The complexity of theorem-proving procedures. In J.J. Grefenstette, editor, *In Proceedings of Third Annual ACM Symposium on Theory of Computing*, pp. 151–158, New York, 1971.
4. K. de Jong and W. Spears. Using genetic algorithms to solve NP-complete problems. In J.D. Schaffer, editor, *Proceedings of the Third International Conference on Genetic Algorithms*, pp. 124–132, San Mateo, CA, 1989. Morgan Kaufmann.
5. R. Duda and P. Hart. *Pattern Classification and Scene Analysis*. John Wiley and Sons, New York, 1973.
6. A.E. Eiben and J.K. van der Hauw. Solving 3-SAT by GAs adapting constraint weights. In *Proceedings of The IEEE Conference on Evolutionary Computation, IEEE World Congress on Computational Intelligence*, 1997.
7. N. Friedman, D. Geiger, and M. Goldsmidt. Bayesian network classifiers. *Machine Learning*, 29(2):131–163, 1997.
8. J. Gottlieb, E. Marchiori, and C. Rossi. Evolutionary Algorithms for the Satisfiability Problem. *Evolutionary Computation*, 10(1):35–50, 2002.
9. J. Gottlieb and N. Voss. Representations, fitness functions and genetic operators for the satisfiability problem. In *Proceedings of Artificial Evolution. Lecture Notes in Computer Science*, volume 1363, pp. 55–68, Springer, Berlin, Germany, 1998.
10. J.-K. Hao. A clausal genetic representation and its evolutionary procedures for satisfiability problems. In D. Pearson et al., editor, *Proceedings of the International Conference on Artificial Neural Nets and Genetic Algorithms*, pp. 289–292, Vienna, Austria, 1995. Springer.
11. V. Herves, P. Larrañaga, V. Robles, J. M. Peña, M. S. Pérez, and F. Rosales. EDA paralelos multipoblación para el problema SAT. In *Proceedings of the III Congreso Español de Metaheurísticas, Algoritmos Evolutivos y Bioinspirados*, Cordoba (Spain), 2004.
12. H.H. Hoos and T. Stützle. SATLIB: An online resource for research on SAT. In I.P.Gent et al., editor, *SAT 2000*, pp. 283–292. IOS Press, 2000. SATLIB is available online at www.satlib.org.

13. I. Inza, P. Larrañaga, R. Etxeberria, and B. Sierra. Feature subset selection by Bayesian network-based optimization. *Artificial Intelligence*, 123(1-2):157–184, 2000.

14. R. Kohavi and G. John. Wrappers for feature subset selection. *Artificial Intelligence*, 97(1-2):273–324, 1997.

15. I. Kononenko. Semi-naïve Bayesian classifiers. In *Proceedings of the 6th European Working Session on Learning*, pp. 206–219, Porto, Portugal, 1991.

16. P. Langley and S. Sage. Induction of selective Bayesian classifiers. In *Proceedings of the 10th Conference on Uncertainty in Artificial Intelligence*, pp. 399–406, Seattle, WA, 1994.

17. H. Liu and H. Motoda. *Feature Selection for Knowledge Discovery and Data Mining*. Kluwer Academic Publishers, Boston, 1998.

18. X. Llorà and D.E. Goldberg. Wise breeding GA via machine learning techniques for function optimization. In Cantú-Paz et al., editor, *Proceedings of the Genetic and Evolutionary Computation Conference GECCO-03, Part I*, Lecture Notes in Computer Science 2723, pp. 1172–1183, Chicago, Illinois, 2003. Springer.

19. E. Marchiori and C. Rossi. A flipping genetic algorithm for hard 3-SAT problems. In W. Banzhaf et al., editor, *Proceedings of Genetic and Evolutionary Computation Conference*, pp. 393–400, San Francisco, CA, 1999. Morgan Kaufmann.

20. R.S. Michalski. Learnable evolution model: Evolutionary processes guided by machine learning. *Machine Learning*, 38:9–40, 2000.

21. M. Minsky. Steps toward artificial intelligence. *Transactions on Institute of Radio Engineers*, 49:8–30, 1961.

22. T. Miquélez, E. Bengoetxea, and P. Larrañaga. Evolutionary computation based on Bayesian classifiers. *International Journal of Applied Mathematics and Computer Science*, 14(3):335–349, 2004.

23. D. G. Mitchell, B. Selman, and H. J. Levesque. Hard and easy distributions for SAT problems. In P. Rosenbloom and P. Szolovits, editors, *Proceedings of the Tenth National Conference on Artificial Intelligence*, pp. 459–465, Menlo Park, California, 1992. AAAI Press.

24. H. Mühlenbein and T. Mahning. FDA - a scalable evolutionary algorithm for the optimization of additively decomposed functions. *Evolutionary Computation*, 7(4):353–376, 1999.

25. M. Pazzani. Searching for dependencies in Bayesian classifiers. In D. Fisher and H.-J. Lenz, editors, *Learning from Data: Artificial Intelligence and Statistics V*, pp. 239–248, New York, NY, 1997. Springer–Verlag.

26. G. Syswerda. Simulated crossover in genetic algorithms. In L.D. Whitley, editor, *Foundations of Genetic Algorithms*, volume 2, pp. 239–255, San Mateo, California, 1993. Morgan Kaufmann.

Feature Ranking Using
an EDA-based Wrapper Approach

Yvan Saeys, Sven Degroeve, and Yves Van de Peer

Department of Plant Systems Biology, Ghent University, Flanders Interuniversity Institute for Biotechnology (VIB), Technologiepark 927, B-9052 Ghent, Belgium
{yvan.saeys,sven.degroeve,yves.vandepeer}@psb.ugent.be

Summary. Feature subset selection is an important pre-processing step for classification. A more general framework of feature selection is feature ranking. A feature ranking provides an ordered list of the features, sorted according to their relevance. Using such a ranking provides a better overview of the feature elimination process, and allows the human expert to gain more insight into the processes underlying the data. In this chapter, we describe a technique to derive a feature ranking directly from the estimated distribution of an EDA. As an example, we apply the method to the biological problem of acceptor splice site prediction, demonstrating the advantages for knowledge discovery in biological datasets with many features.

1 Introduction

Reduction of data dimensionality has become an apparent need in machine learning during the past decades. Examples of large datasets with instances described by many features include problems in image processing, text mining and bioinformatics. To efficiently deal with such data, dimension reduction techniques emerged as a useful pre-processing step in the flow of data analysis. A subset of these techniques is referred to as feature (subset) selection techniques. These techniques differ from other reduction techniques (like projection and compression techniques) in that they do not transform the original input features, but merely select a subset of them.

The reduction of data dimensionality has a number of advantages: attaining good or even better classification performance with a restricted subset of features, faster and more cost-effective predictors, and the ability to get a better insight in the processes described by the data. An overview of feature selection techniques can be found in [10] and [5].

Techniques for feature selection are traditionally divided into two classes: filter approaches and wrapper approaches [12]. Filter approaches usually compute a feature relevance score such as the feature-class entropy, and remove low-scoring features. As such, these methods only look at the intrinsic properties of the dataset, providing a mechanism that is independent of the classification algorithm to be used afterwards. In the wrapper approach, various subsets of features are generated, and evaluated using a specific classification model. A heuristic search through the space of all subsets is then conducted, using the classification performance of the model as a guidance to find promising subsets. In addition to filter and wrapper approaches, a third class of feature selection methods can be distinguished: embedded feature selection techniques [1]. In embedded methods, the feature selection mechanism is built into the classification model, making direct use of the parameters of the induction model to include or reject features. Examples of these methods are the pruning of decision trees, and recursive feature elimination (RFE) using the weight vector of a linear Support Vector Machine [6].

In this chapter, we will focus on the wrapper approach for feature selection. Wrapper based methods combine a specific classification model with a strategy to search the space of all feature subsets. Commonly used search strategies are sequential forward or backward selection [11], and stochastic iterative sampling methods like genetic algorithms (GA, [13]) or estimation of distribution algorithms (EDA, [14]). The EDA approach to feature selection is shown in Fig. 1. In this case, each individual in the population represents a feature subset, coded as a binary string. Each bit represents a feature, a 1 indicating the presence, a 0 the absence of a particular feature. Individuals are evaluated (step 2, Fig. 1) by training a classification model with the features present in the individual (i.e. the ones having a 1), and afterwards validating it, either by cross-validation on the training set, or by using a separate training and holdout set. The feature subset returned by the algorithm is then the best subset found during the search.

Instead of using the traditional crossover and mutation operators, inherent to GA, an EDA explicitly constructs a model of the selected feature subsets (step 4). Depending on the complexity of the model, univariate, bivariate or multivariate interactions between the encoded features are modelled. In a subsequent step (step 5), the new population is created by sampling feature subsets from this model. The new population can either be completely sampled from the distribution, or can partly consist of sampled subsets and subsets retained from the previous population (elitists). The use of EDAs for feature subset selection was pioneered in [8] and the use of EDAs for FSS in large scale domains, was reported to yield good results [9, 22].

2 EDA-based Feature Ranking

2.1 Feature Ranking

As mentioned in the introduction, the standard approach to using EDA for feature subset selection (FSS), is to select the best feature subset encountered in the iterative process as the final solution. However, selecting the single best subset of features provides a rather static view of the whole elimination process. When using FSS to gain more insight in the underlying processes, the human expert has no idea of the context of the specific subset. Questions about how much and which features can still be eliminated before the classification performance drastically drops down provide

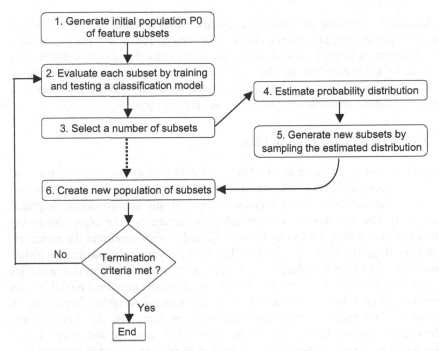

Fig. 1. The general scheme of the EDA approach to wrapper based feature selection

interesting information, yet remain unanswered using a static analysis. Feature ranking is a first step towards a dynamical analysis of the feature elimination process. The result of a feature ranking is an ordering of the features, sorted from the least relevant to the most relevant. Starting from the full/empty feature set, features can then be removed/added and the classification performance for each subset can be calculated, providing a dynamic view.

Traditional sequential wrapper algorithms such as sequential forward/backward search inherently provide a feature ranking. These algorithms either start from the full or empty feature set, and greedily add or discard one feature at the time. If this process is iterated until all features are added or removed, a complete view of the selection process can be obtained. A similar methodology can be applied in the case of most filter methods, where the feature relevance scores can be sorted and provide a feature ranking.

When using stochastic methods like GA or EDA, a hybrid approach can be used to yield a dynamical view of the selection process. The solution found by the evolutionary algorithm is then used as the starting point for a sequential forward or backward wrapper method. However, such practice may result in a large, sometimes unfeasible, number of additional calculations, depending on the number of features selected, or the range of the dynamic view.

Instead of combining an evolutionary method with a sequential method into a hybrid, we present an EDA-based technique that directly results in a feature ranking. Instead of using a single best solution, we use the estimated probability distribution as a basis for the feature ranking. As a consequence, this technique does not require any additional calculations and, as all features are modelled in the estimated distribution, it provides a dynamic view of the whole selection process.

2.2 Deriving a Ranking from an EDA

The main action to be taken in an EDA-based evolutionary algorithm is the construction of the probability distribution that models the variables and their dependencies. In general, most EDAs can be represented graphically as probabilistic graphical models [19]. The structure of the graphical model determines the expressive power of the EDA to model dependencies between variables, and constitutes the major criterion to distinguish subclasses of EDAs. The most common subclassification distinguishes between EDAs modelling univariate, bivariate and multivariate dependencies between the variables. A second aspect of the probabilistic graphical model is a set of generalized probability distributions, associated with the variables. Depending on the domain of the variables, these distributions can be either discrete or continuous. In the case of feature subset selection, all variables are discrete and binary. Fig. 2 shows a few examples of probabilistic graphical models for the three major classes of EDAs in the case of a feature selection problem with eight features $(X_1, \cdots, X_8)$. The notation $p(x_i^j)$ denotes the probability of feature i having value j. As features are either present or absent, j can only be 0 or 1.

The Univariate Marginal Distribution Algorithm (UMDA [18]) is a very simple model, assuming variables are independent. This is reflected in the structure of the graphical model, as no arcs between different variables are present, and the probability distributions do not contain conditional probabilities. In the Bivariate Marginal Distribution Algorithm (BMDA [21]), pairwise interactions between variables are modelled, and in the case of multiple dependencies, higher order interactions between the variables are modelled. Examples of these include the Bayesian Optimization Algorithm (BOA [20]) and the Estimation of Bayesian Networks Algorithm (EBNA [4]).

To derive a feature ranking from a probability distribution, some sort of importance or relevance score for each feature needs to be calculated. Evidently, a feature i having a higher value for $p(x_i^1)$ could be considered more important than a feature j with a lower value for $p(x_j^1)$. The generalized probabilities $p(x_i^1)$ can thus be considered as feature relevance scores, and a list of features sorted by these probabilities returns a feature ranking. The general algorithm to calculate such a ranking consists of the steps presented in Fig. 3.

The most important step in this algorithm is the extraction of the probabilities $p(x_i^1)$ from the model. For models with univariate dependencies like the UMDA, the extraction of these probabilities is trivial, as they can be directly inferred from the model. For higher order EDAs like BMDA, BOA and EBNA, the probabilities $p(x_i^1)$

UMDA

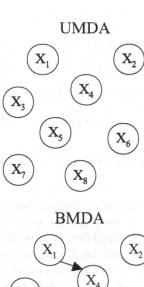

X_i	$p(x_i^j)$ j=0,1
X_1	$p(x_1^j)$
X_2	$p(x_2^j)$
X_3	$p(x_3^j)$
X_4	$p(x_4^j)$
X_5	$p(x_5^j)$
X_6	$p(x_6^j)$
X_7	$p(x_7^j)$
X_8	$p(x_8^j)$

BMDA

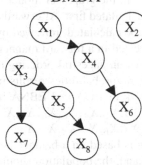

X_i	$p(x_i^j)$ j,k=0,1
X_1	$p(x_1^j)$
X_2	$p(x_2^j)$
X_3	$p(x_3^j)$
X_4	$p(x_4^j \mid x_1^k)$
X_5	$p(x_5^j \mid x_3^k)$
X_6	$p(x_6^j \mid x_4^k)$
X_7	$p(x_7^j \mid x_3^k)$
X_8	$p(x_8^j \mid x_5^k)$

BOA, EBNA

X_i	$p(x_i^j)$ j,k,l=0,1
X_1	$p(x_1^j)$
X_2	$p(x_2^j)$
X_3	$p(x_3^j)$
X_4	$p(x_4^j \mid x_1^k)$
X_5	$p(x_5^j \mid x_3^k, x_4^l)$
X_6	$p(x_6^j \mid x_4^k)$
X_7	$p(x_7^j \mid x_3^k)$
X_8	$p(x_8^j \mid x_5^k)$

Fig. 2. Some examples of probabilistic graphical models for EDAs with varying complexity: univariate dependencies (UMDA), bivariate dependencies (BMDA) and multiple dependencies (BOA, EBNA). The probability distributions are illustrated for a problem with discrete, binary variables, e.g. FSS. The notation x_i^j denotes the instantiation of variable i with value j

EDA-R

1. Select S individuals from the final population $D_{\texttt{final}}$
2. Construct the probability model P from $D^{S_j}_{\texttt{final}}, j \in 1, \cdots, S$
   using an EDA (e.g. UMDA, BMDA, BOA/EBNA)
3. For each variable (feature) X_i, calculate the probability
   $p(x_i^1)$
4. Sort features $X_1, \cdots, X_n$ by their $p(x_i^1)$ probabilities
5. Write out the array of sorted features

Fig. 3. General algorithm to calculate a feature ranking (EDA-R)

need to be calculated in a forward manner, as they may involve conditional probabilities. To enable this, an ancestral ordering of the nodes in the graphical model is needed. The probabilities of nodes without ancestors are calculated first. Afterwards, probabilities for nodes depending on these ancestors can be calculated, followed by the probabilities of their descendants. This process is repeated in a forward manner, until all probabilities are calculated. It has to be noted that an ancestral ordering is not unique [7], yet the forward procedure of calculating the probabilities results in a unique probability distribution. For the example network of BOA and EBNA in Fig. 2, a possible ancestral ordering of the nodes is $X_1, X_2, X_3, X_4, X_5, X_6, X_7, X_8$. Another possible ordering would be for example $X_1, X_3, X_4, X_6, X_5, X_8, X_7, X_2$.

Conceptually, the idea of EDA-based feature ranking is based on a balance between two characteristics of the population. On the one hand, the population should consist of medium to good quality solutions, implying that already some sort of convergence has been accomplished. On the other hand, the population should still preserve some diversity, implying that it has not fully converged yet (e.g. in the ultimate case of convergence all individuals in the populations are the same). Thus, we seek a measure to define how long the iterative process should be continued, resulting in a population that has already converged, but not too much.

To quantify this idea of "convergence" of a population we need a measure of how similar/diverse it is. In the case of feature selection, the individuals are represented by bitstrings, and we can use the Hamming distance as a measure of distance between two individuals [15]. The Hamming distance $HD(x, y)$ between two bitstrings of length N is the number of bits in which the two strings differ. This number thus varies between 0 and N. To normalize this number we calculate the scaled hamming distance as

$$HD_s(x, y) = \frac{HD(x, y)}{N} \tag{1}$$

The convergence of a population can then be calculated as the average scaled hamming distance between all pairs of individuals. For a population P of size S the convergence is calculated as

$$C(P) = \frac{2 \left(\sum_{i=1}^{S-1} \sum_{j=i+1}^{S} HD_s(\mathbf{x}_i, \mathbf{x}_j) \right)}{S(S-1)} \tag{2}$$

where $\mathbf{x}_i$ denotes the ith individual in the population. The parameter $C(P)$ can then be monitored during the stochastic iterative sampling process, which can be stopped when $C(P)$ falls below an a priori specified threshold. The calculation of this threshold can be done using a sub-sample of the data.

A more advanced, yet computationally more expensive way of tuning the parameter $C(P)$ can be thought of, stopping the iterative procedure automatically when the "optimal" convergence point has been reached. This can be done by calculating at each iteration the feature ranking curve, and tracking the area under this curve. A simple greedy heuristic can then be used to halt the iterative process when the area under the ranking curve of the current iteration is less then the area under the ranking curve of the previous iteration.

2.3 Deriving a Feature Weighting Scheme from an EDA

In the previous section, we described how a feature ranking could be derived from the generalized probabilities $p(x_i^1)$. However, these probabilities could also be directly used as feature relevance scores, or feature weights. In this way, we can construct a wrapper based feature weighting mechanism. The derivation of feature weights conveys important additional information, that can be used to gain new insights in the processes that generated the data (knowledge discovery). We will elaborate further on that aspect in the second part of this chapter, where we will discuss the application of the method to a biological classification problem.

The advantage of using EDA-R as a feature weighting mechanism, compared to other feature weighting methods like filter methods, is that it can directly use the feedback (classification performance) of classifiers that allows modelling of higher order dependencies, whereas most filter methods only determine the relevance of each feature by itself. As a direct extension of using EDA-R to rank individual features, it can be easily seen that the method can be generalized to subsets of k features (e.g. weighting of all pairs, triples, of features). Thus the method can be easily extended to *feature subset weighting*. Another possibility for future extensions of this method is by incorporating a more sophisticated weighting scheme, and also taking into account the fitness of the selected individuals when assigning the feature weights.

3 A Real-World Application: Acceptor Splice Site Prediction

Recent advances in genomics have generated large amounts of biological sequence data. An important problem in bioinformatics is to analyse these sequences and predict the location and structure of genes, often referred to as *gene prediction*. Because the problem of correctly predicting genes is quite complex [16], gene prediction systems have a modular structure, combining the outputs of several components that are specialized in recognizing specific structural elements of a gene. An example of such structural elements are the so-called splice sites. These sites are the boundaries between coding and non coding regions in the genomes of higher organisms (eukaryotes), and are of key importance in identifying the correct gene structure. In this chapter we will focus on acceptor splice site prediction, which is the transition from a non coding region (intron) to a coding region (exon). Acceptor splice sites are characterized by the fact that they have a conserved AG subsequence at the intron border side. As a result, acceptor prediction can be formally stated as a two-class classification task: given an AG subsequence, predict whether it is a true acceptor splice site or not. In this chapter we will focus on the prediction of acceptor splice sites in the plant model species *Arabidopsis thaliana*.

The *Arabidopsis thaliana* data set was generated from sequences that were retrieved from the EMBL database, and contained only experimentally validated genes (i.e. no genes that resulted from a prediction). Redundant genes were excluded, and splice site datasets were constructed from 1495 genes. More details on how these datasets were generated can be found in [2].

Because in real sequences, the number of true acceptor sites is largely outnumbered by the number of false acceptor sites, we chose to enforce a *class imbalance* in our datasets for feature selection. We constructed a dataset of 6000 positive instances and 36,000 negative instances. To obtain stable solutions for feature selection, a 10-fold validation of this dataset was used to test all feature selection methods. This was done by doing 5 replications of a two-fold cross-validation, maintaining the same class imbalance of 1 positive versus 6 negative instances in every partition. For the EDA-based wrapper approach, the internal evaluation of classification performance was obtained by doing a 5-fold cross-validation on the training set.

As the EDA-R method is a wrapper approach, it is specific to a fixed classification model. In our experiments, we used the Naive Bayes method (NBM [3]). This classification method follows the Bayes optimal decision rule, combining it with the assumption that the probability of the features given the class, is the product of the probabilities of the individual features (conditional independence assumption). The advantages of using NBM in the context of feature selection are its abilities to cope with high-dimensional data, its robustness and its speed. The latter aspect is of particular importance when using population based methods like EDA or GA , because for every individual, a classification model has to be trained, and tested by cross-validation. As a measure of classification performance, we used the F-measure [17] due to its ability to deal well with imbalanced datasets.

3.1 Feature Ranking for Acceptor Prediction

We start from the knowledge that the discrimination between true and false acceptor sites is determined by the part of the sequence where the site is located, more precisely the *local context* around the acceptor site. Therefore, the nucleotides A,T,C and G occurring on either side of the acceptor constitute a basic feature set. A local context of 100 nucleotides (50 to the left, 50 to the right) around the acceptor sites was chosen, having at each position one of the four nucleotides {A,T,C,G}. These features were extracted for the positive and negative instances, resulting in a dataset of 100 4-valued features, which were converted into binary format using

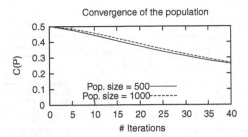

Fig. 4. Evolution of the population convergence as a function of the number of iterations

sparse vector encoding (A = 1000, T = 0100, C = 0010, G = 0001). This results in a dataset described by 400 binary features.

To this dataset, we applied the EDA-R feature ranking method. For different population sizes, ranging from 100 to 1000 individuals, we ran the experiments for 40 iterations. For each iteration i, we monitored the value of $C(P)$ and derived a feature ranking F_i. Afterwards, we compared the evaluations for each feature ranking.

The convergence of the population was calculated at each iteration using equation 2, and its evolution for population sizes of 500 and 1000 is shown in Fig. 4. The x-axis shows the number of iterations, while the y-axis shows the convergence value $C(P)$ of the population. At the beginning of the iterative process, the initial population consists of randomly generated feature subsets, where, for every feature, $p(x_i^1) = p(x_i^0) = 0.5$. As a result, feature subsets will have, on average, half of the features in common, and $C(P)$ will be approximately equal to 0.5. When the iterative process would be repeated ad infinitum, all individuals in the population would converge to the same individual, resulting in $C(P) = 0$. The figure shows that for 40 iterations, convergence will be roughly half way between 0.5 and 0.

As mentioned earlier, the ideal value of $C(P)$ is achieved when the population has already converged, yet not too much. To explore the effect of the number of iterations (and thus $C(P)$) on the feature ranking, we compared the evaluation of the feature ranking during the course of evolution. For a particular iteration number, we derived a feature ranking from the population at that time. This was done by starting with the full feature set, and iteratively eliminating the least relevant feature, according to the feature ranking. The results for a few iterations (iteration 1, 20 and 40) are shown in Fig. 5.

The left part of the figure shows the results for a population of 500 individuals, the right part for a population of 1000 individuals. The results after the first iteration are shown as a baseline result. As soon as the first iterations have passed, the feature ranking improves quickly, until at some point a good feature ranking is obtained (iteration 20). If the iterative process is then continued, populations that are too specific are obtained (iteration 40), characterized by the fact that classification performance drops down earlier when smaller feature sets are evaluated. Furthermore it can be observed that the results for a population size of 1000 individuals are only marginally better than the results using a population of 500 individuals. Gradually worse results are obtained when populations smaller than 500 individuals are used.

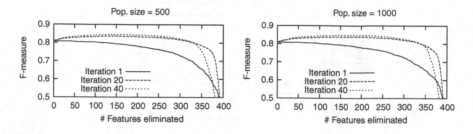

Fig. 5. Evaluation of a feature ranking for a number of iterations (1, 20 and 40). The left part shows the results for a population size of 500 individuals, the right part for a population of 1000 individuals. The origin represents the full feature set. The x-axis represents the number of features that have been eliminated thus far, while the y-axis shows the classification performance on the test set (F-measure)

It can be observed that, e.g. for the case of 20 iterations, many features can be eliminated before the classification performance drops down, showing that many irrelevant or redundant features are present in the dataset. The advantage of a feature ranking is the identification of a "break-point" region. This is the part of the graph where the classification performance drastically drops down, indicating the removal of strongly relevant features. It should be noted that the observation of a break-point region is strongly dependent on the dataset. For many biological processes, it is not completely known which features are relevant for the classification task at hand. Therefore, many potentially useful features are included in the dataset, hoping that the relevant features are included as well. As a result, many irrelevant or redundant features will be present, and a clear break-point can be observed. For other datasets with little or no redundant features, this phenomenon will not be observed.

Strictly speaking, the identification of the break-point region should be considered as a part of the training process. Therefore, the identification of this region should be done on the training set, and only thereafter the test set can be used for evaluation. Fig. 6 shows the results for feature selection on both the training and test set. When comparing the results, it can observed that the break-point regions for both data sets are very similar.

3.2 Feature Weighting for Knowledge Discovery in Acceptor Prediction

An important advantage of feature selection techniques is their ability to distinguish between relevant and irrelevant features, providing new insights in complex datasets.

It is known that correlations exist between nucleotides in the vicinity of splice sites. To detect these dependencies, higher-order (i.e. non-linear) classification methods can be used. When combining higher-order classification methods with EDA-based feature selection, this would require the use of higher-order estimation models. An example of such a combination could be a second order polynomial Support Vector Machine , in combination with the BMDA. However, using such higher order

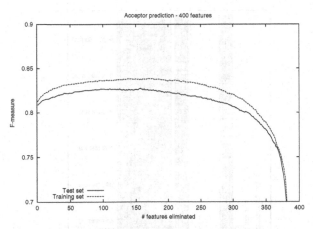

Fig. 6. Evaluation of the feature ranking obtained with a population size of 500 individuals after 20 iterations. Both the evaluation of the ranking on the training and on the test set are shown

classification algorithms and estimation models would make the EDA-R method very slow.

To circumvent the use of higher order models, yet still be able to extract correlations between nucleotides, we applied the following trick. We constructed an additional set of features that captures the nucleotide correlations already at the feature level. This has the advantage that linear models can still be used, while at the same time considering nucleotide dependencies. Another important advantage is that the combination with feature selection techniques allows us to select those dependencies that are of primary importance, and visualize them.

In addition to the simple nucleotide features used in the previous section, we added two layers of more complex features. The first layer captures the idea of compositional sequence information. These type of features extract sequence information that is position invariant. In our experiments, we included position invariant features of length 3, capturing the occurrence of subsequences of length 3 in the sequence neighbouring the splice site. An example of such a feature would be the occurrence of the subsequence "TCA" in the sequence to the left of the acceptor site. For the sequence on either side of the acceptor, including these features results in an additional set of 128 binary features, a 1 indicating the presence, a 0 the absence of the specific subsequence. The second layer of complexity comprises features that capture dependencies between adjacent nucleotides. To this end, we included all position dependent dinucleotides (subwords of length 2) in our analysis, resulting in an additional set of 1568 features. Summing up all features eventually results in a dataset described by 2096 features.

To this dataset we applied the EDA-R feature ranking method, deriving the feature weights from the probabilities $p(x_i^1)$ of the population at iteration 20 (for computational reasons we used the same EDA setting as in the previous experiment). A nice way of visualizing the feature weights is by color coding them using a heat

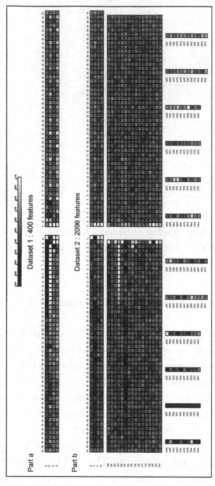

Fig. 7. Color coding of the feature weights using a gradient from black (the feature should be left out) to white (the feature should be included). Part 'a' shows the result for the simplest dataset (400 features), part 'b' shows the result for the complex dataset (2096 features). Features are grouped by their position relative to the acceptor site, which is denoted by the blank space in the middle. For each part of the context the position invariant features of length 3 are grouped according to their composition (A-rich, T-rich, C-rich, G-rich, equally distributed)

map, where a gradient ranging from black (the feature should be left out) to white (the feature should be included) shows the feature weights. This is shown in Fig. 7 where we graphically show the feature weights for both the simple dataset (400 features) and the extended dataset (2096 features). In this figure, features are shown row wise, while the columns indicate positions around the splice site (the gap in the middle). Part 'a' shows the results for the simplest dataset: every row represents one of the nucleotides A,T,C and G. Part 'b' shows the results for the complex dataset, with

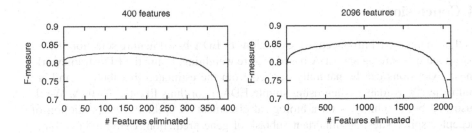

Fig. 8. Evaluation of the classification performance on both datasets, when features are iteratively discarded. The left part of the figure shows the result for the most simple dataset (400 features), while the right part shows the result for an extended version of this dataset, using also position invariant features and position dependent dinucleotide features (2096 features). The x-axis shows the number of features that have been eliminated, the y-axis shows the classification performance (F-measure)

the position invariant features (middle part) and the position dependent dinucleotides (lower part).

Several patterns can be observed. In both datasets, the nucleotides immediately surrounding the acceptor splice site are of key importance. Another pattern can be explained by looking at the right side of the context. In the simplest dataset, a clear periodical pattern is visible for the nucleotides T and G, capturing the fact that the right side of the context is a coding region (exon). In this region, nucleotides are organised in codons (triplets). However, as this is a general characteristic of the sequence, it is observed that the position invariant features in the second dataset better seem to grasp this characteristic.

Another important pattern are the nucleotides T at the left side of the acceptor for the simplest dataset. In the complex dataset, this pattern has completely disappeared, and is replaced by a stretch of AG dinucleotides that now seems to be most important. This is a nice example of the combination of a wrapper based method with NBM. It is known that the classification performance of NBM can be improved by discarding correlated features. Without going into much detail, we here mention that the presence of a poly-pyrimidine stretch (an excess of nucleotides C and T) to the left of the acceptor is correlated with the absence of AG dinucleotides in this part of the sequence. Apparently, NBM chooses the absence of AG dinucleotides as being more informative than the importance of T, and thus discards the T features to the left of the acceptor. The benefit of not including these features is even more apparent when looking at the position invariant features in part 'b', where all T-rich subsequences of length 3 are colored dark. This indicates that NBM strongly benefits from not including these features.

To verify that the features selected for the complex dataset are indeed better at describing the acceptor prediction problem, we compared the classification performance for both datasets (Fig. 8). This figure shows that better classification performance can be obtained using the more complex features.

4 Conclusions

In this chapter, we introduced two extensions of EDA-based feature selection: EDA-based feature ranking and EDA-based feature weighting. Using the EDA framework, these extensions can be naturally derived from the estimated distribution, and are immediately available when using simple EDA algorithms like the UMDA. We illustrated both techniques on a biological classification problem: the prediction of acceptor splice sites, an important subtask of gene prediction. Using the combination of more complex features and feature selection, we were able to extract a new, important feature for acceptor prediction: the inhibition of dinucleotides AG immediately upstream the acceptor site. Using feature selection allows us in this way to gain more insight in the computational modelling of this particular biological classification problem.

References

1. A.I. Blum and P. Langley. Selection of relevant features and examples in machine learning. *Artificial Intelligence*, 97:245–271, 1997.
2. S. Degroeve, B. De Baets, Y. Van de Peer, and P. Rouzé. Feature subset selection for splice site prediction. *Bioinformatics*, 18(2):75–83, 2002.
3. R. Duda and P. Hart. *Pattern Classification and Scene Analysis*. John Wiley and Sons, New York, 1973.
4. R. Etxeberria and P. Larrañaga. Global optimization with Bayesian networks. In *Proceedings of the Second Symposium on Artificial Intelligence. Special Session on Distributions and Evolutionary Optimization*, pp. 332–339, 1999.
5. I. Guyon and A. Elisseeff. An introduction to variable and feature selection. *Journal of Machine Learning Research*, 3:1157–1182, 2003.
6. I. Guyon, J. Weston, S. Barnhill, and V.N. Vapnik. Gene selection for cancer classification using support vector machines. *Machine Learning*, 46(1-3):389–422, 2000.
7. M. Henrion. Propagating uncertainty in Bayesian networks by probabilistic logic sampling. In J. F. Lemmer and L. N. Kanal, editors, *Uncertainty in Artificial Intelligence*, volume 2, pp. 149–163. North-Holland, Amsterdam, 1988.
8. I. Inza, P. Larrañaga, R. Etxeberria, and B. Sierra. Feature subset selection by Bayesian network-based optimization. *Artificial Intelligence*, 123(1-2):157–184, 2000.
9. I. Inza, P. Larrañaga, and B. Sierra. Feature Subset Selection by Estimation of Distribution Algorithms. In P. Larrañaga and J. A. Lozano, editors, *Estimation of Distribution Algorithms. A New Tool for Evolutionary Computation*, pp. 269–294. Kluwer Academic Publishers, 2001.
10. A. K. Jain, R. W. Duin, and J. Mao. Statistical Pattern Recognition. A review. *IEEE Transactions on Pattern Analysis and Machine Intelligence*, 22(1):4–37, 2000.
11. J. Kittler. Feature set search algorithms. In C.H. Chen, editor, *Pattern Recognition and Signal Processing*, pp. 41–60. Sithoff and Noordhoff, 1978.
12. R. Kohavi and G. John. Wrappers for feature subset selection. *Artificial Intelligence*, 97(1-2):273–324, 1997.
13. M. Kudo and J. Sklansky. Comparison of algorithms that select features for pattern classifiers. *Pattern Recognition*, 33:25–41, 2000.

14. P. Larrañaga and J. A. Lozano. *Estimation of Distribution Algorithms. A New Tool for Evolutionary Computation.* Kluwer Academic Press, 2001.

15. S.J. Louis and G.J.E. Rawlins. Predicting convergence time for genetic algorithms. Technical Report TR370, Indiana University, 1993.

16. C. Mathé, M.F. Sagot, T. Schiex, and P. Rouzé. Current methods of gene prediction, their strengths and weaknesses. *Nucleic Acids Research*, 30:4103–4117, 2002.

17. D. Mladenić and M. Grobelnik. Feature selection on hierarchy of web documents. *Decision Support Systems*, 35:45–87, 2003.

18. H. Mühlenbein. The equation for response to selection and its use for prediction. *Evolutionary Computation*, 5:303–346, 1998.

19. J. Pearl. *Probabilistic Reasoning in Intelligent Systems.* Morgan Kaufmann, 1988.

20. M. Pelikan, D.E. Goldberg, and E. Cantú-Paz. BOA: the Bayesian optimization algorithm. In *Proceedings of the Genetic and Evolutionary Computation Conference*, pp. 525–532, 1999.

21. M. Pelikan and H. Mühlenbein. The bivariate marginal distribution algorithm. In *Advances in Soft Computing-Engineering Design and Manufacturing*, pp. 521–535, 1999.

22. Y. Saeys, S. Degroeve, D. Aeyels, Y. Van de Peer, and P. Rouzé. Fast feature selection using a simple estimation of distribution algorithm: a case study on splice site prediction. *Bioinformatics*, 19(2):179–188, 2003.

Learning Linguistic Fuzzy Rules by Using Estimation of Distribution Algorithms as the Search Engine in the COR Methodology

M. Julia Flores, José A. Gámez and José M. Puerta

Departamento de Informática. Universidad de Castilla-La Mancha.
Campus Universitario s/n. Albacete, 02071. Spain.
{Julia.Flores,Jose.Gamez,Jose.Puerta}@uclm.es

Summary. Learning models from data which have the double ability of being predictive and descriptive at the same time is currently one of the major goals of machine learning and data mining. Linguistic (or descriptive) fuzzy rule-based systems possess a good tradeoff between the aforementioned features and thus have received increasing attention in the last few years.

In this chapter we propose the use of estimation of distribution algorithms (EDAs) to guide the search of a *good* linguistic fuzzy rule system. To do this, we integrate EDAs in a recent methodology (COR) which tries to take advantage of the cooperation among rules. Experiments are carried out with univariate and bivariate EDAs over four test functions, and the results show that the exploitation of (pairwise) dependencies done by bivariate EDAs yield to a better performance than univariate EDAs or genetic algorithms.

1 Introduction

It is clear that machine learning and data mining have lately become a focus of attention in the computational scientific world [16]. There is great interest in the automatic construction of a model from a data set, specially for complex domains and/or in the absence of experts capable of collaborating on such modelling. Those models which are at the same time descriptive and predictive are specially interesting, because they allow us to estimate the system outcome and also to obtain insight into the inner structure of the problem domain.

Fuzzy rule-based systems (FRBSs) are a formalism that has gained relevance in data mining because of its dual predictive and modelling capability. These systems are based on fuzzy set theory, proposed by Zadeh [24] and, therefore, they are able to cope with those problems presenting uncertainty and/or vagueness [1, 10]. Among all different types of fuzzy rules existing in literature [1, 18, 21, 22], *linguistic* or *descriptive* fuzzy rules result specially attractive because they permit us to achieve the double-goal of description plus prediction. On this field, also known as, *linguistic modelling of a system* we will focus this work.

M.J. Flores et al.: *Learning Linguistic Fuzzy Rules by Using Estimation of Distribution Algorithms as the Search Engine in the COR Methodology*, StudFuzz **192**, 259–280 (2006)
www.springerlink.com

There are different approaches to the fuzzy rule learning problem. Among them, we are interested in the COR (*cooperative rules*) [7] methodology since it sets the learning problem as a combinatorial optimisation task. The goal of this work is to study the application of *estimation of distribution algorithms* (EDAs) [13,14] to carry out the corresponding search. Therefore, this chapter is structured in five sections, apart from this introduction. Section 2 is a revision of concepts relating to fuzzy logic rule based systems and some algorithms used for learning them. In Sect. 3 we revise estimation of distribution algorithms. In Sect. 4 we specify the search space considered in this work, while in Sect. 5 we present all the material related to the experiments carried out. Finally in Sect. 6 we present our final conclusions.

2 Fuzzy Logic Rule Based System (FLBRS) Learning

A linguistic fuzzy rule has the following structure:

$$\texttt{If } X_1 \text{ is } v_1^j \& \dots \& X_n \text{ is } v_n^j \text{ then } Y \text{ is } v_y^j$$

where, variables $X_1, \dots, X_n$ and Y are *linguistic variables* [25–27], i.e. variables that take values in a set of *linguistic labels* $\{v_i^1, \dots, v_i^{k_i}\}$ that define the semantics of such variables. Each label has a fuzzy set associated to it (see Fig. 1).

Hence, in a FLRBS we find two clearly different components (Fig. 1):

- A domain data base, in which domains would be defined (fuzzy sets) underlying each linguistic variable.
- A linguistic fuzzy rule base.

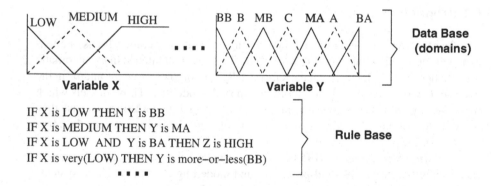

IF X is LOW THEN Y is BB
IF X is MEDIUM THEN Y is MA
IF X is LOW AND Y is BA THEN Z is HIGH
IF X is very(LOW) THEN Y is more–or–less(BB)

Fig. 1. Structure of a FLRBS

In this work we will assume that the domains of the linguistic variables have been obtained from the range of values that the corresponding variable takes in the instances of the training set, and considering a symmetrical fuzzy partition in k labels using triangular membership functions.[1]

For example, the domain for the variable Y in Fig. 1 is obtained by taking 7 linguistic labels. Then, we will focus on the rule base generation process. Although, in literature we find several methods for this kind of system generation, we will only revise the Wang and Mendel algorithm and the COR methodology, because these are the methods used in this work.

2.1 Wang and Mendel Algorithm

The Wang and Mendel algorithm (WM) [23] belongs to the category of *ad hoc* data-driven methods. In this kind of method the learning process is guided by covering criteria of the examples in the *training set*. These methods fall into the category of heuristic and deterministic, and are computationally very efficient.

Let $X_1, \ldots, X_n$ be the variables in the rule antecedent, Y the variable in the rule consequent, $C(X_i) = \{v_i^1, \ldots, v_i^{k_i}\}$ the linguistic label set for X_i, $\mu_A(x)$ the degree of membership of value x to fuzzy set A, and $\mathbf{D} = \{(\mathbf{x}^1, y_1), \ldots, (\mathbf{x}^d, y_d)\}$ the data base (examples) for the learning process, where $\mathbf{x}^i$ is a configuration of (real) values for input variables, $\mathbf{x}^{i \downarrow j}$ is the j-th component of the array $\mathbf{x}^i$, and y_i is the (real) output value for such configuration.

If we assume that the linguistic variables (domain base) have already been specified/constructed, WM algorithm generates the rule set in the following way (for details see [23]):

- A space of possible (antecedents for the) rules is created and it is made up of the Cartesian product $C(X_1) \times C(X_2) \times \cdots \times C(X_n)$.
- For every example $(\mathbf{x}^i, y^i)$ generate the rule R_i: *If X_1 is $v_1^{r_1}$ & ... & X_n is $v_n^{r_n}$ then Y is $v_Y^{r_m}$*, where $r_j = \arg \max_{l=1..k_j} \mu_{X_j^l}(\mathbf{x}^{i \downarrow j})$. Associate the obtained rule with the sub-space $\{v_1^{r_1}, \ldots, v_n^{r_n}\}$.
- For every non-empty sub-space select the rule with the highest degree of importance

$$G(R_i) = \mu_{v_1^{r_1}}(\mathbf{x}^{i \downarrow 1}) \cdot \cdots \cdot \mu_{v_n^{r_n}}(\mathbf{x}^{i \downarrow j}) \cdot \mu_{v_Y^{r_m}}(y^i)$$

and add it to the fuzzy rules system.

In spite of its simplicity, the Wang and Mendel algorithm has clearly demonstrated a high performance in practice.

---

[1] The membership degree for a triangular function defined in the interval [a,c] and maximum value in b is obtained as:

$$\mu_{Triangular}(x) = \begin{cases} \frac{x-a}{b-a}, & \text{if } a \leq x \leq b \\ \frac{c-x}{c-b}, & \text{if } b \leq x \leq c \\ 0, & \text{otherwise} \end{cases}$$

2.2 Wang and Mendel in Practice

As we assume that potential readers of this book are more familiar with EDAs than with fuzzy logic, we introduce an example of WM execution. Suppose that we have a problem with two input variables $\{X_1, X_2\}$ taking values in $[0, 10]$ and one output variable Y taking values in $[0, 20]$. For the sake of simplicity we consider linguistic variables with only three labels ($\{Left, Center, Right\}$ for $\{X_1, X_2\}$ and $\{Low, Medium, High\}$ for Y, depicted on the left part of Fig. 2).

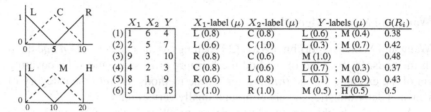

	X_1	X_2	Y	X_1-label (μ)	X_2-label (μ)	Y-labels (μ)	$G(R_i)$
(1)	1	6	4	L (0.8)	C (0.8)	L (0.6) ; M (0.4)	0.38
(2)	2	5	7	L (0.6)	C (1.0)	L (0.3) ; M (0.7)	0.42
(3)	9	3	10	R (0.8)	C (0.6)	M (1.0)	0.48
(4)	4	2	3	C (0.8)	L (0.6)	L (0.7) ; M (0.3)	0.37
(5)	8	1	9	R (0.6)	L (0.8)	L (0.1) ; M (0.9)	0.43
(6)	5	10	15	C (1.0)	R (1.0)	M (0.5) ; H (0.5)	0.5

Fig. 2. Example for Wang and Mendel algorithm. *Left*: domains for $\{X_1, X_2\}$ (up) and Y (down); *Middle*: data set; *Right*: execution of WM algorithm

If we run WM algorithm over the data set shown in the central part of Fig. 2, and using the linguistic variables previously specified, we get five non-empty subspaces (as shown in the right part of Fig. 2): $\{(L, C), (R, C), (C, L), (R, L), (C, R)\}$. From them, only (L, C) has more than one candidate rule. Two to be precise:

If X_1 *is* L & X_2 *is* C then Y *is* L and If X_1 *is* L & X_2 *is* C then Y *is* M. As the second rule has a higher degree than the first one, the output of WM algorithm is:

$$\begin{cases} \text{If } X_1 \;\; is \;\; L \;\& \; X_2 \;\; is \;\; C \text{ then } Y \;\; is \;\; M \\ \text{If } X_1 \;\; is \;\; R \;\& \; X_2 \;\; is \;\; C \text{ then } Y \;\; is \;\; M \\ \text{If } X_1 \;\; is \;\; C \;\& \; X_2 \;\; is \;\; L \text{ then } Y \;\; is \;\; L \\ \text{If } X_1 \;\; is \;\; R \;\& \; X_2 \;\; is \;\; L \text{ then } Y \;\; is \;\; M \\ \text{If } X_1 \;\; is \;\; C \;\& \; X_2 \;\; is \;\; R \text{ then } Y \;\; is \;\; H \end{cases}$$

Where we should notice that the consequent of the last rule could be Y is M, because $\mu_M(15) = \mu_H(15) = 0.5$, and we chose H randomly.

2.3 COR Methodology

Despite the advantages exhibited by the WM algorithm, it also has some shortcomings. The WM algorithm looks for the rules with the best individual performance, taking the rule that best covers each sub-space. That implies a local action, without considering that the interaction between all the system rules will actually define its global performance. Furthermore, the local way in which the rules are treated, makes this procedure more sensitive to *noise*.

Because of these problems, Casillas et al. ([6,7]) propose a WM-based method in which they study the cooperation between the different rules of the system. This modification, known as *COR methodology* (from *cooperative rules*) is based on replacing the *greedy* behaviour of WM algorithm in the selection of each rule, by a *combinatorial search of cooperative rules* in the space of all rule candidate sets. As opposed to the *greedy* and local philosophy of the WM algorithm, the use of COR tries to accomplish a global analysis.

If $ant(SE_l)$ is the antecedent corresponding to the sub-space SE_l, and $cons(SE_l^{\mathbf{D}})$ represents the set of possible consequents (depending on $\mathbf{D}$) for the sub-space SE_l, then, as we have seen applying WM the resulting system would be formed by:

$$\bigcup_{l:cons(SE_l^{\mathbf{D}}) \geq 1} \left(ant(SE_l), \underset{c_i \in cons(SE_l^{\mathbf{D}})}{best} \right),$$

whereas applying the COR methodology the best consequent set $(c_{l_1}, \ldots, c_{l_t})$ would be sought in the Cartesian product:

$$cons(SE_{l_1}^{\mathbf{D}}) \times \cdots \times cons(SE_{l_t}^{\mathbf{D}}),$$

where $SE_{l_1}^{\mathbf{D}}, \ldots, SE_{l_t}^{\mathbf{D}}$ is the sub-spaces set with $cons(\cdot) \geq 1$. Thus, in the example of our previous section, each subspace will have all the labels (with $\mu(value) > 0$) as possible consequents instead of directly taking the best one (underlined in Fig. 2) as WM does.

In [6, 19] the authors propose to use simulated annealing or ant colony optimization as the method to guide the search and therefore to select the *best* rule set as that which minimises the *mean square error* when we apply it to the training set $\mathbf{D}$.

3 Estimation of Distribution Algorithms

Estimation of distribution algorithms (EDAs) family [14] is a new metaheuristics which have attained interest during the last 5 years. EDAs are evolutionary algorithms based on populations as well as genetic algorithms[2](GAs), but in which genetics has been removed and replaced by the estimation/ learning and sampling of a probability distribution which relates the variables or genes forming an individual or chromosome. In this way the dependence/independence relations between these variables are explicitly modelled in the EDAs framework. To learn these (in)dependencies (some of the individuals of) the population are used as training set.

Figure 3 shows the general outline of EDAs evolution process. As we can see, steps (b) and (c) replace the classical selection+crossover+mutation used in genetic algorithms. Step (b) is the key point in EDAs algorithms, because working with joint probability distribution is not useful even in small problems, so a simpler model has to be estimated/learned. Depending on the complexity of the model considered,

---

[2] In this chapter we assume basic knowledge of genetic algorithms. Good descriptions can be found in [15]

EDA Approach

---

1. $D_0 \leftarrow$ Generate the initial population (m individuals)
2. Evaluate the population D_0
3. $k = 1$
4. Repeat
   a) $D_{tra} \leftarrow$ Select $N \leq M$ individuals from D_{k-1}
   b) Estimate/learn a new model $\mathcal{M}$ from D_{tra}
   c) $D_{aux} \leftarrow$ Sample M individuals from $\mathcal{M}$
   d) Evaluate D_{aux}
   e) $D_k \leftarrow$ Select M individuals from $D_{k-1} \cup D_{aux}$
   f) $k = k + 1$
   Until stop condition

Fig. 3. Description of EDAs operation mode

different models of EDAs arise. Thus, the more complex this model is the better collection of dependencies between variables it will show, but the more complex/time consuming its estimation will be. In literature we can find several proposals that can be grouped into: *univariate* models (no dependencies are allowed), *bivariate* models (pairwise dependencies are allowed), and *n-variate* models. The first and second group are algorithms that provides fast model estimation, while the last one allows for a great capability of modelling, using Bayesian networks [11] as a probabilistic model in most of the cases.

In [9] the authors started the investigation into using EDAs to learn linguistic fuzzy rule based systems. In that work the results were not too strong, mainly because few experiments were carried out. In this work, we have extended and improved our previous approach to the FLRBS learning problem using EDAs. Because of the good complexity-accuracy trade-off shown by bivariate models, we focus our research on this family, although univariate models are also considered in our study.

3.1 Univariate Models

In this case it is supposed that the n-dimensional joint probability distribution is factorised as,

$$p(x_1, x_2, \ldots, x_n) = \prod_{i=1}^{n} p(x_i)$$

That is, no structural learning is needed, and only marginal probabilities are required during parameter learning. We have experimented with two classic approaches to the univariate case:

UMDA

The *univariate marginal distribution algorithm* (UMDA) [17] fits in the general description of univariate models given above. Marginal probabilities for each variable are estimated by using frequencies found in D_{tra}. In our case, Laplace correction has been used to smooth the resulting probabilities.

PBIL

As an alternative to UMDA we have considered *population-based incremental learning* (PBIL) [3], or more concretely a variant of this algorithm that can be viewed as an *incremental* UMDA [17]. In this case, we start with an initial model (a uniform one) and at each generation we refine this reference model by using the estimated one. The degree of importance of the estimated model with respect to the model of reference is given by a constant $\alpha \in (0, 1]$ known as the *learning ratio*. Specifically, if $\mathcal{M}$ is the reference model and $\mathcal{M}^i$ is the model learned in this generation, $\mathcal{M}$ is refined as follows:

$$\mathcal{M}(X_j) = (1 - \alpha) \cdot \mathcal{M}(X_j) + \alpha \cdot \mathcal{M}^i(X_j)$$

Notice that if $\alpha = 1$ this algorithm coincides with UMDA.

3.2 Bivariate Models

In this case it is supposed that the n-dimensional joint probability distribution is factorised as

$$p(x_1, x_2, \ldots, x_n) = p(x_r) \prod_{x_i \neq x_r} p(x_i | pa(x_i)),$$

where X_r is the root variable in the model (the only one without parents) and the remaining variables have a unique parent $(pa(X_i))$.

In this case, structural and parametric learning is needed. Below we briefly describe the approaches considered in this work.

MIMIC

In *mutual information maximizing input clustering* algorithm [5] the probabilistic model has the shape of a chain $(X_1 \rightarrow X_2 \rightarrow \cdots \rightarrow X_l)$. The chain is learned by using a greedy algorithm as follows:

1. Select as root node the variable X_i with minimum entropy $H(X_i)$.
2. For the remaining nodes, if X_p is the last variable included in the chain, then choose the variable X_i which maximizes the mutual information with respect to X_p, that is, $I(X_p, X_i)$.

Again we use Laplace correction when estimating the probabilities.

TREE

Now the structure of the probabilistic model has the shape of a tree. The learning method, described in [4], is based on Chow and Liu algorithm [8], in which mutual information $I(X_i, X_j)$ is used as weight for link (X_i, X_j) and the tree is obtained as the spanning tree of maximum weight. As in previous methods we have used Laplace correction to smooth the estimated probabilities.

Normalizing Mutual Information

Finally, as in our problem the genes or variables representing positions in the individual take different numbers of states, we have decided to use $[0, 1]$-normalized mutual information when applying MIMIC and TREE, yielding the algorithms MIMIC_{01} and TREE_{01}. Mutual information has been normalized as follows

$$I_{01}(X_i, X_j) = \frac{2 \cdot I(X_i, X_j)}{H(X_i) + H(X_j)}$$

4 The Search Space

In Sect. 2.3 we have seen that the search space for COR methodology is made up of the Cartesian product of every set $Cons(SE_i^{\mathbf{D}})$, for all sub-space SE_i so that $Cons(SE_i^{\mathbf{D}}) \geq 1$. But actually, the condition could be strictly greater than 1, seeing that for the one-consequent sub-spaces the election is unique.

For example, in the example used in Sect. 2.2 we get the following distribution of examples and possible consequents among the different sub-spaces (see the right part of Fig. 2):

		X_2	
	L	C	R
L		(1)(2) $\{L, M\}$	
X_1 C	(4) $\{L, M\}$		(6) $\{M, H\}$
R	(5) $\{L, M\}$	(3) $\{M\}$	

Our individual would then have four variables

$SE_{L,C}$	$SE_{C,L}$	$SE_{C,R}$	$SE_{R,L}$
$\{L, M\}$	$\{L, M\}$	$\{M, H\}$	$\{L, M\}$

and moreover to every learned system we should add the rule: If $X_1 = R$ and $X_2 = C$ then $Y = M$

Nevertheless, in this work we will also consider the proposal presented in [19], in which seeking rule cooperation, we admit that some of them can be removed. For that we must add a new value $\aleph$ to the set of possible consequents. We do this in a way that in those individuals where the involved variable has such a value, the rule corresponding to the sub-space represented by this variable will not be in the resulting FLRBS. Coming back to our example we would have:

		X_2	
	L	C	R
L		$\{\aleph, L, M\}$	
X_1 C	$\{\aleph, L, M\}$		$\{\aleph, M, H\}$
R	$\{\aleph, L, M\}$	$\{\aleph, M\}$	

And, therefore, the individuals would have the following representation:

$SE_{L,C}$	$SE_{C,L}$	$SE_{C,R}$	$SE_{R,L}$	$SE_{R,C}$
$\{\aleph, L, M\}$	$\{\aleph, L, M\}$	$\{\aleph, M, H\}$	$\{\aleph, L, M\}$	$\{\aleph, M\}$

As we can see, this consideration makes the problem more difficult, that is because the cardinality of the search space increases significantly. In this example we will go from $2 \cdot 2 \cdot 2 \cdot 2 = 16$ to $3 \cdot 3 \cdot 3 \cdot 3 \cdot 2 = 162$.

5 Experimental Study

In this section we will describe the experiments carried out in order to evaluate the suitability of using EDAs as a search engine inside the COR methodology.

5.1 Test Suite

In our experiments we have used four problems borrowed from the FMLib repository[3]: two laboratory problems and two real-world modelling problems.

Function F_1

By F_1 we refer to the mathematical function (Fig. 4.a):

$$F_1(x_1, x_2) = x_1^2 + x_1^2 \qquad x_1, x_2 \in [-5, 5], F_1(\cdot) \in [0, 50]$$

Therefore, we have two predictive variables and one output variable. The goal is to model F_1 by learning a FLRBS. The downloaded training set has 1681 tuples uniformly distributed in the two-dimensional input space ($x_1 \times x_2$). The test set contains 168 tuples randomly sampled.

Function F_4

By F_4 we refer to the mathematical function (Fig. 4.b):

$$F_4(x_1, x_2) = x_1^2 + x_1^2 - cos(18x_1) - cos(18x_2)$$
$$x_1, x_2 \in [-1, 1], F_4(\cdot) \in [-2, 3.383]$$

As in F_1 we have two predictive variables and one output variable, and again the goal is to model F_4 by learning a FLRBS. The cardinality of the training and test sets are the same as in the previous function, as well as the method used to generate them.

Problem E_1

By E_1 we refer to the following real-world problem taken from the engineering field.

---

[3] http://decsai.ugr.es/fmlib

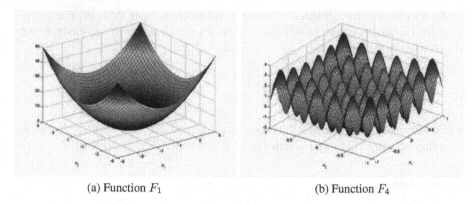

(a) Function F_1 (b) Function F_4

Fig. 4. Graphical representation of F_1 and F_4

The problem involves finding a model that relates the *total length of low voltage line* installed in a rural town with the *number of inhabitants in the town* and the *mean of the distances from the centre of the town to the three furthest clients in it*. The goal is to use the model to estimate the total length of line being maintained.

Therefore, we have two predictive variables ($x_1 \in [1, 320]$ and $x_2 \in [60, 1673.33]$) and one output variable defined in $[80, 7675]$. The cardinality of the training and test sets are 396 and 99 respectively (we have used the first partition available in FMLib).

Problem E_2

By E_2 we refer to another real-world problem also taken from the electrical engineering field.

The problem involves finding how to estimate the minimum maintenance costs. There are four input variables: *sum of the lengths of all streets in the town, total area of the town, area that is occupied by buildings*, and *energy supplied to the town*.

The domains for the four predictive variables are: $[0.5, 11], [0.15, 8.55], [1.64, 142.5]$ and $[1, 165]$. The output variable takes its value in $[64.47, 8546.03]$. Again we have considered the first partition available in FMLib, having 844 and 212 tuples in the training and test set respectively.

5.2 Evaluation/Fitness Function

To evaluate the quality of an individual I, it is translated to the corresponding FLRBS $Rules(I)$ and used for processing every d case within the training set **D**. If for a given tuple $(\mathbf{x}^i, y)$, $\hat{y}$ represents the obtained value when processing $\mathbf{x}^i$ with $Rules(I)$, then the fitness of I is calculated by using the *mean square error* (MSE) as:

$$fitness(I) = MSE_{Rules(I)} = \frac{1}{|\mathbf{D}|} \sum_{i=1}^{|\mathbf{D}|} (\hat{y}_i - y_i)^2.$$

It is evident that our goal is to minimise *fitness(I)*, because the smaller the MSE is, the better the system will be.

Several inference models can be used, which can give rise to different results. In this work we have taken the following decisions:

- The input values are *fuzzyfied* using punctual *fuzzyfication* (that is, as *singleton* fuzzy sets).
- *Max-Min* inference is used.
- *Defuzzyfication* is made using the *gravity centre*.
- The FITA-Weighted[4] approach is used for obtaining a single result when several rules are fired for an example (tuple).

Finally, let us to say that our search algorithms have been written in Java and for the definition and evaluation of the fuzzy rule systems, we interact with FuzzyJess [2, 20] also written in Java.

5.3 Algorithms and Parameters

In the four problems, we have considered two different options: to associate 5 or 7 labels to each linguistic variable, defined to give rise to a symmetric partition of the real domain with triangular fuzzy sets. Of course, the optimization problem is more complex when 7 labels are considered. Table 1 shows some data related to the optimization task of each problem: the cardinality of the search space, the individual length, and the number of values taken by each individual's position (gene).

Table 1. Some data about the considered problems

	5 labels			7 labels		
Prob.	cardinality	length	values(variables)	cardinality	length	values(variables)
F_1	4.22E13	25	2(1),3(9),4(15)	1.99E27	49	2(1),3(16),4(31),5(1)
F_4	6.25E16	25	4(7),5(18)	1.98E36	49	5(24),6(24),7(1)
E_1	1.38E06	12	2(3),3(3),4(4),5(2)	6.37E09	22	2(11),3(4),4(4),5(2),6(1)
E_2	1.27E22	56	2(39),3(17)	6.74E34	104	2(84),3(20)

In the experiments, we have worked with the algorithms previously described: Wang and Mendel (WM) algorithm (that will be taken as a reference), GA, UMDA, PBIL, MIMIC, TREE, MIMIC01 and TREE01. The following parameters have been considered.

---

[4] In FITA (First Infer Then Aggregate), the fuzzy set obtained by each rule is defuzzified and the final value is obtained as a combination of such values weighted by the matching degree of the example with respect to the antecedents (see [12] for details)

- The *population size* (**PopSize**) that has been fixed to 500 and 1000 for the cases of 5 and 7 labels respectively.
- The initial population has been initialised by giving 10% of probability to the null consequent and distributing the remaining 90% uniformly among the rest of possible consequents.
- In GA the crossover probability (**cp**) has been fixed to 1.0 and the mutation probability (**mp**) to 0.01. That is, at each generation **PopSize**/2 pairs of individuals are selected by using rank-based selection, they are crossed and we apply the bit-to-bit mutation to the resulting children with **mp** = 0.01 Classical one-point crossover has been used.
- $\alpha = 0.5$ has been used in PBIL.
- The population D_k is obtained by truncation, that is, the best **PopSize** individuals of $D_{k-1} \cup D_{aux}$ are selected, D_{aux} being the sampled population in the case of EDAs and the genetically-generated population in GAs.
- In EDAs, to estimate the model at the k-th generation we use the best **PopSize**/2 individuals from D_{k-1}^{Se}.
- *Stop condition*: the algorithm will be stopped when the best individual in D_{aux} does not improve any of the best **PopSize**/2 individuals of D_{k-1}, that is, the best **PopSize**/2 individuals of D_k coincide with the best **PopSize**/2 individuals of D_{k-1}. Moreover, we have fixed a maximum number of 100 generations.

5.4 Results

Each algorithm has been run 20 times in each of the eight cases (4 problems with 5 or 7 labels). In Tables 2 to 9 we report mean (first row) and standard deviation (second row) for the following statistics:

- MSE over the training set.
- MSE over the test set.
- Number of evaluations. This figure accounts for the number of individuals actually evaluated, because a hash table is used during the search process.
- Number of generations carried out before the algorithm stops.
- Number of fuzzy rules included in the generated system.

5.5 Analysis

In order to be in a position to analyse our results, we have carried out a series of statistical tests among the seven stochastic algorithms used in the experiments. We have performed t-tests ($\alpha = 0.05$) with respect to the five parameters reported in the previous section: Training error, test error, evaluations, generations and number of rules in the resulting system. Tables 10 to 13 show the results of the statistical analysis.

Table 2. Results for F1 problem with 5 labels and population size 500

Algorithm	Training error	Test error	Evaluations	Generations	Rules
WM	19.97	19.38	–	–	25
GA	7.2404	5.3403	8742.20	23.75	25.00
	0.1100	0.2377	1104.84	2.21	0.00
UMDA	7.1955	5.3188	12045.30	33.75	25.00
	0.0898	0.2682	3438.45	6.78	0.00
PBIL	7.1618	5.2160	13350.85	44.35	25.00
	0.0489	0.0935	4200.88	6.78	0.00
MIMIC	7.1506	5.1946	5665.95	17.10	25.00
	0.0000	0.0000	794.23	1.81	0.00
MIMIC01	7.1506	5.1946	5819.95	17.40	25.00
	0.0000	0.0000	720.73	1.11	0.00
TREE	7.1506	5.1946	5925.30	18.00	25.00
	0.0000	0.0000	1073.65	2.97	0.00
TREE01	7.1506	5.1946	5599.50	17.30	25.00
	0.0000	0.0000	778.56	1.49	0.00

Table 3. Results for F1 problem with 7 labels and population size 1000

Algorithm	Training error	Test error	Evaluations	Generations	Rules
WM	5.4067	5.8188	–	–	49
GA	2.8965	2.3886	32664.35	40.80	49.00
	0.0757	0.1435	4867.68	5.21	0.00
UMDA	2.8451	2.3205	31847.80	40.35	49.00
	0.0280	0.0379	3912.91	6.97	0.00
PBIL	2.8387	2.3204	49189.95	62.15	49.00
	0.0000	0.0185	6057.39	5.96	0.00
MIMIC	2.8415	2.3104	22405.70	29.20	49.00
	0.0122	0.0236	4317.12	4.24	0.00
MIMIC01	2.8451	2.3419	21683.70	29.15	49.00
	0.0280	0.0857	3345.78	4.27	0.00
TREE	2.8387	2.3213	20775.30	28.30	49.00
	0.0000	0.0174	2921.23	2.97	0.00
TREE01	2.8387	2.3223	22789.15	29.45	49.00
	0.0000	0.0125	3541.41	4.02	0.00

Table 4. Results for F4 problem with 5 labels and population size 500

Algorithm	Training error	Test error	Evaluations	Generations	Rules
WM	1.2836	1.4957	–	–	25
GA	1.0016	1.0861	13502.60	32.95	20.00
	0.0003	0.0034	2737.86	5.37	0.95
UMDA	1.0015	1.0850	8145.70	20.80	21.00
	0.0000	0.0000	358.10	0.75	0.00
PBIL	1.0015	1.0850	12653.20	31.90	20.95
	0.0000	0.0000	382.28	0.89	0.22
MIMIC	1.0015	1.0850	10185.75	25.75	20.90
	0.0000	0.0000	1363.60	3.31	0.44
MIMIC01	1.0015	1.0850	10284.40	26.80	20.70
	0.0000	0.0000	1393.07	3.92	0.56
TREE	1.0015	1.0850	10981.20	28.15	20.65
	0.0000	0.0000	1375.46	3.54	0.57
TREE01	1.0015	1.0850	10063.40	26.85	20.75
	0.0000	0.0000	740.49	2.39	0.54

Table 5. Results for F4 problem with 7 labels and population size 1000

Algorithm	Training error	Test error	Evaluations	Generations	Rules
WM	3.4074	3.9503	–	–	49
GA	1.0068	1.0841	51629.35	59.30	47.85
	0.0025	0.0273	6780.26	6.28	0.79
UMDA	1.0092	1.0982	27755.00	34.05	48.95
	0.0007	0.0100	1787.83	1.47	0.22
PBIL	1.0094	1.0932	44828.25	53.25	48.85
	0.0004	0.0077	3130.93	1.73	0.36
MIMIC	1.0059	1.0970	37130.15	44.30	48.20
	0.0026	0.0153	7411.94	7.48	0.75
MIMIC01	1.0071	1.0920	34983.90	41.20	48.55
	0.0021	0.0161	4268.03	5.62	0.59
TREE	1.0058	1.0973	39939.25	46.45	48.25
	0.0020	0.0127	8983.31	9.65	0.70
TREE01	1.0062	1.0999	38732.80	46.30	48.25
	0.0027	0.0201	6775.04	6.42	0.77

Table 6. Results for E1 problem with 5 labels and population size 500

Algorithm	Training error	Test error	Evaluations	Generations	Rules
WM	560417.6055	548392.0448	–	–	12
GA	397270.1750	384926.6230	3780.05	13.95	9.70
	0.0000	0.0124	875.16	1.99	0.46
UMDA	397270.1750	384926.6230	3029.70	16.40	10.00
	0.0000	0.0124	580.28	3.76	0.00
PBIL	397270.1750	384926.6230	4389.65	20.30	9.85
	0.0000	0.0124	727.23	2.37	0.36
MIMIC	397270.1750	384926.6230	3333.75	12.30	9.95
	0.0000	0.0124	527.26	1.31	0.22
MIMIC01	397270.1750	384926.6230	3173.10	12.25	9.95
	0.0000	0.0124	710.36	1.34	0.22
TREE	397318.8916	388147.7117	2994.75	12.15	9.80
	212.3506	14040.4003	727.15	1.35	0.51
TREE01	397270.1750	384926.6230	3217.20	12.70	9.90
	0.0000	0.0124	1037.48	2.69	0.30

Table 7. Results for E1 problem with 7 labels and population size 1000

Algorithm	Training error	Test error	Evaluations	Generations	Rules
WM	467587.4102	417365.2408	–	–	22
GA	313178.5100	349513.1792	13306.30	20.10	17.55
	467.2229	1308.1185	2123.61	2.02	0.92
UMDA	313334.2509	349077.1397	10636.90	15.95	17.90
	0.0000	0.0055	1488.49	1.07	0.30
PBIL	313334.2509	349077.1397	15318.65	22.75	17.95
	0.0000	0.0055	2109.85	0.54	0.22
MIMIC	313334.2509	349077.1397	11207.55	17.95	17.80
	0.0000	0.0055	2116.06	1.72	0.40
MIMIC01	313502.6448	349295.1594	10847.75	17.30	17.60
	1142.7346	950.3260	2221.21	1.73	0.97
TREE	312729.1387	350033.7056	12578.10	18.95	16.60
	744.8612	3496.8269	2247.71	2.50	1.46
TREE01	313118.4911	348943.6069	12129.65	18.95	17.45
	517.7311	2893.3934	2747.53	2.56	1.07

Table 8. Results for E2 problem with 5 labels and population size 500

Algorithm	Training error	Test error	Evaluations	Generations	Rules
WM	381649.0877	383753.3989	–	–	56
GA	189959.4115	196317.5345	29291.35	66.95	37.60
	314.2939	3075.5425	4102.90	8.35	1.50
UMDA	190958.1161	196141.4127	19356.95	44.80	39.10
	783.2563	2497.3381	2796.77	5.85	1.04
PBIL	190945.1776	196187.6407	28475.95	66.05	39.40
	725.1868	2990.8605	4337.75	9.64	1.07
MIMIC	190103.5788	199283.7667	18936.35	44.50	38.00
	813.6191	2077.8690	3189.97	6.42	1.67
MIMIC01	189837.1496	198392.2004	19051.80	44.85	37.85
	546.5796	2061.4484	2807.63	5.83	1.74
TREE	189881.8106	198416.6743	18776.05	43.80	37.75
	469.8954	2309.9994	3211.51	6.60	1.41
TREE01	189759.6925	197657.9061	19475.95	45.85	37.55
	304.9269	2375.1651	4335.95	8.78	1.66

Table 9. Results for E2 problem with 7 labels and population size 1000

Algorithm	Training error	Test error	Evaluations	Generations	Rules
WM	116843.85	122701.0448	–	–	104
GA	83690.2321	94923.8170	55835.45	66.65	80.55
	505.9616	2377.8009	7568.60	7.09	1.75
UMDA	85092.6175	91147.7307	26117.05	31.80	84.30
	631.0082	2735.7334	3459.33	2.80	1.62
PBIL	85081.1355	90970.5983	42678.00	50.00	84.25
	674.4282	2312.6308	5112.42	5.11	1.73
MIMIC	83769.0245	94639.7871	32139.90	38.90	81.35
	483.8130	1279.4342	5559.05	5.76	1.53
MIMIC01	83541.2357	94635.8839	35906.55	42.40	80.40
	260.9488	610.4732	7890.08	8.06	1.50
TREE	83640.1622	94417.7968	33971.60	40.35	81.05
	476.2109	1208.2838	5431.16	5.47	1.47
TREE01	83443.5197	94408.1091	34355.45	41.25	80.65
	142.7718	337.1590	3457.35	4.12	1.06

Table 10. Results of t-tests for function F1: upper diagonal (5 labels) and lower diagonal (7 labels). T/t/e/g/r stands for statistical difference (significance level $\alpha = 0.05$) in Training error/test error/evaluations/generations/rules

	GA	UMDA	PBIL	MIMIC	MIMIC01	TREE	TREE01	
GA		eg	Tteg	Tteg	Tteg	Tteg	Tteg	GA
UMDA	Tte		eg	Tteg	Tteg	Tteg	Tteg	UMDA
PBIL	Tteg	eg		eg	eg	eg	eg	PBIL
MIMIC	Tteg	eg	eg					MIMIC
MIMIC01	Teg	eg	eg					MIMIC01
TREE	Tteg	eg	eg	e				TREE
TREE01	Tteg	eg	eg	t				TREE01
	GA	UMDA	PBIL	MIMIC	MIMIC01	TREE	TREE01	

Table 11. Results of t-tests for function F4: upper diagonal (5 labels) and lower diagonal (7 labels). T/t/e/g/r stands for statistical difference (significance level $\alpha = 0.05$) in Training error/test error/evaluations/generations/rules

	GA	UMDA	PBIL	MIMIC	MIMIC01	TREE	TREE01	
GA		egr	er	gr	gr	eg	gr	GA
UMDA	Tegr		eg	eg	eg	egr	eg	UMDA
PBIL	Tegr	eg		eg	eg	egr	eg	PBIL
MIMIC	tegr	Tegr	Tegr			eg		MIMIC
MIMIC01	egr	Tegr	Tegr			e		MIMIC01
TREE	eg	Tegr	Tegr	r	Tgr		e	TREE
TREE01	tegr	Tegr	Tegr		g	r		TREE01
	GA	UMDA	PBIL	MIMIC	MIMIC01	TREE	TREE01	

Table 12. Results of t-tests for function E1: upper diagonal (5 labels) and lower diagonal (7 labels). T/t/e/g/r stands for statistical difference (significance level $\alpha = 0.05$) in Training error/test error/evaluations/generations/rules

	GA	UMDA	PBIL	MIMIC	MIMIC01	TREE	TREE01	
GA		eg	eg	egr	egr	eg	e	GA
UMDA	g		egr	eg	g	gr	g	UMDA
PBIL	egr	eg		eg	eg	eg	eg	PBIL
MIMIC	eg	eg	eg			e		MIMIC
MIMIC01	g	eg	eg	e				MIMIC01
TREE	Ter	egr	egr	er	Tegr			TREE
TREE01	e	egr	egr		g	Ter		TREE01
	GA	UMDA	PBIL	MIMIC	MIMIC01	TREE	TREE01	

Table 13. Results of t-tests for function E2: upper diagonal (5 labels) and lower diagonal (7 labels). T/t/e/g/r stands for statistical difference (significance level $\alpha = 0.05$) in $\underline{T}$raining error/$\underline{t}$est error/$\underline{e}$valuations/$\underline{g}$enerations/$\underline{r}$ules

	GA	UMDA	PBIL	MIMIC	MIMIC01	TREE	TREE01	
GA		Tegr	Tr	teg	teg	teg	Teg	GA
UMDA	Ttegr		eg	Tter	Tter	Tter	Tter	UMDA
PBIL	Ttegr	eg		Ttegr	Ttegr	Ttegr	Tegr	PBIL
MIMIC	eg	Ttegr	Ttegr				Tt	MIMIC
MIMIC01	eg	Ttegr	Ttegr	Tr				MIMIC01
TREE	eg	Ttegr	Ttegr	e			e	TREE
TREE01	Teg	Ttegr	Ttegr	T		Te		TREE01
	GA	UMDA	PBIL	MIMIC	MIMIC01	TREE	TREE01	

By taking into account the results shown in the previous section and the hypothesis testing carried out, we can reach some general and (sometimes) relevant conclusions depending on the parameter considered. First, we take into account the fitness value, that is, a quality measure of the estimated model. In this aspect, there are four main observations:

- It seems that there are different set of rules that yield the same error over the training set but have different behaviour over the test set. As we have used the error over the training set as fitness measure, we will base our conclusions on this error, though we are considering taking the error using a cross validation over the training set as fitness in future works.

- In all cases WM is surpassed by evolutionary algorithms. That is clear, since these methods pursued an improvement of the first one. Figure 5 shows a graphical example of the error committed by the system obtained by WM and that obtained by EDAs (MIMIC) in function F_1.

- Taking the MSE (evaluation/fitness function) as reference, we find that univariate EDAs behave quite similarly to the genetic algorithm. There is no statistical difference only in problems F1 (in favour of univariate EDAs) and E2 (in favour of GAs). Finally, there is no statistical difference between PBIL and UMDA in any of the eight cases.

- Finally, bivariate EDAs provide in general the best results, being always, at least, as good as the others. In fact, when we consider bivariate EDAs globally, there is statistical difference in 4 out of the 8 cases with respect to univariate EDAs, and in 2 out of the 8 cases with respect to the genetic algorithm. However, if we focus our attention on tree-shaped algorithms, then there is statistical difference in 5 out of the 8 cases with respect to GAs. It is worth mentioning that TREE01 is the algorithm obtaining the best results in problem E2 (with 5 and 7 labels), having statistical difference with respect to GAs and univariate EDAs.

The second, but not less important parameter to study, is the number of generations. That is directly related to the number of evaluated individuals. Let us remember that the evaluation of an individual (a FLBRS) takes a large amount of time (i.e. up to

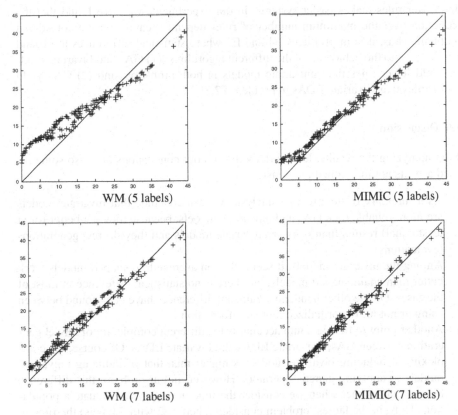

WM (5 labels)   MIMIC (5 labels)

WM (7 labels)   MIMIC (7 labels)

Fig. 5. A plot of predicted vs actual value for the test set in function F1

one minute in some of the problems studied here). Therefore, the fewer the generations, the quicker the method results, this being a considerable difference. Regarding this factor, the main conjectures we can draw from the previous experiments (and supported by the statistical hypothesis testing) are:

- PBIL and GA need more iterations than the other algorithms. Normally this number of generations seems related to the accuracy of the obtained system (error).
- When the population size is 1000, UMDA shows the tendency to stop after fewer generations. That would be nice if the obtained system was as good as the rest. Unfortunately, this is not the case.
- Bivariate models are almost always those running few iterations. This fact indicates that real and significant dependencies among the considered variables (i.e. fuzzy rules) exist. And that is why they obtained the better results for the MSE. Furthermore, there is no (statistically supported) ranking among them.

The last value to analyse might be the number of rules. For the purpose of this work, this parameter is not as important as the other two parameters, but it is clear that fewer rules makes easier systems. In our experiments we have found that the reduction over the maximum number of rules depends heavily on the considered problem. Thus, it is in problems F4 and E2 where statistical differences are found with respect to the behaviour of the different algorithms: (1) GAs and bivariate models yield fewer rules than univariate models in both problems, and (2) GAs yield fewer rules than bivariate EDAs in problem F2.

5.6 Discussion

After analyzing the results, let us briefly discuss our conclusions and also some related aspects not mentioned previously:

- As a summary of the previous analysis, we can conclude that bivariate models are more reliable than GAs and univariate models, because they get better (or at least equal) results than GAs or univariate models and they do few generations (evaluations).
- Among the bivariate models, it seems that on average tree-shaped models work better than chain-shaped models, but there is no statistical difference in most of the cases. On the other hand, no significant differences have been found between using or not $[0, 1]$-normalized mutual information.
- Another point to be taken into account is the different complexity of model generation between GAs, univariate EDAs and bivariate EDAs. Of course, the complexity of inducing bivariate models is higher than that of inducing univariate models or applying genetic operators. However, in our problem this difference has a small impact when we consider the time necessary to evaluate a population. Thus, in the largest problem considered here (E2 with 7 labels) the time to induce+sample a model is 0.037s for GAs, 0.055s for univariate EDAs and 0.35s for bivariate EDAs. Therefore, bivariate models need ten times more CPU than GAs, but if we take into account that evaluating a population takes 55s, then the differences between 55.037s, 55.055s and 55.35s are almost insignificant.
- With respect to comparison with similar works, we can refer to [7, 19], where the COR methodology is instantiated by using simulated annealing [7] or ant colonies [19] as search engine. A fair comparison is not possible because there are some degrees of freedom in the implementation (specially) of the fuzzy inference engine, i.e., which value should be returned by the fuzzy inference engine when an example does not fire any rule?. As an example, in [7] the training error yielded by WM algorithm in problem F1 is 4.0963[5], while in our case we get 5.4067, and WM is (almost[6]) deterministic.

Because of this, we focus our comparison on the reduction of the errors made by COR over the results obtained by WM. Thus, for problem F1 COR with simulated

---

[5] We have multiplied by 2 the errors reported in [7, 19] because in such works the MSE is multiplied by 1/2.

[6] Notice that in the example of Sect. 2.2 we choose arbitrarily/randomly the consequent for the rule produced by instance number 6

annealing or ant colonies has only 78%(54%) of the training(test) error committed by WM, while COR with TREE get a training(test) error which is 52.5%(40%) of the error committed by WM. For problem E2, COR with simulated annealing has a training(test) error of about the 74%(88%) of WM error, while COR with ACO and TREE has, respectively, an error of 74%(70%) and 67%(84%) over the error committed by WM. From this (approximate) comparison, we can see that bivariate EDAs are a very good choice to instantiate the COR methodology, although they seem to need more evaluations than simulated annealing or ant colonies [7, 19].

6 Concluding Remarks

In this work we have presented an application of EDAs in quite an interesting and hard problem which is the machine learning or data mining field. Concretely, we have shown how EDAs are a clear alternative to other metaheuristic algorithms in the problem of learning linguistic fuzzy rule systems. Our experiments show that simple instances of EDAs as is the case of bivariate models offer good behaviour, being quite reliable in both important points: optimisation of the fitness function and number of required evaluations (generations).

Furthermore, the superior behaviour of bivariate models with respect to univariate models, is a clear indication of existing dependence relations among the variables (rules) of the system. This fact constitutes an important starting point for future works.

Acknowledgments

This work has been partially supported by the Junta de Comunidades de Castilla - La Mancha (Consejería de Ciencia y Tecnología) and FEDER, under project PBC-02-002.

References

1. B. A. Bárdossy and Duckstein. *Fuzzy Rule-based Modeling with Application to Geophysical, Biological and Engineering Systems*. CRC Press, 1995.
2. Integrated Reasoning Group at NCR.   Fuzzyj toolkit & fuzzyjess url. http://www.iit.nrc.ca/IR_public/fuzzy/fuzzyJToolkit2.html.
3. S. Baluja. Population-based incremental learning: A method for integrating genetic search based function optimization and competitive learning. Technical Report TR CMU-CS-94-163, Carnegie Mellon University, 1994.
4. S. Baluja and S. Davies. Combining multiple optimization runs with optimal dependency trees. Technical Report TR CMU-CS-97-157, Carnegie Mellon University, 1997.
5. J. S. De Bonet, C. L. Isbell, and P. Viola. Mimic: Finding optima by estimating probability densities. In *Proceedings of Neural Information Processing Systems*, pp. 424–430, 1996.

6. J. Casillas, O. Cordón, and F. Herrera. Improving the Wang and Mendel's fuzzy rule learning method by inducing cooperation among rules. In *8th International Conference on Information Processing and Management of Uncertainty in Knowledge-Based Systems*, pp. 1681–1688, 2000.

7. J. Casillas, O. Cordón, and F. Herrera. COR: A methodology to improve ad hoc data-driven linguistic rule learning methods by inducing cooperation among rules. *IEEE Transactions on Systems, Man, and Cybernetics, Part B: Cybernetics*, 32(4):526–537, 2002.

8. C. Chow and C. Liu. Approximating discrete probability distributions with dependence trees. *IEEE Transactions on Information Theory*, 14:462–467, 1968.

9. M. J. Flores and J. A. Gámez. Applicability of estimation of distribution algorithms to the fuzzy rule learning problem: A preliminary study. In *Proceedings 4th International Conference on Enterprise Information Systems*, pp. 350–357, 2002.

10. K. Hirota. *Industrial Applications of Fuzzy Technology*. Springer-Verlag, 1993.

11. F. V. Jensen. *Bayesian Networks and Decision Graphs*. Springer-Verlag, 2001.

12. G. J. Klir and B. Yuan. *Fuzzy Sets and Fuzzy Logic. Theory and Applications*. Prentice Hall, 1995.

13. P. Larrañaga, R. Etxeberría, J. A. Lozano, and J. M. Peña. Combinatorial optimization by learning and simulation of Bayesian networks. In *Proceedings of the 16th Conference on Uncertainty in Artificial Intelligence*, pp. 343–352, 2000.

14. P. Larrañaga and J. A. Lozano(Eds.). *Estimation of Distribution Algorithms. A New Tool for Evolutionary Computation*. Kluwer Academic Press, 2002.

15. Z. Michalewicz. *Genetic Algorithms + Data Structures = Evolution Programs*. Springer-Verlag, 1996.

16. T. M. Mitchell. *Machine Learning*. McGraw-Hill, 1997.

17. H. Mühlenbein. The equation for response to selection and its use for prediction. *Evolutionary Computation*, 5(3):303–346, 1997.

18. K. Nozaki, H. Ishibuchi, and H. Tanaka. A simple but powerful heuristic method for generating fuzzy rules from numerical data. *Fuzzy Sets and Systems*, 86:251–270, 1997.

19. J. Casillas O., Cordón F., and Herrera. Learning cooperative fuzzy linguistic rules using ant colony optimization algorithms. Technical Report TR DECSAI-00119, University of Granada, 2000.

20. R. Orchand. Fuzzy reasoning in Jess: The fuzzyj toolkit and fuzzyJess. In *Proceedings of the ICEIS 2001, Third International Conference on Enterprise Information Systems*, pp. 533–542, 2001.

21. M. Sugeno and G. T. Kang. Structure identification of fuzzy models. *Fuzzy Sets and Systems*, 28:15–33, 1988.

22. T. Takagi and M. Sugeno. Fuzzy identification of systems and its application to modeling and control. *IEEE Transactions on Systems, Man, and Cybernetics*, 15:116–132, 1985.

23. L. X. Wang and J. M. Mendel. Generating fuzzy rules by learning from examples. *IEEE Transactions on Systems, Man, and Cybernetics*, 22(6):1414–1427, 1992.

24. L. A. Zadeh. Fuzzy sets. *Information and Control*, 8:338–353, 1965.

25. L. A. Zadeh. The concept of a linguistic variable and its application to approximate reasoning. part i. *Information Science*, 8:199–249, 1975.

26. L. A. Zadeh. The concept of a linguistic variable and its application to approximate reasoning. part ii. *Information Science*, 8:301–357, 1975.

27. L. A. Zadeh. The concept of a linguistic variable and its application to approximate reasoning. part iii. *Information Science*, 9:43–80, 1975.

Estimation of Distribution Algorithm with 2-opt Local Search for the Quadratic Assignment Problem

Qingfu Zhang, Jianyong Sun, Edward Tsang and John Ford

Department of Computer Science, University of Essex
Wivenhoe Park, Colchester, CO4 3SQ, U.K.
{qzhang,jysun,edward,fordj}@essex.ac.uk

Summary. This chapter proposes a combination of estimation of distribution algorithm (EDA) and the 2-opt local search algorithm (EDA/LS) for the quadratic assignment problem (QAP). In EDA/LS, a new operator, called *guided mutation*, is employed for generating new solutions. This operator uses both global statistical information collected from the previous search and the location information of solutions found so far. The 2-opt local search algorithm is applied to each new solution generated by guided mutation. A restart strategy based on statistical information is used when the search is trapped in a local area. Experimental results on a set of QAP test instances show that EDA/LS is comparable with the memetic algorithm of Merz and Freisleben and outperforms estimation of distribution algorithm with guided local search (EDA/GLS). The proximate optimality principle on the QAP is verified experimentally to justify the rationale behind heuristics (including EDA/GLS) for the QAP.

1 Introduction

The Quadratic Assignment Problem (QAP) is a combinatorial optimization problem introduced by Koopmans and Beckmann [1] to formulate and solve the situation where a set of facilities have to be assigned in an optimal manner to given locations. The problem can model a variety of applications in scheduling, manufacturing, statistical data analysis, etc. Çela [2] gives a good overview of theory and algorithms for the QAP.

Given $\mathcal{N} = \{1, 2, \cdots, n\}$ and two $n \times n$ matrices $A = (a_{ij})$ and $B = (b_{kl})$, the QAP can be stated as follows:

$$\min_{\pi \in S_n} c(\pi) = \sum_{i=1}^{n} \sum_{j=1}^{n} a_{\pi(i)\pi(j)} b_{ij} \tag{1}$$

where π is a permutation of $\mathcal{N}$ and S_n is the set of all possible permutations of $\mathcal{N}$. In the facility location context, A is the distance matrix, so that a_{ij} represents the distance between locations i and j. B is the flow matrix, so that b_{kl} represents the flow between facilities k and l. π represents an assignment of n facilities to n locations. More specifically, $\pi(i) = k$ means that facility i is assigned to location k.

Q. Zhang et al.: *Estimation of Distribution Algorithm with 2-opt Local Search for the Quadratic Assignment Problem*, StudFuzz **192**, 281–292 (2006)
www.springerlink.com

The QAP is one of the most difficult $\mathcal{NP}$-hard combinatorial problems. Solving QAP instances with $n > 30$ to optimality is computationally impractical for exact algorithms such as the branch-and-bound method [18]. Therefore, a variety of heuristic algorithms for dealing with large QAP instances have been developed, e.g. simulated annealing [4], threshold accepting [5], neural networks [6], tabu search [7], guided local search [8], evolution strategies [9] , genetic algorithms [10], ant colony optimization [11], memetic algorithms [12], and scatter search [13]. These algorithms cannot be guaranteed to produce optimal solutions, but they are able to produce fairly good solutions at least some of the time.

Estimation of Distribution Algorithms (EDAs) [3] are a new class of evolutionary algorithms (EAs). Unlike other EAs, EDAs do not use crossover or mutation. Instead, they explicitly extract global statistical information from the previous search and build a posterior probability model of promising solutions, based on the extracted information. New solutions are sampled from the model thus built. Like other EAs, EDAs are good at identifying promising areas in the search space, but lack the ability of refining a single solution. A very successful way to improve the performance of EAs is to hybridize them with local search techniques. In fact, combinations of genetic algorithms and local search heuristics, often called memetic algorithms in the literature, have been applied successfully to a number of combinatorial optimization problems. Recently, we have combined an EDA with guided local search (EDA/GLS) [19] for the QAP and obtained some encouraging preliminary experimental results.

A combination of an EDA with a very simple local search (EDA/LS) for the QAP is proposed and studied in this chapter. EDA/LS maintains a population of potential solutions and a probability matrix at each generation. The offspring generation scheme in EDA/LS is guided mutation [19] [20]. Guided by the probability matrix, guided mutation randomly mutates a selected solution to generate a new solution. Each new solution is improved by the 2-opt local search. A novel restart strategy is used in EDA/LS to help the search escape from areas where it has been trapped. The experimental results show that EDA/LS is comparable to the memetic algorithm (MA) of Merz and Freisleben [12] and outperforms EDA/GLS on a set of QAP instances.

The rest of the chapter is organized as follows. In Sec. 2, EDA/LS is introduced. Section 3 presents the comparison of EDA/LS, EDA/GLS and the memetic algorithm [12]. The proximate optimality principle, the underlying assumption in heuristics including EDA/LS, has been experimentally verified in Sec. 3. Section 4 concludes the chapter.

2 Algorithm

At each generation t, EDA/LS maintains a population $Pop(t) = \{\pi^1, \pi^2, \ldots, \pi^N\}$ of N solutions and a probability matrix:

$$p(t) = \begin{pmatrix} p_{11}(t) & \cdots & p_{1n}(t) \\ \vdots & & \vdots \\ p_{n1}(t) & \cdots & p_{nn}(t) \end{pmatrix},$$

where $p(t)$ models the distribution of promising solutions in the search space. More precisely, $p_{ij}(t)$ is the probability that facility i is assigned to location j in a promising assignment.

2.1 2-opt Local Search

The local search used in this chapter is the 2-opt local search [16]. Let π be a solution for the QAP. Then its 2-opt neighborhood $\mathcal{N}(\pi)$ is defined as the set of all possible solutions resulting from π by swapping two distinct elements. The 2-opt local search algorithm searches the neighborhood of a current solution for a better solution. If such a solution is found, it replaces the current solution and the search continues. Otherwise, a local optimum has been reached. In our experiments, the first better solution found is accepted and used to replace the current solution. In other words, we use the first-improvement principle.

2.2 Initialization

EDA/LS randomly chooses N solutions and then applies the 2-opt local search to improve them. The N resultant solutions $\{\pi^1, \pi^2, \ldots, \pi^N\}$ constitute the initial population $Pop(0)$. The initial probability matrix $p(0)$ is set as

$$p_{ij} = \frac{1}{n}.$$

2.3 Update of Probability Matrix

Assume that the population at generation t is $Pop(t) = \{\pi^1, \pi^2, \ldots, \pi^N\}$. Then the probability matrix $p(t)$ can be updated (as in PBIL [14]) as follows:

$$p_{ij}(t) = (1 - \beta)\frac{1}{N}\sum_{k=1}^{N} I_{ij}(\pi^k) + \beta p_{ij}(t-1), \quad (1 \le i, j \le n), \tag{2}$$

where

$$I_{ij}(\pi) = \begin{cases} 1 & \text{if } \pi(i) = j, \\ 0 & \text{otherwise}. \end{cases}$$

$0 \le \beta \le 1$ is a learning rate. The bigger β is, the greater the contribution of the solutions in $Pop(t)$ is to the probability matrix $p(t)$.

GuidedMutation(π, p, α) *Input: a permutation*
$\pi = (\pi(1), \ldots, \pi(n))$, *a probability matrix* $p = (p_{ij})$ *and a positive parameter* $\delta < 1$. *Output:* $\sigma = (\sigma(1), \ldots, \sigma(n))$, *a permutation.*

Step 1 Randomly pick $[\alpha n]$ integers uniformly from $\{1, 2, \ldots n\}$ and
      let these integers constitute a set $K \subset I$. Set $V = I \backslash K$ and
      $U = I$.

Step 2 For each $i \in K$, set $\sigma(i) = \pi(i)$ and $U = U \backslash \{\pi(i)\}$.

Step 3 While$(U \neq \emptyset)$ do:

      Select a i from V, then randomly pick up a $k \in U$ with
        probability
$$\frac{p_{ik}}{\sum_{j \in U} p_{ij}}.$$
      Set $\sigma(i) = k$, $U = U \backslash \{k\}$ and $V = V \backslash \{i\}$.

Step 4 Return σ.

Fig. 1. Guided Mutation for creating offspring with permutation representation

2.4 Generation of New Solutions: Guided Mutation

Guided by a probability matrix $p = (p_{ij})_{n \times n}$, guided mutation [19] [20] mutates an existing solution to generate a new solution. This operator also needs a control parameter $0 < \alpha < 1$. It works as shown in Fig. 1.

The goal of guided mutation is to generate a solution σ. Step 1 randomly divides the facilities into two groups. The first group has $[\alpha n]$ facilities and the second one has $n - [\alpha n]$ facilities. In Step 2, facility i in the first group is assigned to location $\pi(i)$, which is the location for this facility in solution π. Step 3 arranges the facilities in the second group sequentially, based on the probability matrix p.

2.5 Restarting Strategy

In EDA/LS, if the average cost of the population does not decrease for successive L generations, EDA/LS will re-initialize its population. New initial solutions should be as far from the current population as possible, since EDA/LS has intensively exploited the current area. Let $p = (p_{ij})$ be the current probability matrix. Then EDA/LS generates a new initial solution as shown in Fig. 2.

Obviously, the larger p_{ij} is, the smaller the probability that $\pi(i) = j$ is in the above procedure. Therefore, the resultant π should be far from the current population.

Two other commonly-used restart strategies are the random restart and the mutation restart. The random restart generates the new initial population randomly. It does not take into consideration any information from the previous search. In the mutation restart [12], each solution except the best one in the current population is mutated to yield a new initial solution. Mutation restart does not explicitly utilize global statistical information in the current population.

REstart(p)
Input: $p = (p_{ij})$: *a probability matrix.*
Output: $\pi = (\pi(1), \ldots, \pi(n))$, *a solution.*

Step 1 Set $U = \{1, 2, \ldots, n\}$
Step 2 For $i = 1, 2, \ldots, n$

    Randomly pick a $k \in U$ with probability

$$\frac{[1 - p_{ik}]}{\sum_{j \in U}[1 - p_{ij}]}.$$

    Set $\pi(i) = k$ and $U = U \backslash \{k\}$.

Step 3 2-opt Local Search: use the 2-opt local search to improve π.
Step 4 Return π.

Fig. 2. The probability-based restart strategy

2.6 Structure of EDA/LS

The framework of EDA/LS is described in Fig. 3.

3 Computational Experiments and Analysis

3.1 Experimental Comparison with EDA/GLS and MA

EDA/LS has been compared with the memetic algorithm (MA) of Merz and Freisleben [12] and EDA/GLS [19] on a set of QAPLIB test instances [15]. EDA/LS was implemented in C++. All the experiments reported in this chapter were performed on identical PCs (AMD Athlon 2400MHZ) running Linux. The parameter settings for EDA/LS were as follows:

- Population size $N = 10$;
- The number of new solutions generated at each generation: $M = \frac{N}{2}$;
- The number of generations used in the restart condition: $L = 30$;
- The control parameter in Guided Mutation α and the learning rate β used in the update of probability matrix. We have used two different settings: $(\alpha, \beta) = (0.3, 0.3)$ and $(\alpha, \beta) = (0.5, 0.1)$.

The experimental results are given in Table 1.

In this table, the MA results are from one of the best MA variants with the diversification rate $R = 1$ and CX recombination operator (please see [12] for details). The *instance* column lists the QAPLIB instances (the number in the name is the problem size). The cost of the best-known solution for each instance is given in the *best known*

Step 0 Parameter Setting Population Size: N. The number of new solutions generated at each generation: M. The control parameter in GuidedMutation: α. The learning rate used in the update of the probability matrix: β. The number of generations used in the restart strategy: L.

Step 1 Initialization Set $t := 0$. Do initialization as described in subsection 2.2. Set π^* to be the solution with the lowest cost in $Pop(0)$.

Step 2 Guided Mutation For $j = 1, 2, \ldots, M$, do:

    pick up a solution π from $Pop(t)$, do

    guided mutation

        $\sigma = \mathbf{GuidedMutation}(\pi, p(t), \alpha)$

    2-Opt local search

        improve σ by the 2-opt local search.

Step 3 New Population Choose the N best solutions from $\{\sigma^1, \ldots \sigma^M\} \cup Pop(t)$ to form $Pop(t+1)$. Set $t := t+1$. Set π^* to be the solution with the lowest cost in $Pop(t)$. Update the probability matrix using (2).

Step 4 Stopping Condition If the stopping condition is met, stop. Return π^*.

Step 5 Restart Condition If the restart condition is not met, go to Step 2.

Step 6 Restart For $j = 1, 2, \ldots, N$, set $\pi^j = \mathbf{REstart}(p(t))$. Set $Pop(t) = \{\pi^1, \pi^2, \ldots, \pi^N\}$. Find the lowest cost solution σ^* in $Pop(t)$. If $c(\pi^*) > c(\sigma^*)$, set $\pi^* = \sigma^*$. Update the probability matrix using (2). Go to Step 2.

Fig. 3. The framework of EDA/LS

column. The average percentage excess over the best-known solution obtained over 10 runs for MA, EDA/LS and EDA/GLS is listed under $avg\%$ for each algorithm. t/s is the time in seconds used in each run. A number in bold type indicates the result is the best among the three algorithms.

The one tailed t-test results at the 0.05 significance level are also presented in Table 1 for the alternative hypothesis that the mean best solutions obtained by EDA/LS have lower costs than those obtained by EDA/GLS or MA. Column t-$test_1$ lists the t-test values between EDA/LS and EDA/GLS and column t-$test_2$ lists the values between EDA/LS and MA, where t is the absolute value of the t statistic. $sig < 0.05$ suggests that EDA/LS is better than EDA/GLS or MA in terms of solution quality.

In Table 1, the better results obtained by the two sets of parameters in EDA/LS are listed. The respective results of EDA/LS with the two sets of parameters on these test QAP instances are listed in Table 2. In Table 2, "$*$" denotes that the version of the algorithm with the parameter set $(\alpha, \beta) = (0.3, 0.3)$, while "$+$" denotes the version with the parameter set $(\alpha, \beta) = (0.5, 0.1)$. The numbers in bold in these two tables indicate the better result of the two obtained

The results in Table 1 show that in 5 QAP instances (tai60a, tai80a, tai100a, tho150, and tai256c), the results obtained by EDA/LS are better than those of MA, whereas they are worse in 3 instances (sko100a, tai100b, and tai150b). Based on the t-test, EDA/LS is significantly better than MA in 2 instances (with $sig < 0.05$). In 7 instances EDA/LS is significantly better than EDA/GLS. Therefore, we can claim

Table 1. Comparison of EDA/GLS, MA and EDA/LS

		EDA/GLS	MA	EDA/LS	t-test$_1$		t-test$_2$		
instance	best known	avg%	avg%	avg%	t	sig	t	sig	t/s
els19	17212548	**0.000**	**0.000**	**0.000**	-	-	-	-	5
chr25a	3796	2.391	**0.000**	**0.000**	6.129	0.000	-	-	15
bur26a	5426670	**0.000**	**0.000**	**0.000**	-	-	-	-	20
nug30	6124	**0.000**	0.001	**0.000**	-	-	1.000	0.172	20
kra30a	88900	**0.000**	**0.000**	**0.000**	-	-	-	-	20
ste36a	9526	0.041	0.087	**0.000**	2.475	0.018	1.769	0.056	30
tai60a	7208572	**1.209**	1.517	1.320	0.555	0.296	0.989	0.174	90
tai80a	13557864	**0.887**	1.288	1.138	0.827	0.215	0.850	0.204	180
tai100a	21125314	**0.779**	1.213	1.158	3.137	0.006	0.627	0.273	300
sko100a	152002	0.066	**0.027**	0.034	2.395	0.020	0.786	0.226	300
tai60b	608215054	0.132	**0.000**	**0.000**	1.906	0.045	-	-	180
tai80b	818415043	0.513	**0.000**	**0.000**	4.419	0.001	-	-	300
tai100b	1185996137	0.135	**0.000**	0.005	3.966	0.001	1.000	0.176	300
tai150b	498896643	0.351	**0.180**	0.357	0.989	0.348	3.422	0.004	600
tho150	8133484	**0.091**	0.187	0.169	1.899	0.045	0.485	0.319	600
tai256c	44759294	**0.042**	0.096	0.074	1.695	0.062	2.077	0.034	1200
Avg. %		0.414	0.287	**0.265**					

"-" indicates that the t-test has not been carried out for these instances since the corresponding algorithms found the optimal solutions.

Table 2. Comparison of EDA/LS and MA

	EDA/LS*	EDA/LS$^+$
instance	avg.%	avg.%
els19	**0.000**	**0.000**
chr25a	**0.000**	1.713
bur26a	**0.000**	**0.000**
nug30	**0.000**	0.039
kra30a	**0.000**	0.728
ste36a	**0.000**	0.075
tai60a	1.522	**1.320**
tai80a	1.206	**1.138**
tai100a	2.080	**1.158**
sko100a	0.222	0.034
tai60b	**0.000**	**0.000**
tai80b	0.034	**0.000**
tai100b	0.142	0.005
tai150b	0.508	0.357
tho150	0.364	**0.169**
tai256c	0.120	**0.074**

that EDA/LS outperforms EDA/GLS and is comparable with MA in terms of solution quality within a given time limit.

3.2 The QAP and the Proximate Optimality Principle

The proximate optimality principle (POP) assumes that good solutions have similar structure [17]. It is an underlying assumption in most heuristics including EDA/LS. In fact, only when the POP holds, a probability model used in EDA/LS approximates a promising area.

To verify the POP on the QAP instances, we have conducted the following experiments: 500 different local optima $\pi^1, \ldots, \pi^{500}$ are generated by applying the 2-opt local search on randomly generated solutions, then we sort all the 500 obtained local optima with respect to their costs in ascending order. For each local optimum π^k, we generate 1000 distinct local optima $\sigma_k^1, \ldots, \sigma_k^{1000}$ by applying the 2-opt local search on randomly generated solutions in a neighborhood of π^k (the set of all the solutions differing from π^k on at most $0.1n$ items in our experiments). We compute the average cost and the average Hamming distance to π^k of the local optima $\sigma_k^1, \ldots, \sigma_k^{1000}$. Figures 4 and 5 plot these average costs and average distances.

From these figures we can observe the following:

- The average of local optima around a better local optimum is lower.
- The better π^k, the shorter the average distance of $\sigma_k^1, \ldots, \sigma_k^{1000}$ to π^k is.

These observations verify the POP in these QAP instances. Therefore, it is reasonable to use statistical information collected from the better local optima visited in the previous search to build the probability model.

4 Conclusion

In this chapter, we have proposed EDA/LS, a hybrid evolutionary algorithm for the QAP. In EDA/LS, a new operator, guided mutation, is used to produce new solutions. Guided by a probability model which characterizes the distribution of promising solutions in the search space, guided mutation alters a parent solution randomly to generate a new solution. Every new solution is then improved by the 2-opt local search. The search is re-initialized when it gets trapped in a local area. EDA/LS has been compared with MA and EDA/GLS on a set of QAP instances. The comparison results show that EDA/LS is comparable with MA, and outperforms EDA/GLS.

Most, if not all, meta-heuristics implicitly or explicitly use the proximate optimality principle The preliminary experiments in this chapter have verified the POP on several QAP instances. We believe that a deep understanding of the POP will be helpful in designing efficient algorithms for hard optimization problems.

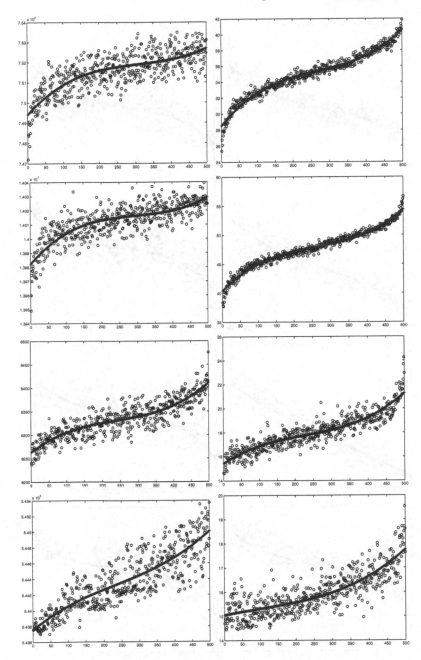

Fig. 4. The POP verification for four QAP instances **tai60a**, **tai80a**, **nug30** and **bur26a** (from top to bottom). On the x-axis is given the order of $\pi^1, \ldots, \pi^{500}$ w.r.t their costs. The left figures plot the average costs of $\sigma_k^1, \ldots, \sigma_k^{1000}$ while the right figures show their average distances to π^k. The continuous lines are the interpolatory curves

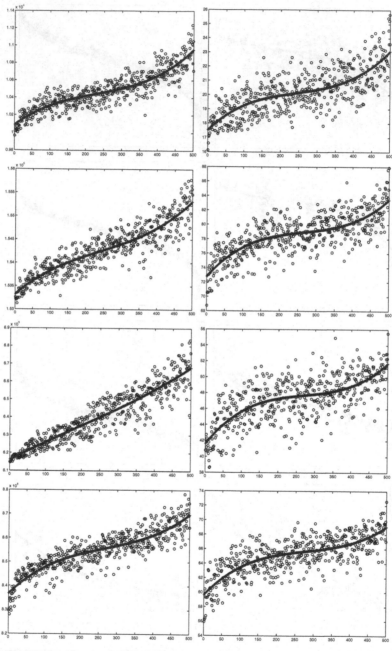

Fig. 5. The POP verification for the QAP instances **ste36a**, **sko100a**, **tai60b** and **tai80b** (from top to bottom). On the x-axis is given the order of $\pi^1, \ldots, \pi^{500}$ w.r.t their costs. The left figures plot the average costs of $\sigma_k^1, \ldots, \sigma_k^{1000}$ while the right figures show their average distances to π^k. The continuous lines are the interpolatory curves

References

1. T.C. Koopmans and M.J. Beckmann, "Assignment Problems and the Location of Economic Activities," *Econometrica*, vol. 25, pp. 53–76, 1957.

2. E. Çela, "The Quadratic Assignment Problem: Theory and Algorithms", Kluwer Academic Publishers, 1998.

3. Larranaga, P. and Lozano, J.A., "Estimation of Distribution Algorithms: A New Tool for Evolutionary Computation," Kluwer Academic Publishers, 2001.

4. D.T. Connolly, "An Improved Annealing Scheme for the Quadratic Assignment Problem," *European Journal of Operational Research*, vol. 46, pp. 93–100, 1990.

5. V. Nissen and H. Paul, "A Modification of Threshold Accepting and its Application to the Quadratic Assignment Problem," *OR Spektrum,* vol. 17, pp. 205–210, 1995.

6. S. Ishii and M. Sato, "Constrained Neural Approaches to Quadratic Assignment Problems," *Neural Networks*, vol. 11, pp. 1073–1082, 1998.

7. V. Bachelet, P. Preux, and E.-G. Talbi, "Parallel Hybrid Meta-Heuristics: Application to the Qudratic Assignment Problem," in *Proceedings of the Parallel Optimization Colloquium,* (Versailles, France), 1996.

8. P. Mills, E.P.K. Tsang and J.A. Ford, "Applying an Extended Guided Local Search on the Quadratic Assignment Problem," *Annals of Operations Research*, Kluwer Academic Publishers, vol. 118, pp. 121–135, 2003.

9. V. Nissen, "Solving the Quadratic Assignment Problem with Clues from Nature," *IEEE Transactions on Neural Networks,* vol. 5, no. 1, pp. 66–72, 1994.

10. D.M. Tate and A.E. Smith, "A Genetic Approach to the Quadratic Assignment Problem," *Computers and Operations Research,* vol. 22, no. 1, pp.73–83, 1995.

11. L. Gambardella, É. Taillard and M. Dorigo, "Ant Colonies for the QAP," *Journal of the Operations Research Society,* 1999.

12. P. Merz and B. Freisleben, "Fitness Landscape Analysis and Memetic Algorithms for the Quadratic Assignment Problem," *IEEE Transactions on Evolutionary Computation*, vol. 4, no. 4, pp. 337–352, 2000.

13. V.-D. Cung, T. Mautor, P. Michelon and A. Tavares, "A Scatter Search Based Approach for the Quadratic Assignment Problem," in *Proceedings of the 1997 IEEE International Conference on Evolutionary Computation (ICEC)*, (T. Bäck, Z. Michalewicz, and X. Yao, eds.), (Indianapolis, USA), pp. 165–170, IEEE Press, 1997.

14. S. Baluja, "Population-based Incremental Learning: A Method for Integrating Genetic Search Based Function Optimization And Competitive Learning," Technical Report, Carnegie Mellon University, 1994.

15. R.E. Burkard, S. Karisch and F. Rendl, "QAPLIB - A Quadratic Assignment Problem Library," *European Journal of Operational Research*, vol. 55, pp. 115–119, 1991. Updated Version: http://www.imm.dtu.dk/ sk/qaplib.

16. E.S. Buffa, G.C. Armour and T.E. Vollmann, "Allocating Facilities with CRAFT," *Harvard Business Review*, pp. 136–158, March 1964.

17. F. Glover and M. Laguna, *Tabu Search*, Kluwer, 1997.

18. P.M. Hahn, T. Grant and N. Hall, "A Branch-and-Bound Algorithm for the Quadratic Assignment Problem based on the Hungarian Method," *European Journal of Operational Research,* vol. 108, pp. 629–640, 1998.

19. Q. Zhang, J. Sun, E.P.K. Tsang and J.A. Ford, "Combination of Guided Local Search and Estimation of Distribution Algorithm for Solving Quadratic Assignment Problem," *Proceedings of the Bird of a Feather Workshops, Genetic and Evolutionary Computation Conference*, pp. 42–48, 2004.

20. Q. Zhang, J. Sun, and E. Tsang, "An evolutionary algorithm with guided mutation for the maximum clique problem. *IEEE Trans. Evolutionary Computation*", vol. 9, no. 2. pp. 192–200, 2005.

Index